Student Solutions Manual
for Moore, McCabe, Duckworth, and Sclove's

The Practice of Business Statistics

Michael A. Fligner
William I. Notz
Both of Ohio State University

W. H. Freeman and Company
New York

Printed in the United States of America

ISBN: 0-7167-9860-3

First Printing 2003

CONTENTS

CHAPTER 1

EXAMINING DISTRIBUTIONS

SECTION 1.1

OVERVIEW

Section 1.1 introduces several methods for exploring data. These methods should be applied only after clearly understanding the background of the data collected. The choice of method depends to some extent upon the type of variable being measured. The two types of variables described in this section are:

- **Categorical variables** – variables that determine the group or category to which an individual belongs. Hair color and gender are examples of categorical variables. Although we might count the number of people in the group with brown hair, we would not compute an average hair color for the group, even if numbers were used to represent the hair color categories.

- **Quantitative variables** – variables that have numerical values and with which it makes sense to do arithmetic. Height, weight, and GPA are examples of quantitative variables. It makes sense to talk about the average height or GPA of a group of people.

To summarize the **distribution** of a variable, for categorical variables use **bar charts** or **pie charts,** while for numerical data use **histograms** or **stemplots.** Also, when numerical data are collected over time, in addition to a histogram or stemplot, a **timeplot** can be used to look for interesting features of the data. When examining the data through graphs, we should be on the alert for

- unusual values that do not follow the pattern of the rest of the data
- some sense of a central or typical value of the data
- some sense of how spread out or variable the data are
- some sense of the shape of the overall pattern.

In addition, when drawing a timeplot, be on the lookout for **trends** occurring over time. Although many of the graphs and plots may be drawn by computer, it is still up to you to recognize and interpret the important features of the plots and the information they contain.

APPLY YOUR KNOWLEDGE

1.1 The "individuals" are the objects described, and the "variables" are the characteristics being measured. The individuals in this exercise are the different make and model cars, and the variables are vehicle type (categorical), transmission type (categorical), number of cylinders (probably can be considered as categorical), city MPG (quantitative) and highway MPG (quantitative). Although the variable cylinders is recorded as a number, this variable divides the cars into only a few categories. For a group of cars, we would not necessarily compute the average number of cylinders, but instead the number or percentage of 6- or 8-cylinder cars.

1.3 a) The bar chart below was produced using MINITAB.

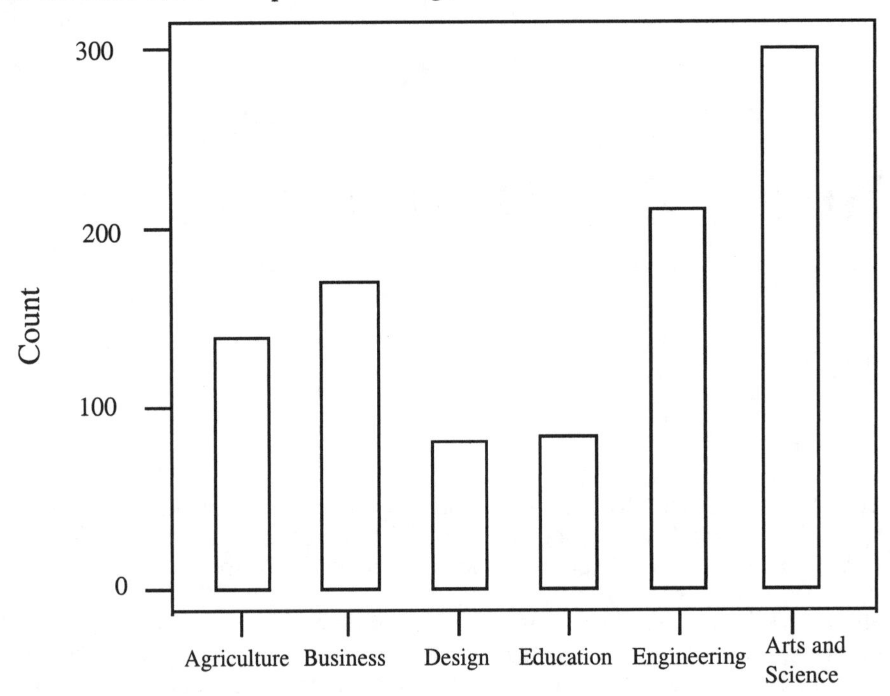

b) The percentages add to 98.5%. This seems to be more than a roundoff error, so if using a pie chart it would be necessary to include a category "other."

1.5 The histogram below was produced using MINITAB.

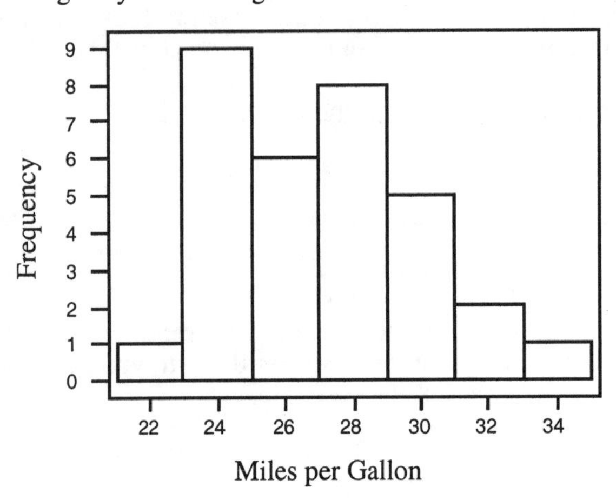

The classes are $21 \leq \text{MPG} < 23$, $23 \leq \text{MPG} < 25$, etc.

1.7 a) The distribution of highway mileage is somewhat skewed to the right, with a spread from 22 MPG to 33 MPG. There are no real outliers (observations far from the rest of the data), and the center of the data appears to be around 27 MPG, with about half the cars getting higher mileage and half getting lower mileage.

b) There are no obvious "gas guzzlers" in the sense of being a low outlier, so any cutoff would be somewhat arbitrary. The lower third of the cars (10 of the 32) have gas mileages of 24 or less, so a gas guzzler tax could be imposed on these cars, or the three cars with gas mileages of 23 or less could be singled out for the gas guzzler tax.

1.9 a) The distribution of monthly returns for all stocks over the 600 month period looks fairly symmetric with one low outlier. Excluding the low outlier, the spread is from about –15% to 16%.

b) Visually, the center of the distribution seems to lie in the tallest bar corresponding to the class from 0% to 2%. If the center corresponds to the value with approximately half the returns above it and half below it, you would need to keep adding the heights of the bars, until you found the bar containing the 300th smallest return, because there are 600 observations in total. About 200 months have returns below zero (the sum of the heights of the nine bars below 0 — see part [d]) and about 125 months have returns between 0% and 2%, so the middle return falls in the interval that goes from 0% to 2%. Taking the center of this interval could be used to approximate the center of the distribution at 1%. (Using the original data, the center is 1.33%.)

c) The smallest return is about –15% (exact value is –15.55%), and the largest return is about 17% (exact value is 16.56%). It is not possible to pick out the exact values using the graph alone because for observations in a class, you cannot determine their exact value because each class contains a range of values.

d) Adding up the nine bars to the left of zero (with the first bar corresponding to the low outlier), there are approximately

$$1 + 1 + 2 + 4 + 4 + 12 + 35 + 70 + 92 = 221$$

returns below zero, or about 37%. Your answers should be close to these as it is difficult to determine the exact count in some classes using the graph alone. The original data show exactly 223 monthly returns are less than zero.

1.11 The stemplot on the left has split stems so that the expenses in a \$10 range are split into two groups. For example, expenses between \$30 and \$39 are split into two groups — from \$30 to \$34 and from \$35 to \$39. The stemplot on the right does not have split stems. Using either stemplot, the center is \$28 and spread is from \$3 to \$93. There are no obvious outliers. Examination of either stemplot shows the distribution is clearly right skewed. In this particular example, little information is gained by splitting the stems and the plot is getting a little jagged. For these reasons, the stemplot on the right would probably be preferred.

```
0 | 3
0 | 99
1 | 134
1 | 5677889
2 | 0001234
2 | 55668888
3 | 2
3 | 5699
4 | 134
4 | 5579
5 | 03
5 | 59
6 | 1
6 |
7 | 0
7 |
8 | 3
8 | 66
9 | 3
```

```
0 | 399
1 | 1345677889
2 | 000123455668888
3 | 25699
4 | 1345579
5 | 0359
6 | 1
7 | 0
8 | 366
9 | 3
```

SECTION 1.1 EXERCISES

1.13 a) The different public mutual funds are the individuals in this data set.

b) Four variables are in the data set in addition to the fund's name. The variables "category" and "largest holding" are both categorical, while the variables "net assets" and "year-to-date return" are quantitative.

c) The units of measurement for net assets is millions of dollars and year-to-date return is given as a percentage.

1.15 Some variables that could be measured for each location include cost of property (expense to build a new facility), taxes (operational costs), utility costs (operational costs), cost of living (salaries that would need to be paid), number of similar facilities in the area (competition), average age of residents in the location (availability of workers or customers), percentage of land available for recreation (quality of life), average rainfall per year or other climate measures (desirability of living in location), transportation access (operational costs), and crime rates (quality of life).

1.17 The histogram below of exam scores is skewed to the left.

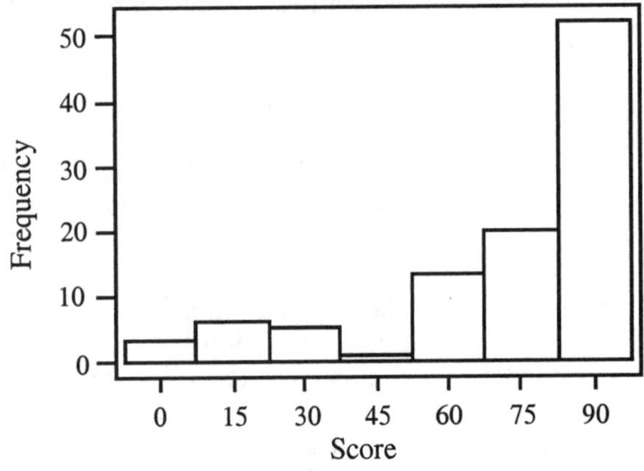

The distribution of dates of coins would be skewed to the left because as the dates go further back there are fewer of these coins currently in circulation.

1.19 a) The first thing to notice is that many more readers who completed the survey owned brand A than brand B. Even if both products had the same reliability, more owners of brand A than brand B would require service calls. The count is not a good measure of reliability.

b) Because different numbers of readers reported owning each brand, instead of looking at the count of service calls it would be better to consider the proportion of readers owning each brand who required service calls. In this case, 2942 / 13,376 = 22% of the brand A owners required a service call while 192 / 480 = 40% of the brand B owners required a service call. In terms of percentages of owners requiring a service call, brand A appears to be much more reliable. Of course, that may be why more people have chosen to purchase brand A!

1.21 Other variables that could be used to measure the "size of a company" include number of employees, assets, sales, market value, profits, and cash flow.

1.23 The stemplot on the next page uses split stems to reveal more of the pattern present in the data. Without splitting, almost the entire sample would be included in two stems — the tens and twenties. The stemplot shows the costs per month are skewed slightly to the right, with a center of $20. The spread goes from $8 (a provider with few services) to $50, which probably corresponds to a provider of fast Internet access. America Online and its largest competitors were probably charging around $20, and the members of the sample corresponding to early adopters of fast Internet access probably correspond to the monthly costs of more than $30.

```
0|89
1|00234
1|55558899
2|000000000000011111222222223
2|59
3|0
3|5
4|00
4|
5|0
```

1.25 Both timeplots below use the same data as in Example 1.7 of the text on page 21. The plot on the left has stretched the vertical axis in addition to adding a fair amount of white space and gives the impression of fairly slow changes in the price of oranges. The plot on the right has compressed the time axis and gives an impression of very rapid change in the price. For example, consider the extreme spike in the prices occurring in 1999 that is very apparent in the plot on the right. If you look for this same spike in the plot on the left, you will see it is barely noticeable.

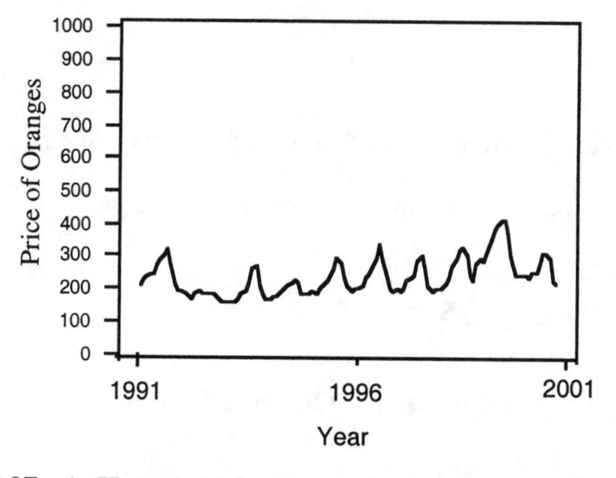

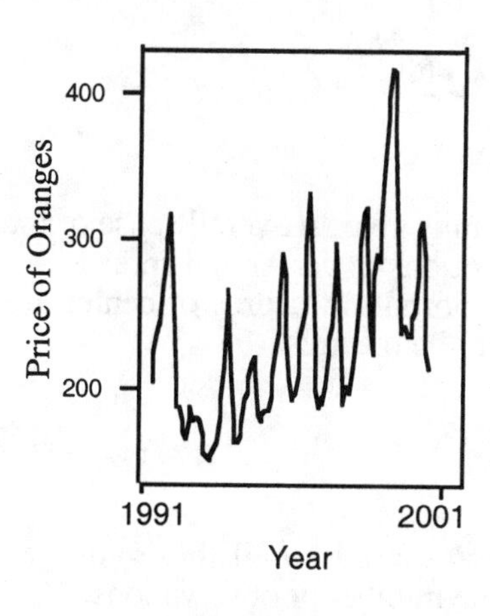

1.27 a) Household income is the total income of all persons in the household, so it will be higher than the income of any individual in the household. If a person doesn't have an income, he will not contribute to the household income, but he will also not be included in those with personal income.

b)

```
   Mean Personal        Median Household
      Income                 Income
              9|1|
        1111000|2|
      333222222|2|
  5555555554444|2|
         776666|2|67
       99988888|2|9
             00|3|0111
             32|3|2333
              5|3|445555
               |3|66666777
             88|3|8899999
               |4|0000111
               |4|233
               |4|45
               |4|6677
               |4|9
               |5|00
```

The units of the stems are $10,000. The back-to-back stemplot is using split stems, where the stems have been split into five. The first stem with a 2 includes states with incomes of $20,000 and $21,000, the second stem with a 2 includes states with incomes of $22,000 and $23,000, etc. Although incomes were rounded and recorded to the nearest hundred dollars, the information about the hundreds place is lost in this stemplot. To include this information, the units of the stems would have to be thousands of dollars (19, 20, 21, 22, etc.) and the leaves hundreds. This would have required twice as many stems and made the general shape of the distribution less apparent.

c) Both distributions are fairly symmetric, although the distribution of mean personal income has two high outliers, corresponding to Connecticut and the District of Columbia. The distribution of median household incomes has larger spread and a larger center. The center of the distribution of mean personal income is about $25,000, while the center of the distribution of median household income is about $37,000. Note also that personal income is using means while household income uses medians. The mean is pulled up by extremes, which is the reason the mean personal income is higher than the median household income for the District of Columbia, although this reversal of order does not occur in any of the other states.

SECTION 1.2

OVERVIEW

Although graphs give an overall sense of the data, numerical summaries of features of the data make more precise the notions of center and spread.

Two important measures of center are the **mean** and the **median.** If there are n observations, x_1, $x_2,\ldots,x_n$, then the mean is

$$\bar{x} = \frac{x_1 + x_2 + \cdots + x_n}{n} = \frac{1}{n}\sum x_i$$

where $\sum$ means "add up all these numbers". Thus, the mean is just the total of all the observations divided by the number of observations.

While the median can be expressed by a formula, it is simpler to describe the rules for finding it.

How to find the median.

1. List all of the observations from smallest to largest.

2. If the number of observations is odd, then the median is the middle observation. Count from the bottom of the list of ordered values up to the $(n + 1) / 2$ largest observation. This observation is the median.

3. If the number of observations is even, then the median is the average of the two center observations.

The most important measures of spread are the **quartiles,** the **standard deviation,** and **variance.** For measures of spread, the quartiles are appropriate when the median is used as a measure of center. In general, the median and quartiles are more appropriate when outliers are present or when the data are skewed. In addition, the **five-number summary,** which reports the largest and smallest values of the data, the quartiles and the median, provides a compact description of the data that can be represented graphically by a **boxplot.** Computationally, the first quartile, Q_1, is the median of the lower half of the list of ordered observations and the third quartile, Q_3, is the median of the upper half of the list of ordered values.

If you use the mean as a measure of center, then the standard deviation and variance are the appropriate measures of spread. Remember that means and variances can be strongly affected by outliers and are harder to interpret for skewed data.

If we have n observations, $x_1, x_2, \ldots, x_n$, with mean $\bar{x}$, then the variance s^2 can be found using the formula

$$s^2 = \frac{(x_1 - \bar{x})^2 + (x_2 - \bar{x})^2 + \cdots + (x_n - \bar{x})^2}{n-1} = \frac{1}{n-1}\sum (x_i - \bar{x})^2$$

The standard deviation is the square root of the variance, i.e., $s = \sqrt{s^2}$, and is a measure of spread in the same units as the original data. If the observations are in feet, then the standard deviation is in feet as well.

APPLY YOUR KNOWLEDGE

1.29 In practice, you will probably enter the data into a spreadsheet on a computer and have a software package do the calculations, or you would use a calculator that computes means automatically. The text has illustrated the use of the formula for black female bank workers to get a mean amount earned of \$17,528.93. For the other categories, you get

$$\bar{x}_{\text{BM}} = \frac{x_1 + x_2 + \cdots + x_n}{n} = \frac{18{,}365 + 17{,}755 + \cdots + 19{,}028}{12}$$
$$= \frac{237{,}650}{12}$$
$$= \$19{,}804.17$$

$$\bar{x}_{\text{WF}} = \frac{429{,}696}{20} = \$21{,}484.80$$

$$\bar{x}_{\text{WM}} = \frac{319{,}259}{15} = \$21{,}283.93$$

The black workers of either gender are making less on the average than their white counterparts. There is little difference in average salaries of white males and females, although black females are making less on average than black males. Because we have not taken into account the type of jobs performed by individuals in each category or years employed, we cannot make claims of discrimination without first adjusting for these factors.

1.31 In practice calculation of medians is best left to statistical software, particularly for large data sets. The text has illustrated the calculations for black female workers to get a median amount earned of \$17,516. For black males, there are 12 observations and the location of M_{BM} is (12 + 1) / 2 = 6.5. A location of 6.5 means halfway between the sixth and seventh observations. The ordered earning are:

16576 16853 16890 17147 17755 18365 18402 19028 20972 21565 24750 29347

with the sixth smallest observation equal to \$18,365 and the seventh equal to \$18,402. Averaging these,

$$M_{\text{BM}} = \frac{18{,}365 + 18{,}402}{2} = \$18{,}383.50$$

For the white females, there are 20 observations and the location of M_{WF} is (20 + 1) / 2 = 10.5. The ordered earnings are

14698 15904 17233 17576 17757 18002 18863 19029 19102 19308 20612 21596 22477
24497 24780 25249 26606 26885 28346 31176

Averaging the 10th and 11th smallest observations,

$$M_{\text{WF}} = \frac{19,308 + 20,612}{2} = \$19,960$$

Finally, their are 15 white males and the locations of M_{WM} is (15 + 1) / 2 = 8. The ordered earnings are

15100 16411 17194 18245 18364 19007 19268 19977 22049 22078 22346 23531 26970 28336 30383

The eighth smallest observation is

$$M_{\text{WM}} = \$19,977$$

The relative orderings of the groups are similar to the ordering when computing means, although the order of white males and white females reverses. Because the medians are less affected by extremes (in this case, the higher salaries within each group), the medians are smaller than the means.

1.33 When a distribution is skewed to the right, the mean is pulled toward the larger observations and will be larger than the median. With strong skewness, as in this example, the mean and median can be quite different. The median is the lower number of \$330,000 and the mean is \$675,000.

1.35 a) There are 12 Asian countries with Japan excluded. The ordered growth of per capita consumption for these countries is

1.3 2.3 2.9 3.9 4.1 **4.2** **5.1** 5.6 5.9 6.2 6.4 8.8

The location of the median is (12 + 1) / 2 = 6.5, so the median is M = (4.2 + 5.1) / 2 = 4.65, the average of the two numbers in bold. The first quartile is the median of the six observations below the overall median and the third quartile is the median of the six observations above the overall median. The quartiles are Q_1 = (2.9 + 3.9) / 2 = 3.4 and Q_3 = (5.9 + 6.2) / 2 = 6.05. The minimum is 1.3 and the maximum is 8.8 giving a five-number summary

1.3 3.4 4.65 6.05 8.8

There are 13 Eastern European countries. The ordered growth of per capita consumption for these countries is

–12.1 –5.2 –1.7 –1.5 –1.0 1.0 **1.4** 3.2 3.5 3.7 4.9 6.0 7.0

The location of the median is (13 + 1) / 2 = 7, so the median is M = 1.4, the number in bold. The first quartile is the median of the six observations below the overall median of 1.4. This is the average of –1.7, the third smallest observation, and –1.5, the fourth smallest observation, so Q_1 = –1.6. The third quartile is the median of the six observations above the overall median of 1.4. This is the average of 3.7, the third smallest of these observation, and 4.9, the fourth smallest, so Q_3 = 4.3. The minimum is –12.1 and the maximum is 7.0 giving a five number summary

–12.1 –1.6 1.4 4.3 7.0

b) The side-by-side boxplot on the next page was produced by MINITAB. Basically, the boxplot is a picture of the information in the five-number summary, with outliers identified. The key differences between the two groups of countries are that the growth of per capita consumption tends to be much higher for the Asian countries than the Eastern European countries and the growth of per capita consumption for the Eastern European countries is much more spread out.

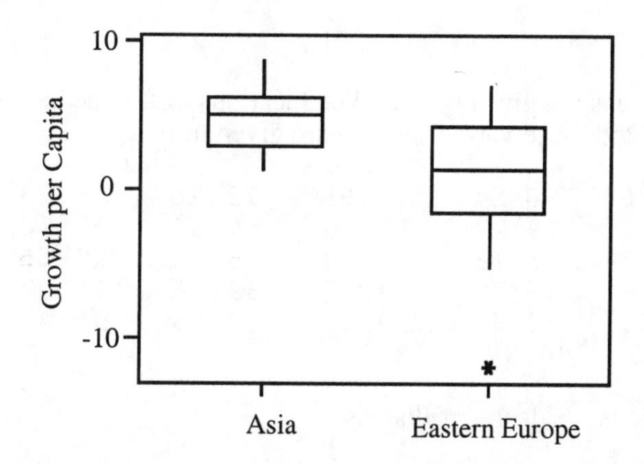

1.37 a) Using the definition, the mean is computed as

$$\bar{x} = \frac{x_1 + x_2 + \cdots + x_n}{n} = \frac{1600+1157+937+800+707+700}{6}$$

$$= \frac{5901}{6}$$

$$= \$983.5 \text{ million}$$

b) In practice, you will be using software or your calculator to obtain the mean and standard deviation from keyed-in data. However, we illustrate the step-by-step calculations to help you understand how the standard deviation works. Be careful not to round off the numbers until the last step, as this can sometimes introduce fairly large errors when computing *s*.

Observation	Difference	Difference squared
x_i	$x_i - \bar{x}$	$(x_i - \bar{x})^2$
1600	616.5	380072.25
1157	173.5	30102.25
937	-46.5	2162.25
800	-183.5	33672.25
707	-276.5	76452.25
700	-283.5	80372.25
		602833.50

Adding the last column in the table gives the sum of the differences squared. The formula for the variance is

$$s^2 = \frac{(x_1 - \bar{x})^2 + (x_2 - \bar{x})^2 + \cdots + (x_n - \bar{x})^2}{n-1} = \frac{602833.5}{6-1} = 120566.7$$

Taking the square root gives the standard deviation as $s = \sqrt{120566.7} = \$347.2272$ million.

c) Computer software gives the mean as 983.5 and the standard deviation as 347.23. Although different software packages use different conventions for the number of digits printed out in the final answer, the intermediate computations in all packages are done retaining many more digits. Just because a package prints a standard deviation out to eight decimal places does not mean that it is doing the computations more accurately than one that prints the result to two decimal places.

SECTION 1.2 EXERCISES

1.39 The 50 ordered observations are given below. The locations of the median is (50 + 1) / 2 = 25.5, or midway between the 25th and 26th observations, which are given in bold.

3.11	8.88	9.26	10.81	12.69	13.78	15.23	15.62	17.00
17.39	18.36	18.43	19.27	19.50	19.54	20.16	20.59	22.22
23.04	24.47	24.58	25.13	26.24	26.26	**27.65**	**28.06**	28.08
28.38	32.03	34.98	36.37	38.64	39.16	41.02	42.97	44.08
44.67	45.40	46.69	48.65	50.39	52.75	54.80	59.07	61.22
70.32	82.70	85.76	86.37	93.34				

Averaging these two observations yields the median as

$$M = \frac{27.65 + 28.06}{2} = \$27.855$$

The mean was \$34.70 and is larger than the median. This is because the distribution is skewed to the right as can be seen from the histogram below.

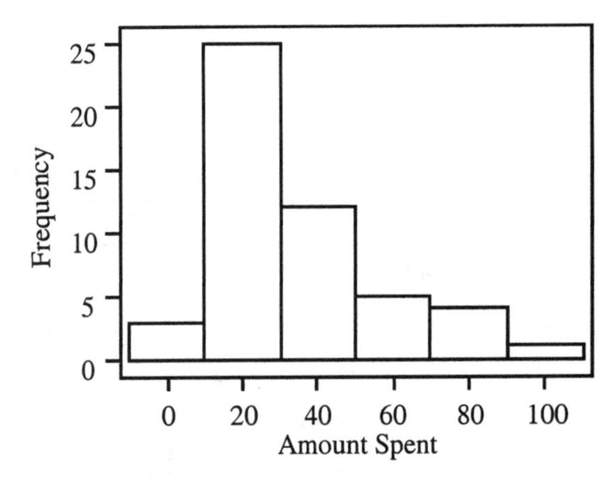

1.41 a) Using statistical software, the mean change is $\bar{x}$ = 28.767% and the standard deviation is s = 17.766%.

b) Ignoring the outlier, there are 14 remaining observations. Recalculating the mean gives the mean change as $\bar{x}$ = 31.707% and the standard deviation as s = 14.150%. The low outlier has pulled the mean down toward it and increased the variability in the data.

c) Identical would usually mean the same make and model for the 5-liter vehicles.

1.43 Go to the applet available at www.whfreeman.com/pbs to learn more about the effects of outliers on the mean and median interactively.

1.45 a), b) The boxplot is given on the next page. Although this boxplot was produced by MINITAB, it cannot be done automatically from the information given in the spread sheet. The clear pattern is that as the level of education increases, the incomes tend to increase and also become more spread out. The lower 5% of the incomes are almost independent of level of education and correspond to those working part-time, and the distributions of incomes for those with a high school diploma and those with some college but no degree are quite similar. However, once a bachelor's degree has been obtained, there is a marked increase in incomes and this increase continues with master's and professional degrees.

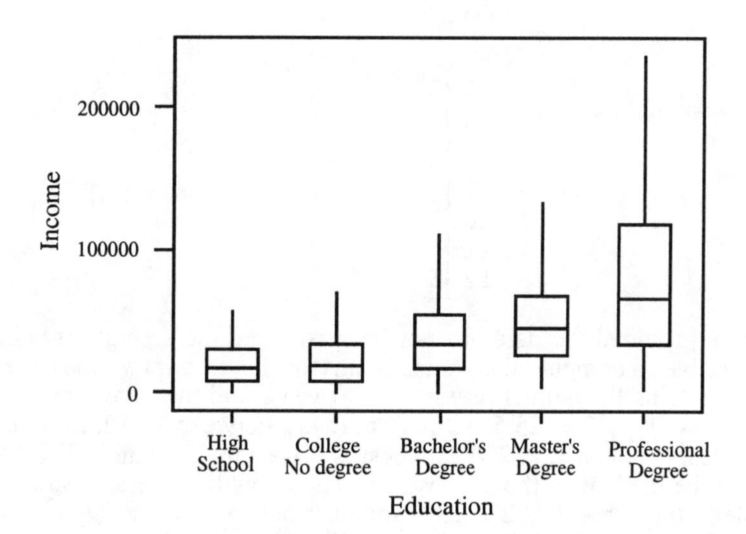

1.47 The histogram on the left includes all states and the District of Columbia. The District of Columbia is an extreme high outlier, and this has reduced the number of classes for the remaining 50 states. This can result in a loss of information about the overall shape. For this reason the histogram was redrawn excluding the District of Columbia giving the histogram on the right.

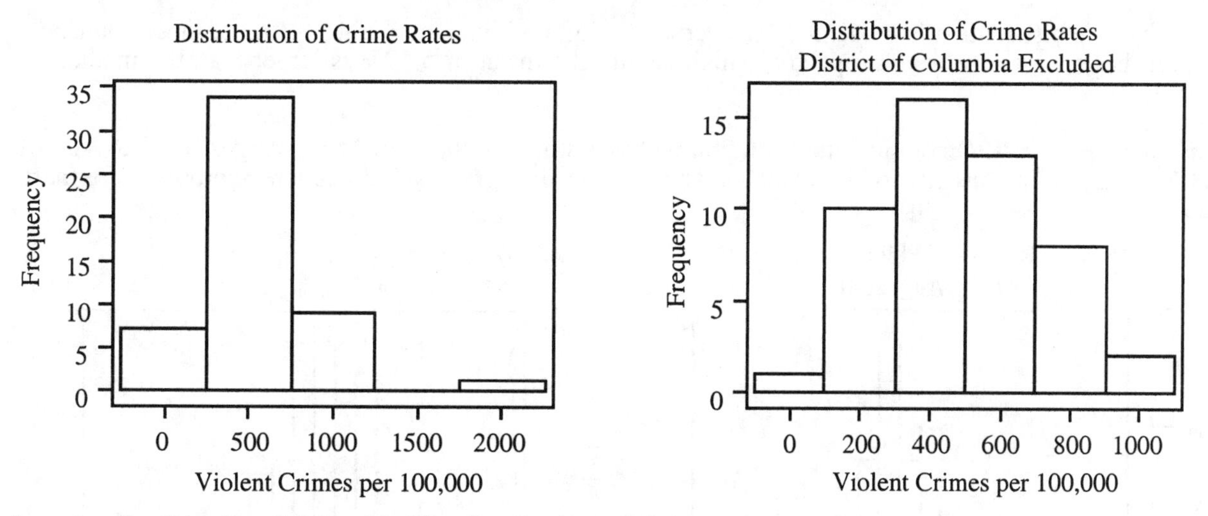

Ignoring the District of Columbia, the distribution of violent crimes is fairly symmetric, with a center of about 450 violent crimes per 100,000. The spread is from about 100 to 1000 violent crimes per 100,000 for the 50 states, with the District of Columbia being a high outlier featuring a rate of slightly more than 2000 violent crimes per 100,000.

1.49 Using statistical software to compute means and variances, we find Data A has a mean of 7.501 and a standard deviation of 2.032. Data B has a mean of 7.501 and a standard deviation of 2.031. Thus these are two data sets of 11 observations with the same means and standard deviations. The mean gives an estimate of center, while the standard deviation gives an estimate of spread. Neither measure is resistant to outliers, and they do not give an indication of the shape of the distribution. On the next page, the stemplot of each data set is provided, where the data has been rounded to the nearest 10th. The stems are ones and the leaves are 10ths. Data A has a distribution that is left skewed while Data B has a distribution that is fairly symmetric except for one high outlier. Thus we see two distributions with quite different shapes, but with the same mean and standard deviation. Data B seems less spread out than Data

A despite the fact that the standard deviations are the same. The reason for this is that the standard deviation is not a resistant measure of spread and its value has been increased by the high outlier.

Data A

```
3 | 1
4 | 7
5 |
6 | 1
7 | 3
8 | 1178
9 | 113
```

Data B

```
 5 | 368
 6 | 69
 7 | 079
 8 | 58
 9 |
10 |
11 |
12 | 5
```

1.51 a) Unlike a histogram, provided the data has not been rounded the stemplot retains the original observations. Thus, it is possible to compute the values of any of the descriptive measures discussed in this section. The stemplot contains the annual returns for 50 years and they are already ordered. The location of the median is (50 + 1) / 2 = 25.5, so it is midway between the 25th and 26th smallest observations. The 25th smallest is 4.8 and the 26th smallest is 5.1, so the median is $M = (4.8 + 5.1) / 2 =$ 4.95%. The first quartile is the median of the 25 observations below the overall median of 4.95. The location of the first quartile is the (25 + 1) / 2 = 13th smallest observation, so Q_1 = 3.0%. The third quartile is the median of the 25 observations above the overall median of 4.95. The location of the third quartile is the (25 + 1) / 2 = 13th smallest observation of the 25 above the median, so Q_3 = 6.6%. The minimum is 0.9% and the maximum is 14.2%, giving a five-number summary

0.9% 3.0% 4.95% 6.6% 14.2%

b) The distribution is moderately skewed to the right with a high outlier. Both of these contribute to pull the mean annual return upward, which is why the mean of 5.19% is larger than the median of 4.95%.

1.53 a) There are 134 monthly returns on Philip Morris stock for the period from August 1990 to August 2001. Stemplots are usually used for smaller data sets so a histogram would be more appropriate for these data.

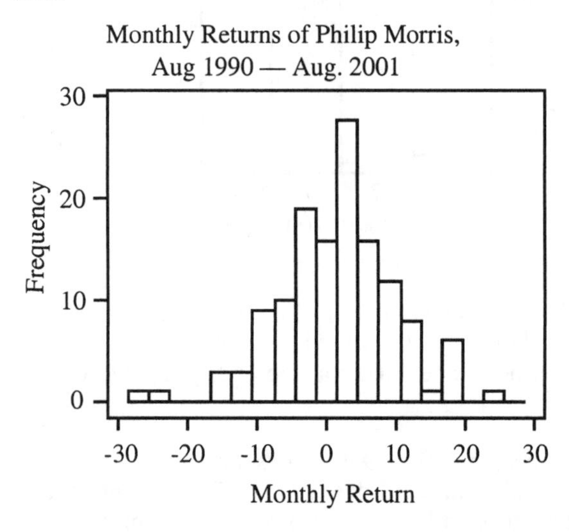

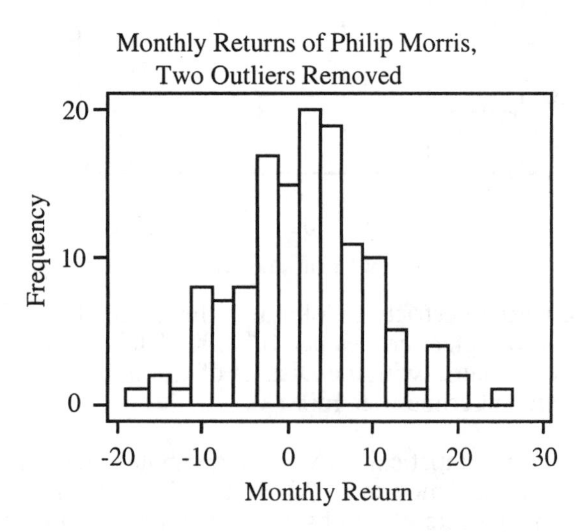

b) The two clear outliers are low outliers of –26.6% and –22.9%. The histogram on the right is of the monthly returns excluding these two outliers. The distribution is fairly symmetric, with a median or center of 2.65%. The spread is from –16.5% to 24.4%. Note that the mean is 1.95% and is slightly less than the median.

c) Using all 134 observations and statistical software, the mean is 1.551% and the standard deviation is 8.218%. If you invested $100 and the return was equal to the mean of 1.551%, then at the end of the month you would have (1 + 0.01551) × $100 = $101.55.

d) The worst month in the data corresponds to a return of –26.6%. If you had invested $100 at the beginning of this month, then at the end of the month you would have (1 – 0.266) × $100 = $74.40. Excluding the two outliers and recomputing the mean and standard deviation gives a mean of 1.551% and a standard deviation of 8.218%. Neither of these measures is resistant, and the two low outliers pull the mean downwards and increase the standard deviation. Quartiles and medians are resistant measures and outliers do not have a large effect on their values, particularly in a data set of this size. We wouldn't expect them to change much, and in fact the median is 2.50% with the two low outliers and 2.65% without them. The quartiles remain almost unchanged as well.

1.55 You are told there are 411 players and that only 139 made more than the "league average salary" of $2.36 million. If $2.36 million were the median salary, then half of the players – or 205 players – would have salaries above it. Because only 139 players have salaries above $2.36 million, it must be the mean. In this case, the mean is larger than the median because the distribution of salaries is skewed to the right.

1.57 a) The standard deviation is always greater than or equal to zero. The only way it can equal zero is if all the numbers in the data set are the same. Because repeats are allowed, just choose all four numbers the same to make the standard deviation equal to zero. Examples are 1, 1, 1, 1 or 2, 2, 2, 2.

b) To make the standard deviation large, numbers at the extremes should be selected. So you want to put the four numbers at zero or 10. The correct answer is 0, 0, 10, 10. You might have thought that 0, 0, 0, 10 or 0, 10, 10, 10 would be just as good, but a computation of the standard deviation of these choices shows that two at either end is the best choice.

c) There are many choices for part (a) but only one for part (b).

SECTION 1.3

OVERVIEW

This section considers the use of mathematical models to describe the overall pattern of a distribution. A mathematical model is an idealized description of this overall pattern, often represented by a smooth curve. The name given to a mathematical model that summarizes the shape of a histogram is a **density curve.** The density curve is a kind of idealized histogram. The total area under a density curve is one and the area between two numbers represents the proportion of the data that lie between these two numbers. Like a histogram, it can be described by measures of center, such as the **median** (a point such that half of the area under the density curve is to the left of the point) and the **mean** μ (the balance point of the density curve if the curve were made of solid material), and measures of spread, such as the **quartiles** and the standard deviation σ.

One of the most commonly used density curves in statistics is the **normal curve** and the distributions they describe are called **normal distributions.** Normal curves are symmetric and bell-shaped. The peak of the curve is located above the mean and median, which are equal because the density curve is symmetric. The standard deviation is the distance from the mean to the change-of-curvature points on either side. It measures how concentrated the area is around this peak. Normal curves follow the **68-95-99.7 rule,** i.e., 68% of the area under a normal curve lies within one standard deviation of the mean (illustrated in the figure on the next page), 95% within two standard deviations of the mean, and 99.7% within three standard deviations of the mean.

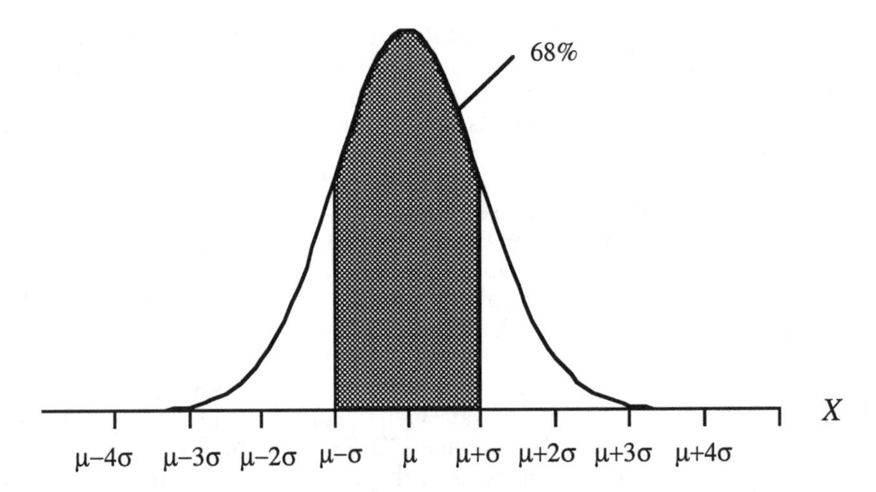

Areas under any normal curve can be found easily if quantities are first **standardized** by subtracting the mean from each value and dividing the result by the standard deviation. This standardized value is sometimes called the ***z*-score.** If data whose distribution can be described by a normal curve are standardized (all values replaced by their *z*-scores), the distribution of these standardized values is called the **standard normal** distribution and they are described by the **standard normal curve.** Areas under standard normal curves are easily computed by using a standard normal table such as that found in Table A in the front inside cover of the text.

If we know the distribution of data is described by a normal curve, we can make statements about what values are likely and unlikely, without actually observing the individual values of the data. Although one can examine a histogram or stemplot to see if it is bell-shaped, the preferred method for determining if the distribution of data is described by a normal curve is a **normal quantile plot.** These are easily made using modern statistical computer software. If the distribution of data is described by a normal curve, the normal quantile plot should look like a straight line.

In general, density curves are useful for describing distributions. Many statistical procedures are based on assumptions about the nature of the density curve that describes the distribution of a set of data. **Density estimation** refers to techniques for finding a density curve that describes a given set of data.

APPLY YOUR KNOWLEDGE

1.59 The sketch on the left is a symmetric density called the uniform density and the sketch on the right is of a density that is strongly skewed to the left.

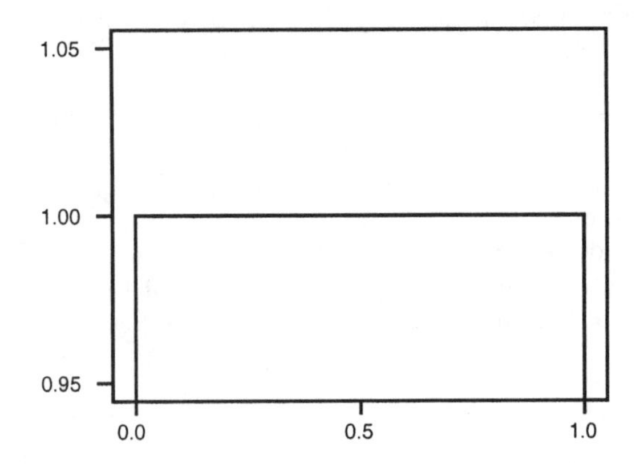

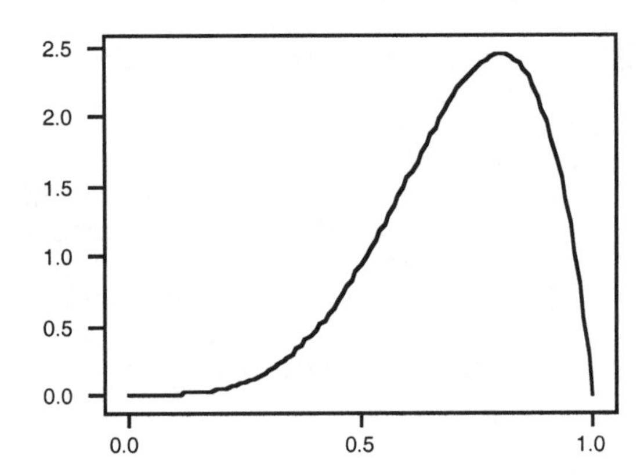

1.61 a) The density curve is right-skewed so the mean is larger than the median (right skew pulls the mean to the right). Point A is too small to be the median, as less than 50% of the area under the density is to the left of point A. Of the two remaining points, point B must be the median and point C must be the mean, because the mean is larger than the median.

b) For a symmetric distribution the mean and median coincide and both are equal to the center of symmetry of the density. Point A corresponds to both the mean and the median.

c) The density curve is left skewed so the mean is smaller than the median (left skew pulls the mean to the left). Point C is too large to be the median, as less than 50% of the area under the density is to the right of point C. Of the two remaining points, point B must be the median and point A must be the mean, because the mean is smaller than the median.

1.63 a) A height of 74 inches is two standard deviations above the mean because 74 = 69 + 2(2.5). The 68–95–99.7 rule says that 95% of the men have heights within two standard deviations of the mean, or between 64 inches and 74 inches. This is equivalent to 5% of the men having heights less than 64 inches or greater than 74 inches. Because the normal curve is symmetric, half of these, or 2.5% of the men, have heights below 64 inches. The other half, or 2.5% of the men, have heights above 74 inches. The answer is 2.5% of the men have heights above 74 inches.

b) The 68–95–99.7 rule says that 95% of the men have heights within two standard deviations of the mean, or within 2 × 2.5 = 5 inches of the mean. Because the mean is 69 inches, two standard deviations below the mean is 64 inches and two standard deviations above the mean is 74 inches, so that the middle 95% of the men have heights between 64 inches and 74 inches.

c) A height of 66.5 inches is one standard deviation below the mean because 66.5 = 69 – 2.5. The 68–95–99.7 rule says that 68% of the men have heights within one standard deviation of the mean, or between 66.5 inches and 71.5 inches. This is equivalent to 32% of the men having heights less than 66.5 inches or greater than 71.5 inches. Because the normal curve is symmetric, half of these, or 16% of the men, have heights below 66.5 inches. The other half, or 16% of the men, have heights above 71.5 inches. The answer is 16% of the men are shorter than 66.5 inches.

1.65 To compare scores from two normal distributions, each can be standardized or converted into z-scores. For example, an SAT score or an ACT score that corresponds to a z-score above two places an individual in the top 2.5% of the distribution of scores for either test, as it corresponds to a score that is at least two standard deviations above the mean. In this example, Eleanor's z-score is $\frac{680-500}{100} = 1.8$, while Gerald's z-score is $\frac{27-18}{6} = 1.5$, so Eleanor scored relatively higher.

1.67 a) *State the problem:* We want the percent of vehicles with $x > 30$.
Standardize:

$$x > 30$$

$$\frac{x-21.22}{5.36} > \frac{30-21.22}{5.36}$$

$$z > 1.64$$

Use the table: The area greater than 1.64 is one minus the area below 1.64, or 1 –.9495 = 0.0505. About 5% of vehicles have ratings greater than 30.

b) *State the problem:* We want the percent of vehicles with $30 < x < 35$.
Standardize:

$$30 < x < 35$$

$$\frac{30-21.22}{5.36} < \frac{x-21.22}{5.36} < \frac{35-21.22}{5.36}$$

$$1.64 < z < 2.57$$

Use the table: The area between 1.64 and 2.57 is the area below 2.57 minus the area below 1.64. This is 0.9949 – 0.9495 = 0.0454, or about 4.5% of vehicles have ratings between 30 and 35.

c) *State the problem:* We want the percent of vehicles with $x < 12.45$.

Standardize:

$$x < 12.45$$

$$\frac{x - 21.22}{5.36} < \frac{12.45 - 21.22}{5.36}$$

$$z < -1.64$$

Use the table: The area less than –1.64 can be read directly from the table and is 0.0505. About 5% of vehicles have ratings less than 12.45.

1.69 a) *State the problem:* We want the percent of scores with $x > 500$.
Standardize:

$$x > 500$$

$$\frac{x - 527}{112} > \frac{500 - 527}{112}$$

$$z > -0.24$$

Use the table: The area greater than –0.24 is one minus the area below –0.24, or 1 –.4052 = 0.5948. About 60% of testtakers have scores above 500.

b) This is an example of a "backward" normal calculation.
State the problem: We want to find the GMAT score x with area 0.25 to its left for a normal distribution with mean $\mu = 527$ and standard deviation $\sigma = 112$.
Use the table: Look in the body of Table A for the entry closest to 0.25. This corresponds to $z = -0.67$, so $z = -0.67$ is the standardized value with area 0.25 to its left.
Unstandardize: The z score of –0.67 needs to be transformed back to the original scale. So x satisfies

$$\frac{x - 527}{112} = -0.67$$

and solving this equation for x gives

$$x = -0.67(112) + 527 = 451.96,$$

or about 25% of the scores are below 452.

c) This is another example of a "backward" normal calculation.
State the problem: We want to find the GMAT score x with area 0.05 to its right for a normal distribution with mean $\mu = 527$ and standard deviation $\sigma = 112$.
Use the table: Look in the body of Table A for the entry closest to 0.95, because if the area to the right is 0.05, the area to the left must be 0.95. This corresponds to $z = 1.65$, so $z = 1.65$ is the standardized value with area 0.95 to its left.
Unstandardize: The z score of 1.65 needs to be transformed back to the original scale. So x satisfies

$$\frac{x - 527}{112} = 1.65$$

and solving this equation for x gives

$$x = 1.65(112) + 527 = 711.8,$$

or the top 5% of the scores are above 712.

1.71 There are no major deviations from normality, although the two lowest returns appear to be off the line of the other points. The smallest return appears to be a low outlier.

SECTION 1.3 EXERCISES

1.73 The 68–95–99.7 rule says that 95% of the trucks have NOX levels within two standard deviations of the mean or between 1.45 – 2(0.4) = 0.65 and 1.45 + 2(0.4) = 2.25 grams per mile driven. The normal curve is sketched below, with the interval containing the middle 95% of NOX levels shaded.

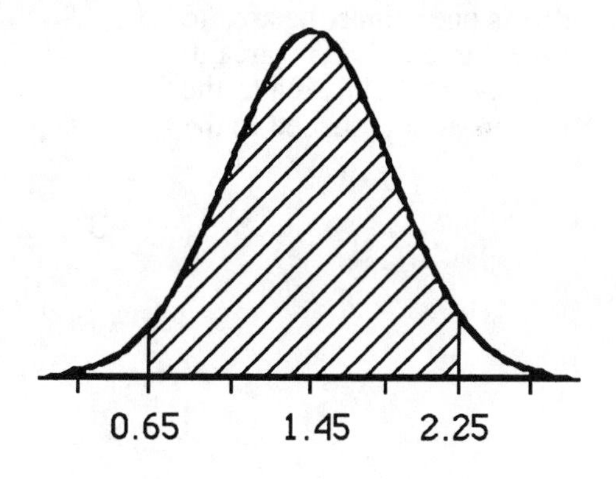

1.75

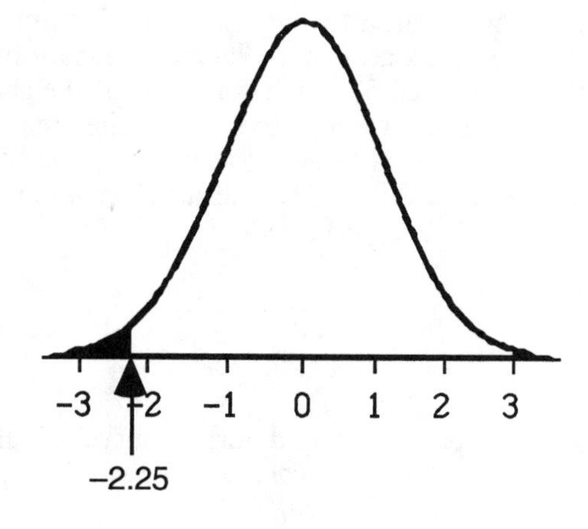

a) This area can be read directly from Table A, which gives the area to the left of any z value. Looking up the value –2.25 in Table A gives a value of 0.0122, which corresponds to the area to the left of –2.25 that is shaded in the figure.

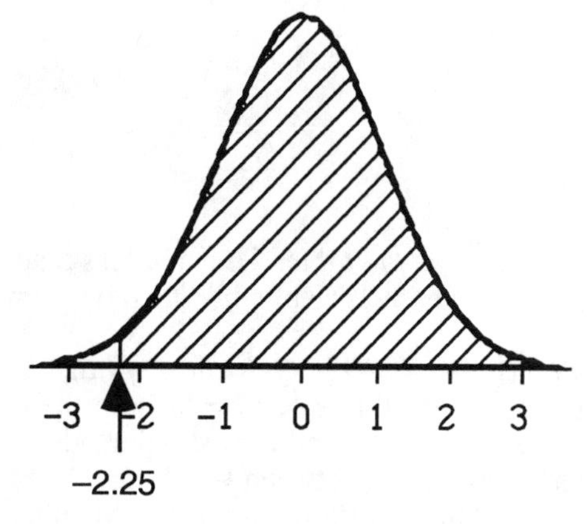

b) The area to the right of any z value is one minus the area to the left because the total area under the curve is 1. The area to the left of –2.25 is 0.0122 from part (a), so the area to the right of –2.25 is 1 – 0.0122 = 0.9878. This area is shaded in the figure.

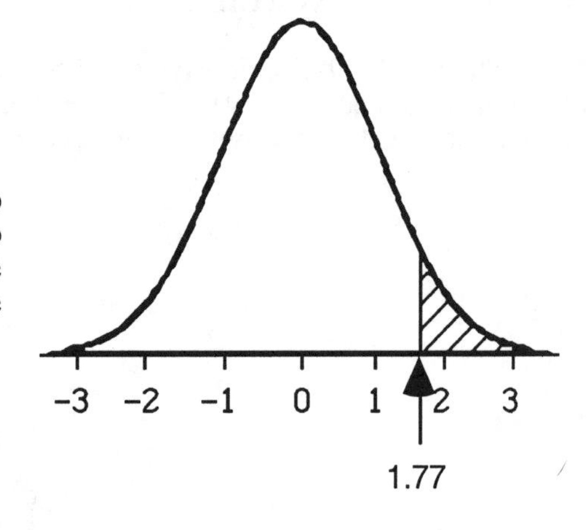

c) The area to the right of any z value is one minus the area to the left because the total area under the curve is 1. The area to the left of 1.77 is given in Table A as 0.9616. The area to the right of 1.77 is $1 - 0.9616 = 0.0384$. This area is shaded in the figure.

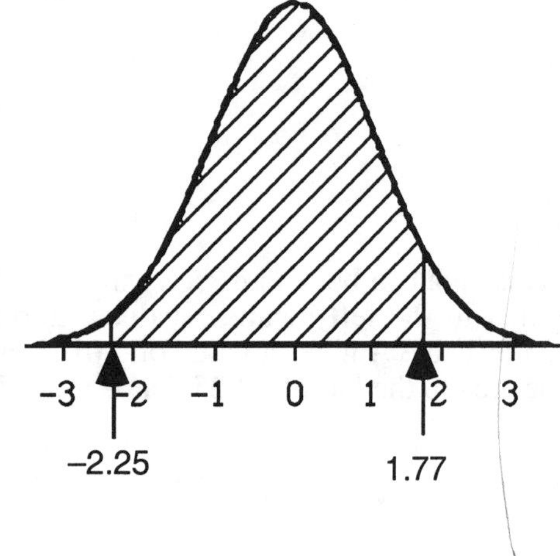

d) The area between –2.25 and 1.77 is obtained by first finding the area to the left of 1.77 from Table A which is 0.9616. This area is too large because it includes the area to the left of –2.25, which isn't part of the shaded region in the figure. Thus we need to subtract the area to the left of –2.25 from 0.9616. The area to the left of –2.25 is obtained from Table A as 0.0122, so the area between –2.25 and 1.77 is $0.9616 - 0.0122 = 0.9494$.

1.77 This is a standard normal curve calculation.

State the problem: We want the percent of scores with $x < 820$.

Standardize:

$$x < 820$$

$$\frac{x-1019}{209} < \frac{820-1019}{209}$$

$$z < -0.95$$

Use the table: The area less than –0.95 can be read directly from the table and is 0.1711. About 17% of students have combined SAT scores less than 820.

1.79 a) The 68–95–99.7 rule says that 95% of the yearly returns are within two standard deviations of the mean, or within $2 \times 17 = 34\%$ of the mean. Because the mean is 13%, two standard deviations below the mean is –21% and two standard deviations above the mean is 47%, so that the middle 95% of the yearly returns are between –21% and 47%.

b) This is a standard normal calculation.

State the problem: We want the percent of years with $x < 0$.

Standardize:

$$x < 0$$

$$\frac{x-13}{17} < \frac{0-13}{17}$$

$$z < -0.76$$

Use the table: The area less than –0.76 can be read directly from the table and is 0.2236. In about 22% of the years the market is down.

c) This is a standard normal calculation.
State the problem: We want the percent of years with $x > 25$.
Standardize:

$$x > 25$$
$$\frac{x-13}{17} > \frac{25-13}{17}$$
$$z > 0.71$$

Use the table: The area to the right of 0.71 is one minus the area to the left of 0.71, or 1 – 0.7661 = 0.2389. In about 24% of the years, the index gains more than 25%.

1.81 a) The area under any density curve to the left of the first quartile is 0.25. To find the z value corresponding to this, you need to look for the value 0.25 in the body of Table A and find the z value, which gives the first quartile of the standard normal is $z = -0.67$. The third quartile has 0.75 of the area to the left of it for any distribution. To find the z value corresponding to this, you need to look for the value 0.75 in the body of Table A and find the z value, which gives the third quartile of the standard normal is $z = 0.67$. Because of the symmetry of the standard normal, the first and third quartiles are the same distance from $\mu = 0$, namely ± 0.67.

b) This problem is basically a "backward" normal calculation that we are doing in two parts (for example, Exercise 1.69b in which we found the first quartile of GMAT scores).

In part (a) of this problem, we have completed the first two steps, which correspond to stating the problem and then using Table A to find the z value. Now what is left is to unstandardize the z values to find the quartiles for the distribution of the lengths of human pregnancies, which has a mean $\mu = 266$ days and a standard deviation of $\sigma = 16$ days. For the first quartile, we have

Unstandardize: The z-score of –0.67 needs to be transformed back to the original scale. So x satisfies

$$\frac{x-266}{16} = -0.67$$

and solving this equation for x gives

$$x = -0.67(16) + 266 = 255.28,$$

or the first quartile is about 255 days.

To find the third quartile of the distribution of the lengths of human pregnancies, we follow the same steps but with the z value corresponding to the third quartile.

Unstandardize: The z-score of 0.67 needs to be transformed back to the original scale. So x satisfies

$$\frac{x-266}{16} = 0.67$$

and solving this equation for x gives

$$x = 0.67(16) + 266 = 276.72,$$

or the third quartile is about 277 days.

1.83 You will need to use statistical software that will make Normal quantile plots. The plot below was made using MINITAB. The two low outliers are quite apparent in the plot, but the remainder of the points look as if they follow a line fairly well. In problem 1.53, the histogram showed the distribution to be fairly symmetric, and the normal quantile plot suggests the distribution should be reasonably well approximated by a normal curve.

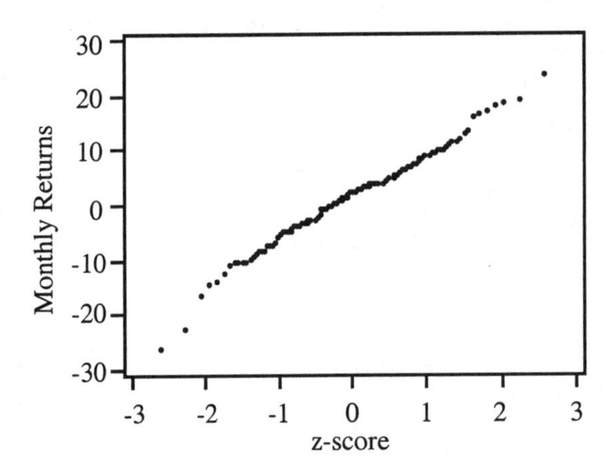

1.85 The histogram of the 100 observations generated from the uniform distribution is given on the left and the normal quantile plot is given on the right. The histogram is very similar in shape to the uniform density as it should be since this is the mathematical model that generated the data. The normal quantile plot does not look linear, but is instead S shaped. This is caused by the small and large observations from the uniform distribution not being as large or small as would be predicted from the normal model.

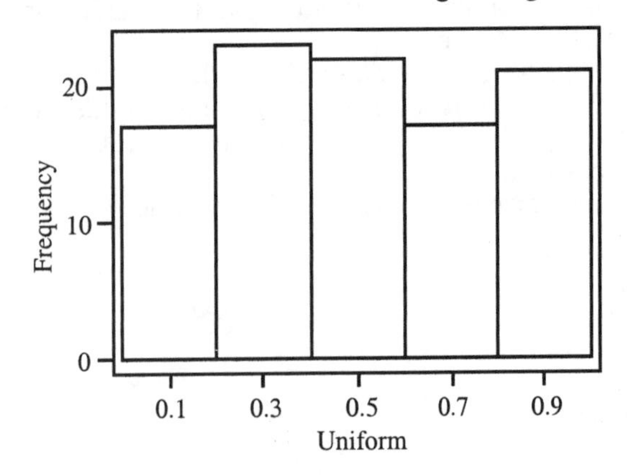

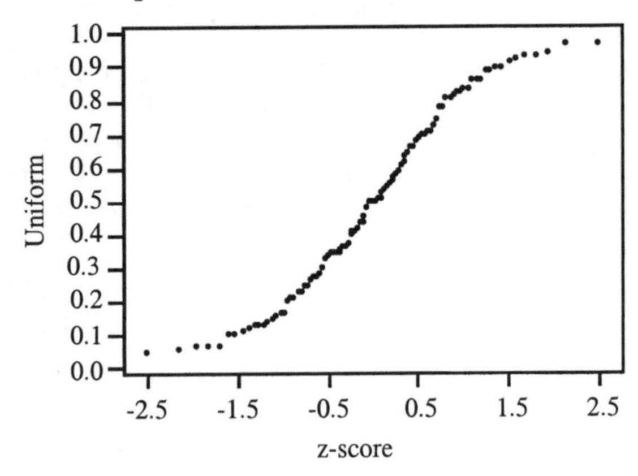

Chapter 1 Review Exercises

1.87 The variables gender and automobile preference are categorical. The variables age and household income are quantitative.

1.89 The salaries of the New York Yankees are given in the histogram on the next page. The data is strongly skewed to the right. Plotting this data is difficult as the salaries cover several orders of magnitude. For example, even the first bar of the histogram, which is the class interval \$0 to \$2000 thousand covers the range of salaries from \$0 to \$2 million, which corresponds to about half of the team. Unfortunately, if the class intervals were made any smaller, most would contain only a single observation. The median salary is \$1.6 million and the mean salary is slightly more than \$3.5 million, being pulled up due to the strong right skewness of the distribution. The spread goes from \$200 thousand to slightly more than \$12 million.

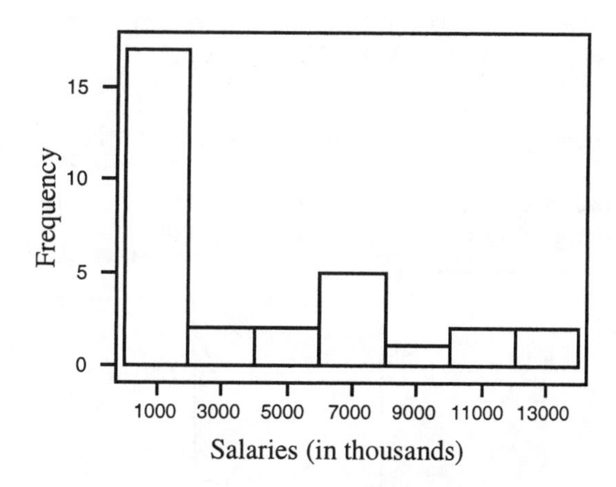

1.91 a) The data from each group need to be ordered to easily compute the five-number summaries. There are 20 observations in each group, so the location of the median is the (20 + 1) / 2 = 10.5, or halfway between the 10th and 11th observations. The first quartile is the median of the 10 observations below the median or midway between the fifth and sixth observations below the median. The third quartile is the median of the 10 observations above the median, or midway between the fifth and sixth observation above the median.

<u>Normal corn</u>

272 283 316 321 329 345 349 350 356 356
360 366 380 384 399 402 410 431 455 462

The median is M = (356 + 360) / 2 = 358 grams, the first quartile is Q_1 = (329 + 345) / 2 = 337 grams, and the third quartile is Q_3 = (399 + 402) / 2 = 400.5 grams. The minimum is 272 grams and the maximum is 462 grams.

<u>New Corn</u>

318 326 339 361 375 392 393 401 403 406
407 410 420 426 427 430 434 447 467 477

The median is M = (406 + 407) / 2 = 406.5 grams, the first quartile is Q_1 = (375 + 392) / 2 = 383.5 grams, and the third quartile is Q_3 = (427 + 430) / 2 = 428.5 grams. The minimum is 318 grams and the maximum is 477 grams.

The comparative boxplots below were produced by MINITAB. From the boxplots, it is apparent that the weight gains for the new corn tend to be larger than for the normal corn.

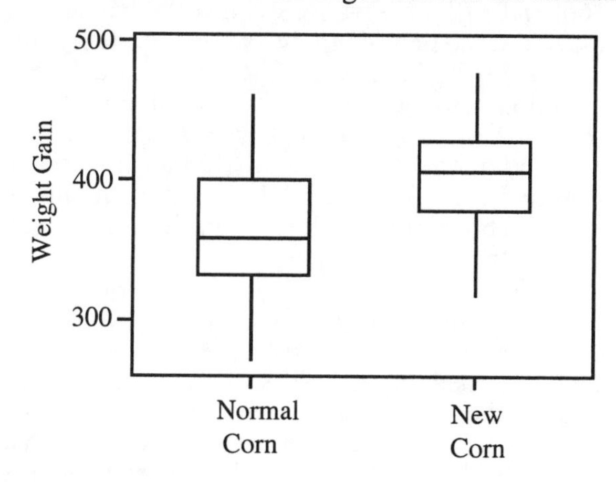

b) The means and standard deviations were computed using statistical software and are given in the table below. N corresponds to the sample size.

Variable	N	Mean	Median	StDev
normal	20	366.30	358.00	50.80
new	20	402.95	406.50	42.73

The mean weight gain for the new corn is 36.65 grams larger than for the normal corn.

1.93 a) The histogram and the boxplot are given below. Using either plot, it is clear the distribution is skewed to the right, with three high outliers.

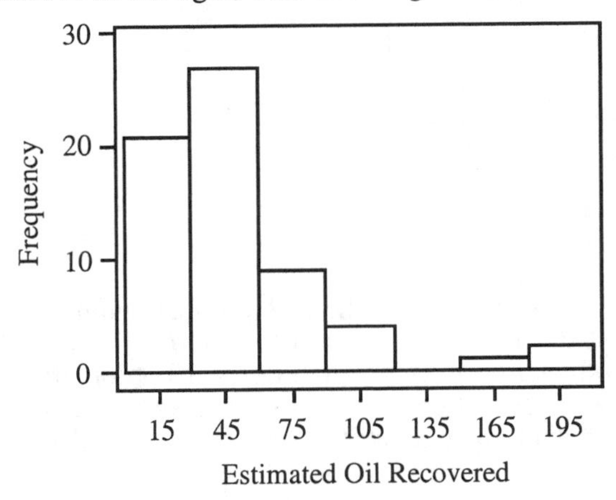

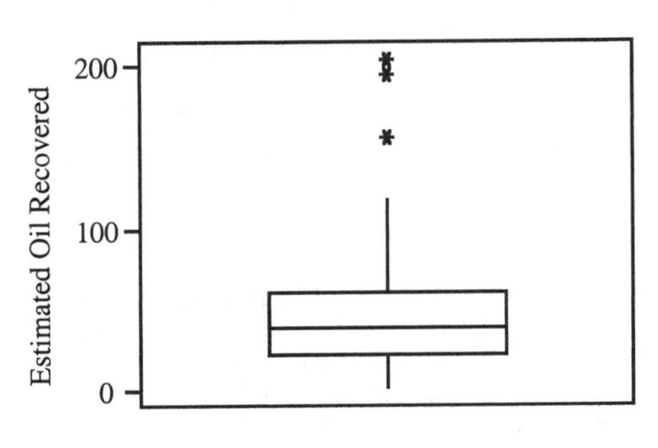

b) There are 64 observations, so the median is midway between the 32nd and 33rd observations which is $M = (37.7 + 37.9) / 2 = 37.8$ thousand barrels (see part [c]). The mean was computed using statistical software as $\bar{x} = 48.25$ thousand barrels. Because the distribution is skewed right and has three high outliers, we would expect the mean to be substantially larger than the median.

c) The 64 observations are ordered below.

2.0	2.5	3.0	7.1	10.1	10.3	12.0	12.1
12.9	14.7	14.8	17.6	18.0	18.5	20.1	21.3
21.7	24.9	26.9	28.3	29.1	30.5	31.4	32.5
32.9	33.7	34.6	34.6	35.1	36.6	37.0	37.7
37.9	38.6	42.7	43.4	44.5	44.9	46.4	47.6
49.4	50.4	51.9	53.2	54.2	56.4	57.4	58.8
61.4	63.1	64.9	65.6	69.5	69.8	79.5	81.1
82.2	92.2	97.7	103.1	118.2	156.5	196.0	204.9

The median was computed in part (b). The first quartile is the median of the 32 observations below the median or midway between the 16th and 17th observations below the median. The third quartile is the median of the 32 observations above the median, or midway between the 16th and 17th observations above the median. The first quartile is $Q_1 = (21.3 + 21.7) / 2 = 21.5$ thousand barrels, and the third quartile is $Q_3 = (58.8 + 61.4) / 2 = 60.1$ thousand barrels. The minimum is 2 thousand barrels and the maximum is 204.9 thousand barrels. The box containing the first quartile, the median and the third quartile is fairly symmetric, with the third quartile being slightly further from the median, indicating some right skewness. The maximum is much further from the median than the minimum, which could suggest right skewness or just a high outlier. The boxplot or the histogram show clearly the right skewness, but it is difficult to be certain of this from only the five-number summary.

1.95 We first need to determine exactly what information is being given to us. Scores less than 25 received a C, and this corresponds to the lowest 10% so 10% of the scores are below 25. Scores above 475 received an A and this corresponds to the highest 10%, so 10% of the scores are above 475. See the plot on the next page. By the symmetry, the mean of the curve must be the average of the 10th percentile and the 90th percentile, or $\mu = (25 + 475) / 2 = 250$. To compute the standard deviation, we know that the z value having 10% of the area to the left of it is -1.28. This is a "backward" normal calculation and is found by looking in the body of Table A for 0.10 and finding the corresponding $z = -1.28$. Thus, the standardized x score is -1.28. The difference in this problem is that usually we know μ and σ and are trying to find the value of x. This time, we know that $x = 25$, $\mu = 250$, and we are trying to find the value of σ. The equation to solve is

$$\frac{25-250}{\sigma}=-1.28\text{, or}$$

$$\sigma=\frac{25-250}{-1.28}=115.68$$

The mean is μ = 250 and the standard deviation is σ = 115.68. The picture below summarizes the information given.

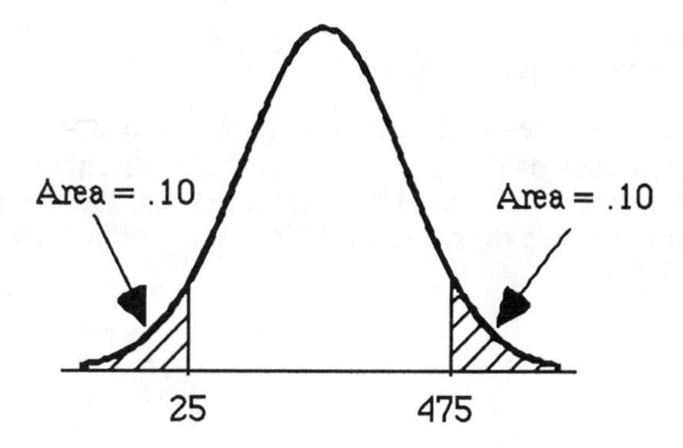

1.97 These data are similar to the Yankee salaries of Exercise 1.89 in that the range of the observations covers several orders of magnitude, although the problem is more severe in this example. When data have this feature, they are often transformed first. Attempts to construct stemplots, boxplots, or histograms for the entire original data set yield plots that show very little of the features of the data, other than the fact that they are strongly skewed to the right with an extreme outlier, as seen in the boxplot below.

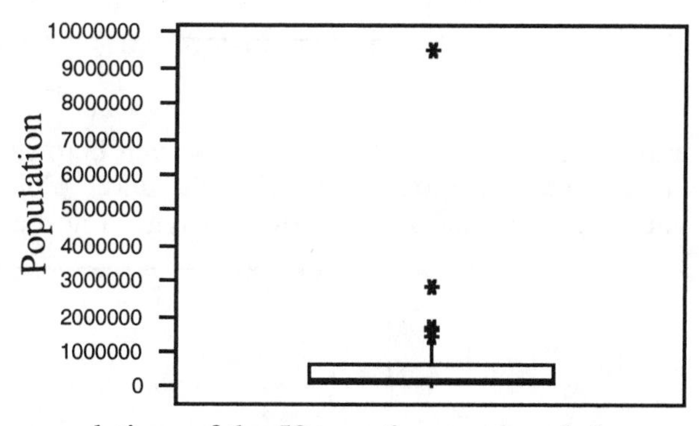

The ordered values of the populations of the 58 counties are given below.

1208	3555	9449	12853	13022	17130	17945
18804	20824	26453	27507	33828	35100	40554
44301	53234	54501	56039	58309	60219	78930
86265	92033	123109	124279	126518	129461	142361
156299	163256	168660	203171	210554	246681	247289
248399	255602	368021	394542	399347	401762	446997
458614	563598	661645	707161	753197	776733	799407
948816	1223499	1443741	1545387	1682585	1709434	2813833
2846289	9519338					

The mean of the populations is 583,994 and is clearly quite far from the center of the data. The median is 159,778. Division into three groups is fairly arbitrary at this point as you have not studied any principles of sampling. A natural thing to do would be to use simple numbers for cutoffs that also correspond to gaps in the data. For example, the large group might consist of the eight counties with populations over 1 million, and we would sample from all of these counties. The next break is not as clear but it could be taken at 100,000. Twenty-seven counties have populations between 100,000 and 1 million, and we could sample households from half of these counties. Finally, there are 23 counties with populations below 100,000 and a smaller fraction of these could be sampled.

1.99 The 25 observations generated are given below.

23.0203	37.2191	22.0657	24.2206	13.3446	20.1758	18.2885
15.4089	13.6384	19.2406	21.2975	22.7387	18.9497	19.0880
24.1798	25.7851	15.1241	14.2617	17.2885	16.8056	19.2017
23.2883	13.0503	21.5958	23.8547			

The mean of these 25 observations is $\bar{x} = 20.13$ and is quite close to μ, and the standard deviation of these 25 numbers is $s = 5.20$, which is quite close to σ.

Repeating this process another 19 times gives 20 values of $\bar{x}$ and 20 values of s. The stemplot and normal quantile plot of the 20 values of $\bar{x}$ are given below. The distribution is fairly symmetric with a center at about 20. Because of the small number of observations, the normal quantile plot is not that smooth, but it does not appear to deviate from a straight line in a systematic way.

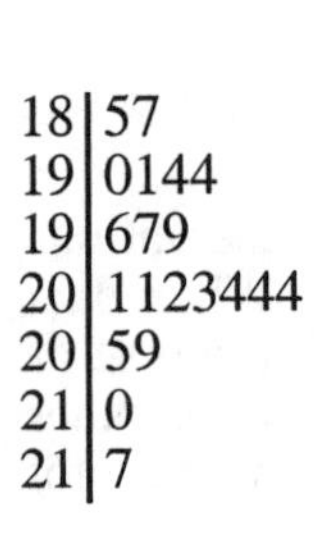

18 | 57
19 | 0144
19 | 679
20 | 1123444
20 | 59
21 | 0
21 | 7

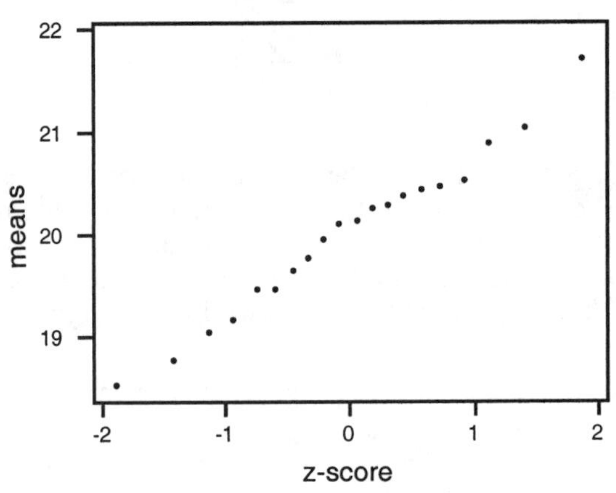

The stemplot and normal quantile plot of the 20 values of s are given below. The distribution looks slightly skewed to the right with a center close to 5. The normal quantile plot does not look nearly as much like a straight line as when plotting the 20 values of $\bar{x}$, suggesting the distribution of s is not normal.

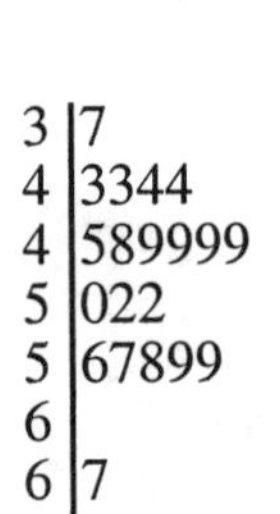

3 | 7
4 | 3344
4 | 589999
5 | 022
5 | 67899
6 |
6 | 7

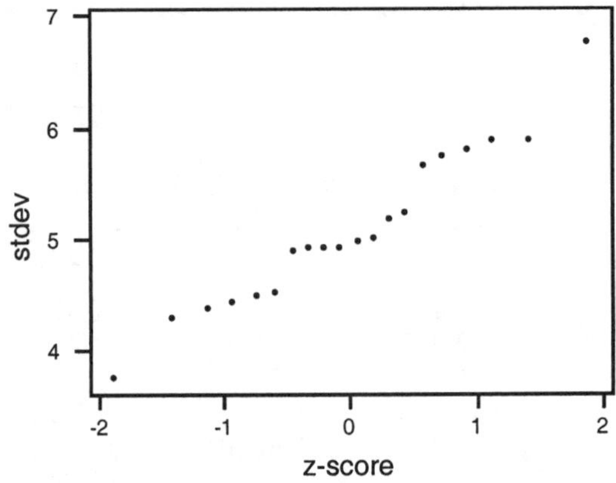

CHAPTER 2

EXAMINING RELATIONSHIPS

SECTION 2.1

OVERVIEW

Chapter 1 provides the tools to explore several types of variables one by one, but in most instances the data of interest are a collection of variables that may exhibit relationships among themselves. Typically, these relationships are more interesting than the behavior of the variables individually. To study these relationships, we must measure the variables on the same group of individuals. The first tool we consider for examining the relationship between variables is the **scatterplot.** Scatterplots show us two quantitative variables at a time, such as the weight of a car and its MPG (miles per gallon). Using colors or different symbols, we can add information to the plot about a third variable that is categorical in nature. For example, if in our plot we wanted to distinguish between cars with manual or automatic transmissions, we might use a circle to plot the cars with manual transmissions and a cross to plot the cars with automatic transmissions.

When drawing a scatterplot, we need to pick one variable to be on the horizontal axis (x axis) and the other to be on the vertical axis (y axis). When there is a **response variable** and an **explanatory variable,** the explanatory variable is always placed on the x axis. In cases where there is no explanatory-response variable distinction, either variable can go on the x axis. After drawing the scatterplot by hand or using a computer, the scatterplot should be examined for an **overall pattern,** which may tell us about any relationship between the variables and for **outliers** or other deviations from it. You should be looking for the **direction, form,** and **strength** of the overall pattern. In terms of form, look for **linear relationships** (the points show a straight-line pattern), curved relationships, or clusters of points. In terms of direction, **positive association** occurs when the variables take on high values together, while **negative association** occurs if one variable takes high values when the other takes on low values. If the plotted values seem to form a line and the line slopes up to the right, the association is positive; if the line slopes down to the right, the association is negative. In terms of strength, see how close the points in the scatterplot lie to a simple form such as a line. The closer to this simple form, the stronger the relationship.

APPLY YOUR KNOWLEDGE

2.1 a) It is natural to think that the amount of time a student spends studying for a statistics exam affects the grade on the exam. Thus, we should probably view the amount of time spent studying as the explanatory variable and grade on the exam as the response variable.

b) We would expect weight and height to be associated but it is not obvious that one "causes" the other. Thus, it is reasonable to simply explore the relationship between the two variables.

c) One generally assumes that the amount of yearly rainfall affects the yield of a crop. Thus, we should probably view the amount of yearly rainfall as the explanatory variable and the yield of a crop as the response variable.

d) It is possible that an employee who takes many sick days is an irresponsible employee and that this will result in a lower salary. However, many factors affect salary and many factors affect the number of sick days one takes. Thus, it may be most reasonable to simply explore the relationship between these two variables.

e) It is not unreasonable to believe that the economic class of a father will have some effect on the economic class of the son. Thus, one should probably take the economic class of the father as the explanatory variable and the economic class of the son as the response variable.

2.3 There are two variables here. One is the type of hand wipe used and the other is whether or not a person's skin appears abnormally irritated. Both are categorical variables. The company believes that hand wipes can irritate the skin, so the type of hand wipe is the explanatory variable and whether or not the skin appears abnormally irritated is the response.

2.5 a) The unusual observation is in the upper right corner of the plot. The city gas mileage is a little larger than 60, approximately 62 miles per gallon. The highway mileage is slightly less than 70, approximately 68 miles per gallon.

b) The pattern is roughly linear. This is not surprising. If you have shopped for a car, you have noticed that highway gas mileage tends to between 5 and 10 miles per gallon greater than city gas mileage. This leads to a roughly linear relationship.

c) It appears to fall roughly on the line defined by the pattern of the remaining points, so it seems to fit the overall relationship portrayed by the other two-seater cars. The only difference is its much higher mileage.

2.7 a) We would expect speed to affect fuel consumption, so speed would be the explanatory variable and would be plotted on the *x* axis. A scatterplot of the data is as follows.

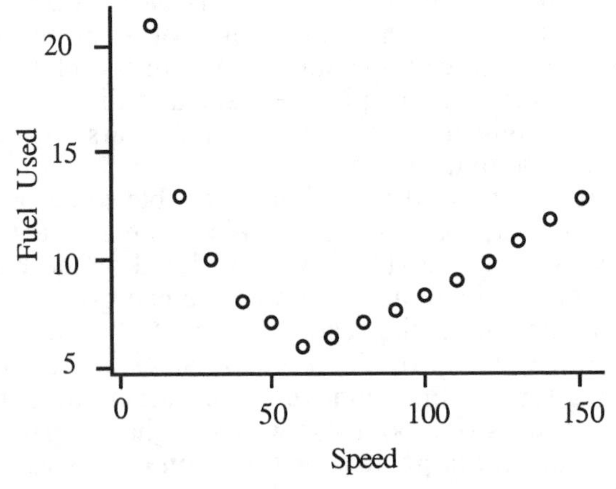

b) The plot shows a curved relationship. Fuel used first decreases as speed increases, and then at about 60 km/h increases as speed increases. This is not surprising. At very slow speeds and at very high speeds, engines are very inefficient and use more fuel, while at moderate speeds engines are more efficient and use less fuel.

c) Variable are positively associated when both take on high values together and both take on low values together. Negative association occurs when high values of one variable are associated with low values of the other. In the scatterplot, both low and high speeds correspond to high values of fuel used, so we cannot say that the variables are positively or negatively associated.

d) The points appear to lie close to a simple curved form, so we would say the relationship is reasonably strong.

SECTION 2.1 EXERCISES

2.9 a) A positive association between decline and duration means that the longer the duration of the bear market, the greater the decline. The plot shows a positive association. The points corresponding to the bear markets with the longer durations also tend to be those with the largest declines.

b) The relationship is roughly linear. A straight line that slopes up from left to right describes the general trend reasonably well. The association is not very strong. The points do not appear to be tightly clustered about a straight line.

c) The bear market with the greatest decline corresponds to the point that is highest in the vertical direction. This bear market appears to have had an approximate decline of 48% and a duration of about 21 months.

2.11 a) Household income is the sum of the incomes of the individuals in the household. If a person has a high income, their household income will also be high, and so personal and household incomes are positively associated. Thus, we would expect mean personal income for each state to be positively associated with median household income per state. Because household income is the sum of the personal incomes of the members of the household, household incomes will always be greater than or equal to the incomes of the individuals in the household. We would therefore expect median household income in a state to be larger than mean personal income in the state.

b) If the distribution of incomes (personal and household) in a state is strongly right skewed, then we know that the mean will be larger than the median. Thus, although mean household income must be larger than mean personal income, and median household income must be larger than median personal income, *median* household income could be smaller than *mean* personal income.

c) The overall pattern of the plot is roughly linear. The variables are positively associated. The strength of the relationship (ignoring outliers) is moderately strong because the points aren't too widely scattered from a line.

d) The two points with the largest mean incomes appear to be outliers and correspond to the District of Columbia and Connecticut. The District of Columbia is a city with a few very wealthy inhabitants and many poorer inhabitants. Many very wealthy people that work in New York City live in Connecticut resulting in a high mean income. We would expect the distribution of both personal and household incomes to be strongly right skewed.

2.13 a) A scatterplot of these data follows.

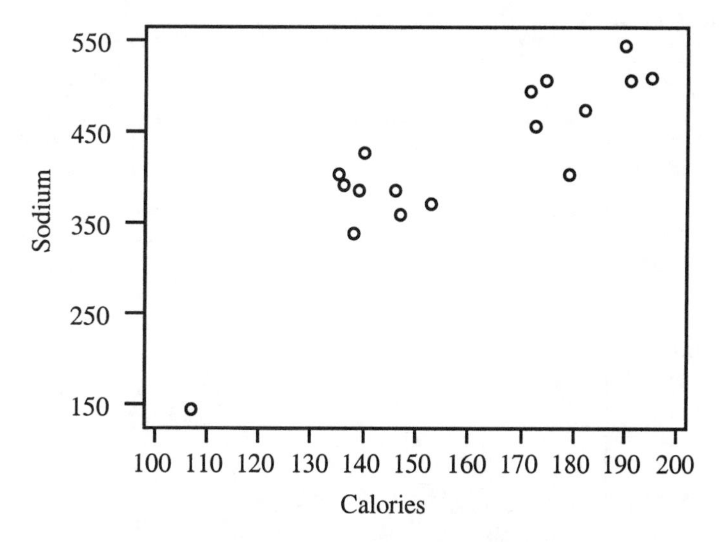

The overall pattern is linear, the association is positive, and the strength is moderately strong. Hot dogs that are high in calories are generally high in sodium.

b) Because this brand positions itself as a diet hot dog, we would expect this brand to be low in calories. Brand 13 has the fewest calories, so this probably corresponds to the brand "Eat Slim Veal Hot Dogs.".

2.15 a) A scatterplot is given below. Because business starts is taken as the explanatory variable, it is represented by the horizontal axis.

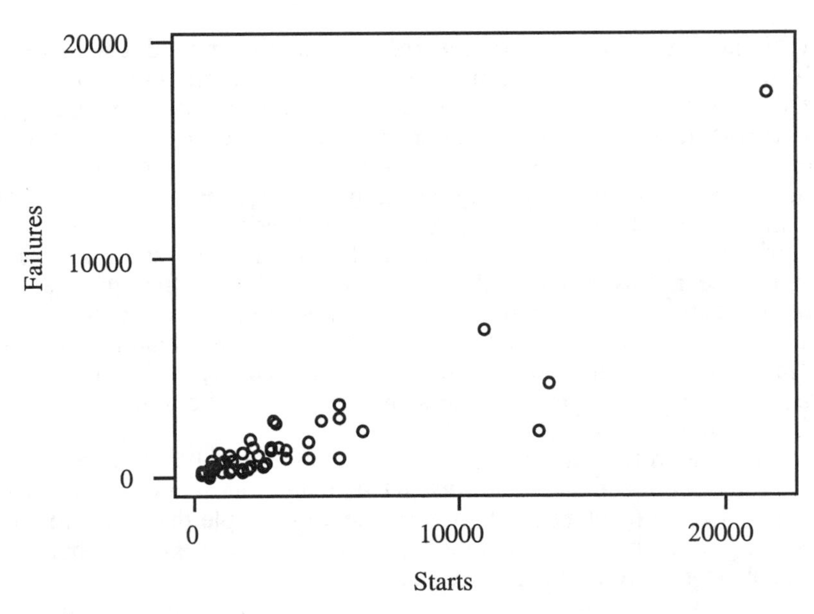

b) The association is positive because states with a large number of business starts tend to have a large number of business failures, and states with a small number of business starts tend to have a small number of business failures.

c)

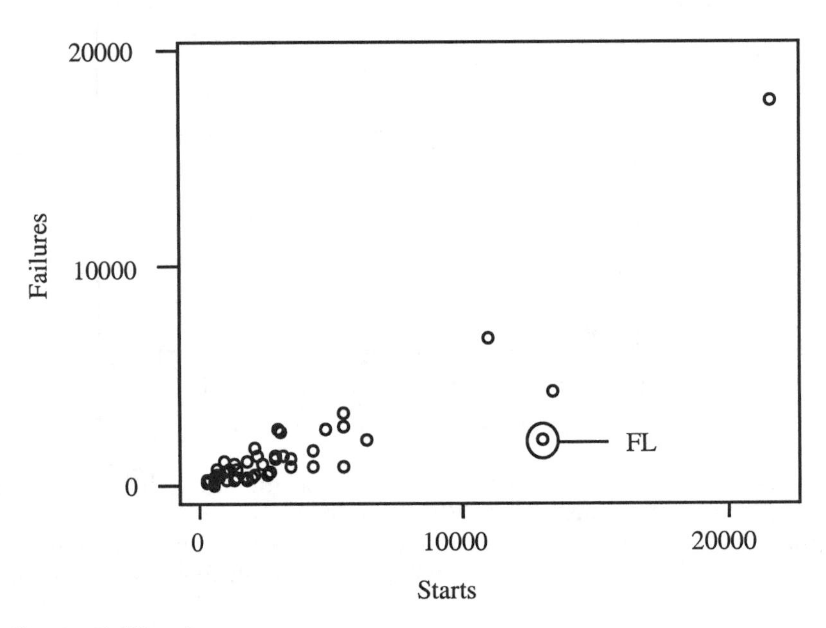

d) The outlier is California.

e) The four states outside the cluster are California, Florida, New York, and Texas. These states are scattered throughout the country. They are the most populous states in the country.

2.17 a) The plot is as follows. Means are represented by the plotting symbol x.

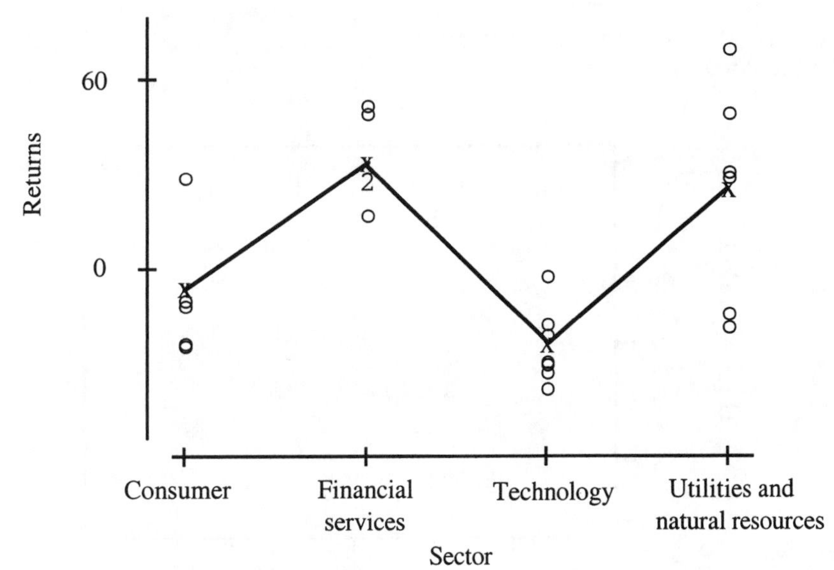

b) Financial services and Utilities and natural resources were good places to invest, based on the data.

c) Market sector is categorical so we cannot speak of positive or negative association between market sector and total return.

SECTION 2.2

OVERVIEW

Scatterplots provide a visual tool for looking at the relationship between two variables. Unfortunately our eyes are not good tools for judging the strength of the relationship. Changes in the scale or the amount of white space in the graph can easily change our judgment as to the strength of the relationship. **Correlation** is a numerical measure we will use to show the strength of **linear association.**

The correlation can be calculated using the formula

$$r = \frac{1}{n-1}\Sigma\left(\frac{x_i - \bar{x}}{s_x}\right)\left(\frac{y_i - \bar{y}}{s_y}\right)$$

where $\bar{x}$ and $\bar{y}$ are the respective means for the two variables X and Y, and s_x and s_y are their respective standard deviations. In practice, you will probably be computing the value of r using computer software or a calculator that finds r from entering the values of the x's and y's. When computing a correlation coefficient, there is no need to distinguish between the explanatory and response variables, even in cases where this distinction exists. The value of r will not change if we switch x and y.

When r is positive, it means that there is a positive linear association between the variables and when it is negative there is a negative linear association. The value of r is always between 1 and –1. Values close to 1 or –1 show a strong association while values near 0 show a weak association. As with means and standard deviations, the value of r is strongly affected by outliers. Their presence can make the correlation much different than what it might be with the outlier removed. Finally, remember that the correlation is a measure of straight line association. There are many other types of association between two variables, but these patterns will not be captured by the correlation coefficient.

APPLY YOUR KNOWLEDGE

2.19 a) In this case, neither variable is necessarily the explanatory variable. We, therefore, arbitrarily put City MPG on the horizontal axis.

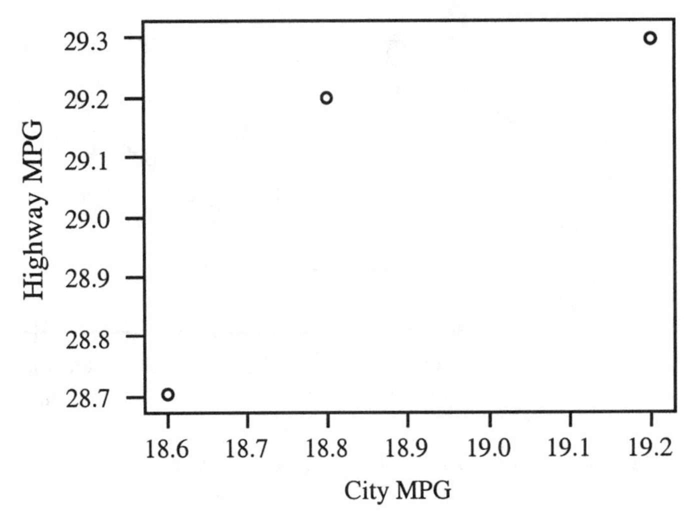

b) We summarize the step-by-step calculations below.

	City(x)	Highway(y)	$x_i - \bar{x}$	$(x_i - \bar{x})^2$	$\frac{x_i - \bar{x}}{s_x}$	$y_i - \bar{y}$	$(y_i - \bar{y})^2$	$\frac{y_i - \bar{y}}{s_y}$	$\left(\frac{x_i - \bar{x}}{s_x}\right)\left(\frac{y_i - \bar{y}}{s_y}\right)$
	18.6	28.7	− 0.27	0.0729	0.8838	−0.37	0.1369	− 1.1509	1.0172
	19.2	29.3	0.33	0.1089	1.0802	0.23	0.0529	0.7154	0.7728
	18.8	29.2	− 0.07	0.0049	− 0.2291	0.13	0.0169	0.4044	− 0.0926
Total	56.6	87.2	− 0.01	0.1867		− 0.01	0.2067		1.6974

From the totals of the x and y columns, we see that $\bar{x}$ = 56.6/3 = 18.87 and $\bar{y}$ = 87.2/3 = 29.07. From the totals of the $(x_i - \bar{x})^2$ and $(y_i - \bar{y})^2$ columns we see that s_x^2 = 0.1867/2 = 0.09335 and s_y^2 = 0.2067/2 = 0.10355. Thus s_x = 0.3055 and s_y = 0.3215. From the entries in the columns labeled $\frac{x_i - \bar{x}}{s_x}$ and $\frac{y_i - \bar{y}}{s_y}$, we finally compute

$$r = \frac{1}{n-1}\sum\left(\frac{x_i - \bar{x}}{s_x}\right)\left(\frac{y_i - \bar{y}}{s_y}\right) = \frac{1}{2}[(-0.8838 \times -1.1509) + (1.0802 \times 0.7154) + (-0.2291 \times 0.4044)]$$

$$= \frac{1.0172 + 0.7728 - 0.0926}{2} = 0.8487.$$

c) The association is positive and is reasonably strong.

2.21 a) The correlation r is clearly positive but not near 1. High values of duration go with high values of decline so the association is positive. The points are not tightly clustered about a line, so the correlation is not near 1. Figure 2.6 in the text looks most like the plot with correlation r = 0.5 in Figure 2.10 of the text.

b) The correlation in Figure 2.2 is closer to 1 than the correlation in Figure 2.6. The points in Figure 2.2 appear to lie more closely along a line than do those in Figure 2.6.

2.23 A scatterplot of mileage versus speed is given below, with speed on the horizontal axis.

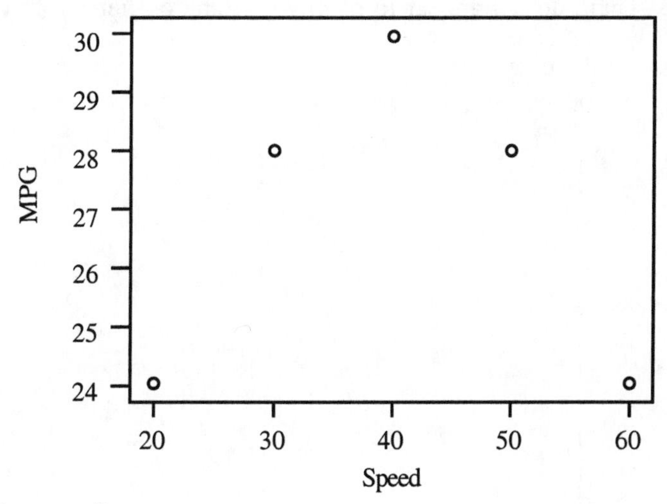

To compute the correlation we first compute

	speed(x)	MPG(y)	$x_i - \bar{x}$	$(x_i - \bar{x})^2$	$\frac{x_i - \bar{x}}{s_x}$	$y_i - \bar{y}$	$(y_i - \bar{y})^2$	$\frac{y_i - \bar{y}}{s_y}$	$\left(\frac{x_i - \bar{x}}{s_x}\right)\left(\frac{y_i - \bar{y}}{s_y}\right)$
	20	24	−20	400	−1.27	−2.8	7.84	−1.04	1.32
	30	28	−10	100	−0.63	1.2	1.44	0.45	−0.28
	40	30	0	0	0.00	3.2	10.24	1.19	0.00
	50	28	10	100	0.63	1.2	1.44	0.45	0.28
	60	24	20	400	1.27	−2.8	7.78	−1.04	−1.32
Totals	200	134		1000			28.80		0.00

Notice that $\bar{x} = 40$, $\bar{y} = 26.8$, $s_x^2 = 250$, $s_y^2 = 7.2$, $s_x = 15.81$, $s_y = 2.68$, and $r = 0$. Even though the plot shows a clear curved relationship between speed and MPG (in fact, they appear to be related by the equation MPG = 30 − |Speed - 40|/10 − [(Speed − 40)/10]^2), it is not a straight line relationship and r measures the strength of a straight line relationship. Looking at the data, high values of speed and low values of speed go with low values of MPG, so there is no linear association.

SECTION 2.2 EXERCISES

2.25 A scatterplot with EAFE on the horizontal axis follows.

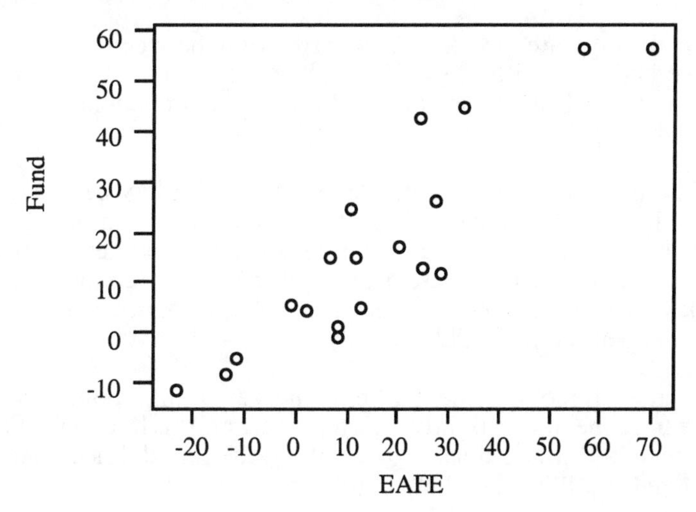

There is a straight line pattern that is fairly strong. The correlation, computed from statistical software, is $r = 0.898$. There do not appear to be any extreme outliers from the straight-line pattern.

2.27 a) The scatterplot of the data follows.

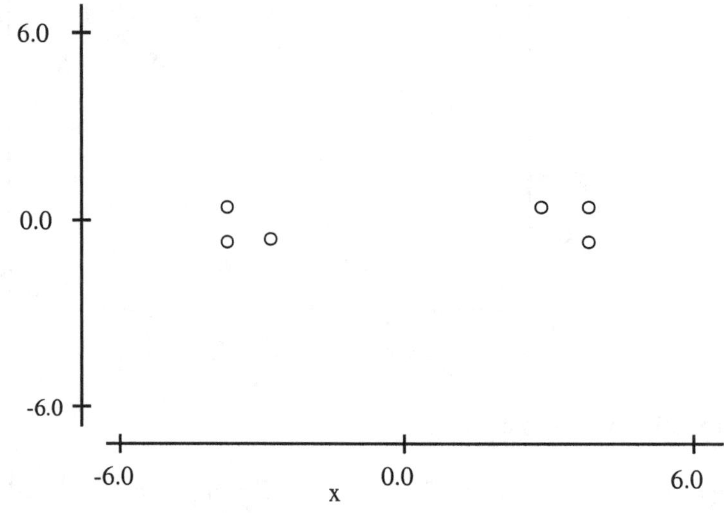

b) The scatterplot follows.

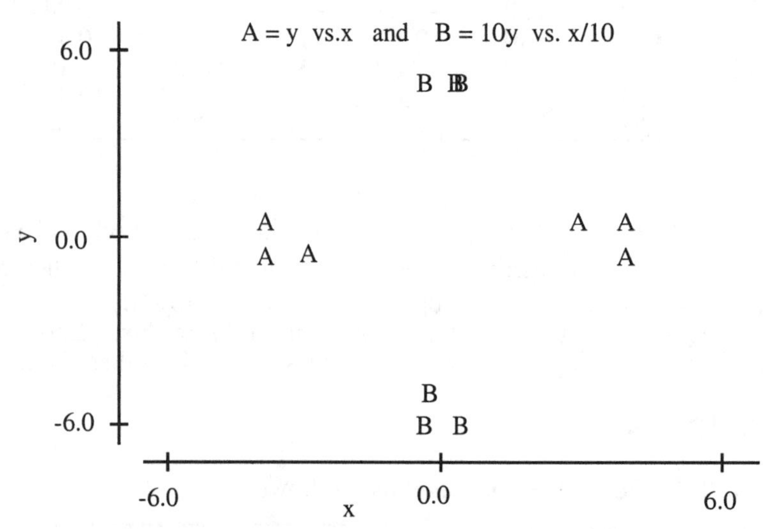

c) Using statistical software, we find the correlation between x and y to be 0.253 and the correlation between x^* and y^* also to be 0.253. They are equal. This is not surprising because (according to Fact 3 on page 110 of the text) r does not change when we change the units of measurement of x, y, or both.

2.29 The magazine's report says that high pay tends to go with low performance and low pay with high performance. This implies that there is a negative association (and hence a negative correlation) between compensation of corporate CEOs and the performance of their company's stock, not that there is no correlation. A more accurate statement might be that in companies that compensate their CEOs highly, the stock is just as likely to perform well as to not perform well. Likewise for companies that do not compensate their CEOs highly.

2.31 In Figure 2.8, after removing the outliers, the remaining points show a stronger linear relationship, i.e., appear to be more tightly clustered around a line. In Figure 2.2, the outlier accentuates the linear trend because it appears to lie along the line defined by the other points. After its removal, the linear trend is still evident but is not as pronounced.

2.33 a) Gender is a categorical variable. Correlation requires both variables to be quantitative

b) A correlation of 1.09 is not possible. The correlation between two quantitative variables is always a number between –1 and +1.

c) The correlation has no unit of measurement (such as bushels). It is just a number.

SECTION 2.3

OVERVIEW

If a scatterplot shows a linear relationship that is moderately strong as measured by the correlation, we would like to draw a line on the scatterplot to summarize the relationship. In the case where there is a response and an explanatory variable, the **least-squares regression** line often provides a good summary of this relationship. A straight line relating y to x has the form $y = a + bx$ where b is the **slope** of the line and a is the **intercept.** The least-squares regression line is the straight line $\hat{y} = a + bx$, which minimizes the sum of the squares of the vertical distances between the line and the observed values y. The formula for the slope of the least squares line is

$$b = r\frac{s_y}{s_x}$$

and for the intercept is $a = \bar{y} - b\bar{x}$, where $\bar{x}$ and $\bar{y}$ are the means of the x and y variables, s_x and s_y are their respective standard deviations and r is the value of the correlation coefficient. Typically, the equation of the least-squares regression line is obtained by computer software or a calculator with a regression function.

Regression can be used to predict the value of y for any value of x. Just substitute the value of x into the equation of the least-squares regression line to get the predicted value for y. Predicting values of y for x values in the range of those x's we observed is called interpolation and is fine to do. However, be careful about extrapolation (using the line for prediction beyond the range of x values covered by the data). Extrapolation may lead to misleading results if the pattern found in the range of the data does not continue outside the range.

Correlation and regression are clearly related as is seen from the equation for the slope, b. However, more important is how r^2, the square of the correlation coefficient, measures the strength of the regression. r^2 tells us the fraction of the variation in y that is explained by the regression of y on x. The closer r^2 is to 1 the better the regression describes the relation between x and y.

APPLY YOUR KNOWLEDGE

2.35 a) Minitab gives the equation of the least-squares regression as $\hat{y} = 1.0892 + 0.188999x$.

b) We find $\bar{x} = 22.31$, $s_x = 17.74$, $\bar{y} = 5.306$, $s_y = 3.368$, and $r = 0.995$. Thus, we compute the slope to be $b = r\frac{s_y}{s_x} = 0.995 \times \frac{3.368}{17.74} = 0.1889$ and the intercept to be $a = \bar{y} - b\bar{x} = 5.306 - 0.1889 \times 22.31 = 1.0916$. The differences from the results in (a) are due to roundoff error.

2.37 a) The percent of the observed variation in yearly changes in the index that is explained by a straight-line relationship with the change during January is 35.5% because $r^2 = (0.596)^2 = 0.355$.

b) We compute the slope to be $b = r\frac{s_y}{s_x} = 0.596\frac{15.35}{5.36} = 1.707$ and the intercept to be $a = \bar{y} - b\bar{x} = 9.07\% - 1.707 \times 1.75\% = 6.08\%$. Thus, the equation of the least-squares line for predicting full-year change from January change is $\hat{y} = 6.08\% + 1.707x$.

c) Using the regression line we would predict the change in the index in a year in which the index rises 1.75% to be $\hat{y} = 6.08\% + 1.707(1.75\%) = 9.08\%$. We could have given this answer (up to roundoff error) without doing the calculation because $\bar{x} = 1.75\%$ and the least-squares regression line passes through the point $(\bar{x}, \bar{y})$. Thus, we would predict $\hat{y} = \bar{y} = 9.07\%$ when $x = \bar{x} = 1.75\%$.

2.39 a) The scatterplot with the regression line included follows.

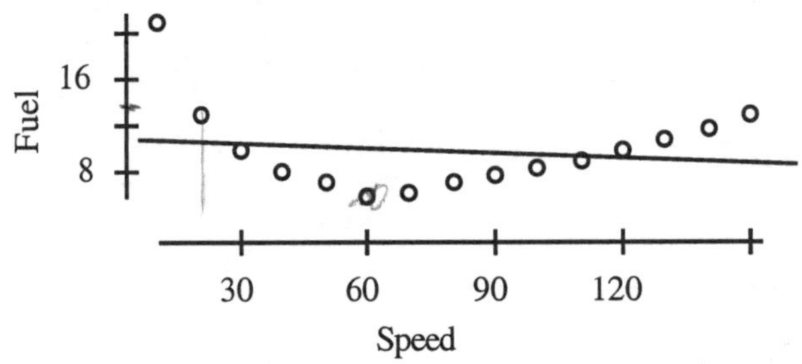

b) We would not use the regression line to predict y from x. The scatterplot suggests that the relation between y and x is a curved relationship, not a straight-line relationship. Thus, the least-squares regression line does not adequately describe the relationship.

c) Using a calculator, we found the sum to be –0.01.

d) A plot of the residuals against x follows.

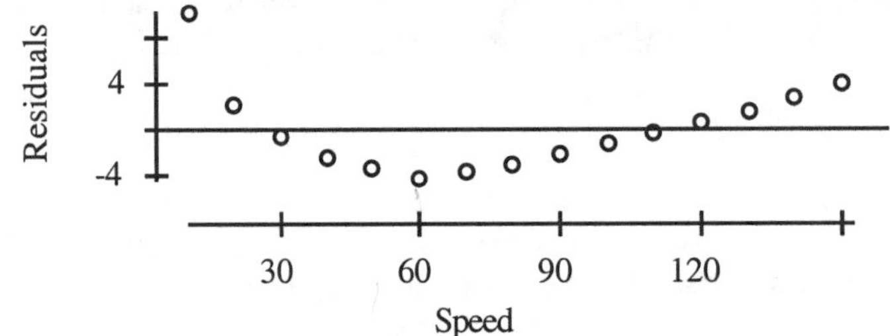

We notice that the residuals show the same curved pattern around the horizontal line at height zero as the data points do around the regression line in the scatterplot in part (a).

2.41 a) The equation of the least-squares regression line, leaving out observation 1 is $\hat{y} = 2.95918 + 6.74776x$. This line and the regression line for all the data are plotted in the scatterplot below.

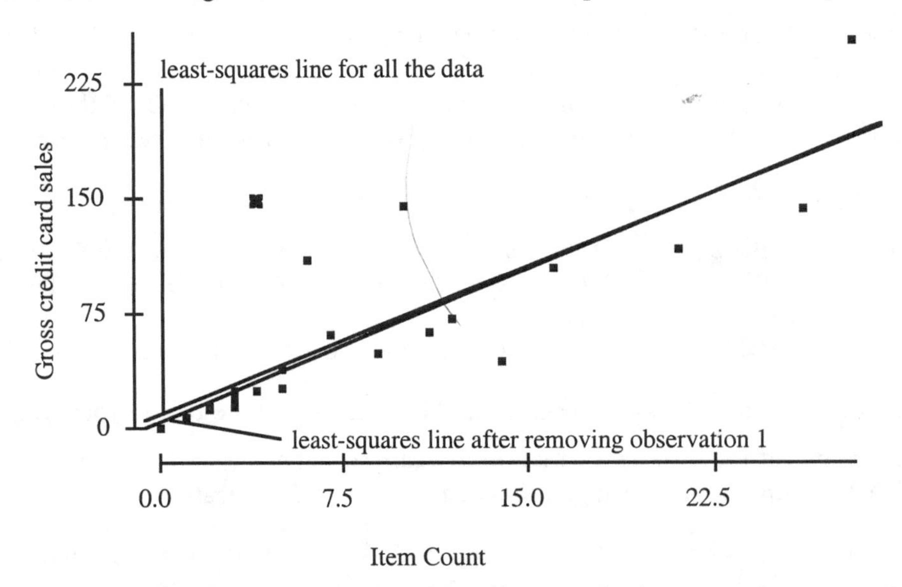

Although removal of observation 1 has a noticeable effect on the intercept (its removal changes the intercept from 14.3908 to 2.95918) it has very little effect on the slope and the two lines are difficult to distinguish in the plot. Thus, I would not call observation 1 very influential.

b) Before removing observation 1 $r^2 = 0.660$, while after removing observation 1 $r^2 = 0.779$. We see that r^2 increases after removing observation 1. This is not surprising. Observation 1 is an outlier and after its removal the remaining points appear more tightly clustered about the least-squares regression line, resulting in a larger value of r^2.

SECTION 2.3 EXERCISES

2.43 a) Presently (at week 0), there are 96 DVD players in inventory. Each week, this number is reduced by exactly 4 (the number sold that week). Thus, the number y in inventory after x weeks must be $y = 96 - 4x$.

b) A graph of this line between week 0 and week 10 follows.

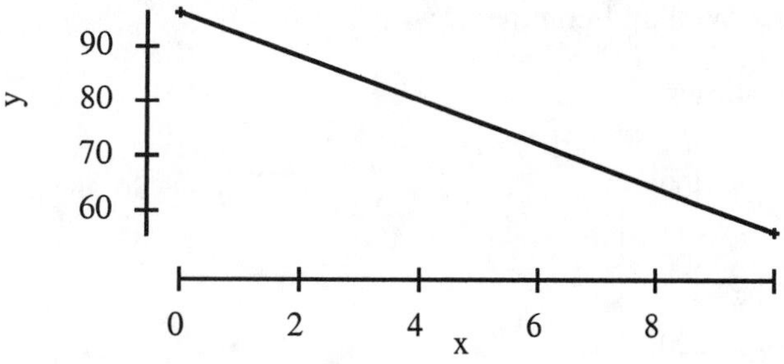

c) No. After 25 weeks the equation predicts there will be $y = 96 - 4(25) = 96 - 100 = -4$. It is not possible to have less than 0 in inventory.

2.45 If the correlation has increased to 0.8, then there is a stronger relationship between American and European stocks. In particular, when American stocks have gone down, European stocks have tended to also go down. Thus, European stocks have not provided much protection against losses in American stocks, because European stocks have had a greater tendency to go down also.

2.47 a) Using the information given, we compute the slope to be $b = r\frac{s_y}{s_x} = 0.6285\frac{11.20}{8.20} = 0.858$ and the intercept to be $a = \bar{y} - b\bar{x} = 24.67 - 0.858 \times 10.73 = 15.464$. Thus, the equation of the least-squares line for predicting decline from duration is $\hat{y} = 15.464 + 0.858x$.

b) The percent of the observed variation in these declines that can be attributed to the linear relationship between decline and duration is determined by $r^2 = (0.6285)^2 = 0.395$ and so is 39.5%.

c) From the equation of the least-squares regression line the predicted decline for a bear market with duration $x = 15$ is $\hat{y} = 15.464 + 0.858(15) = 28.334\%$. The observed decline for this particular bear market is 14% so the residual is observed – predicted = 14% – 28.334% = –14.334%.

2.49 a) A plot of the data follows.

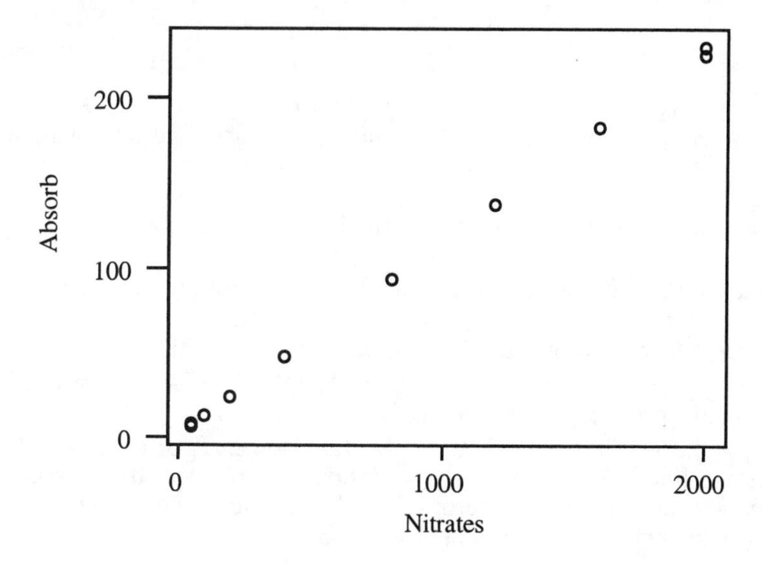

Using software, we compute the correlation to be $r = 0.999993$, so the calibration does not need to be done again.

b) Using software, we find the equation of the least-squares line for predicting absorbance (y) from concentration (x) is $\hat{y} = 1.6571 + 0.113301x$. If the lab analyzed a specimen with 500 milligrams of nitrates per liter, we would expect the absorbance to be (based on the least-squares line) $\hat{y} = 1.6571 + 0.113301(500) = 58.3$. We would expect the predicted absorbance to be very accurate based on the plot (in which the points appear to lie almost perfectly on a straight line) and the correlation (which is 0.999993).

2.51 Do the applet on the Web and comment.

2.53 a) The scatterplot follows.

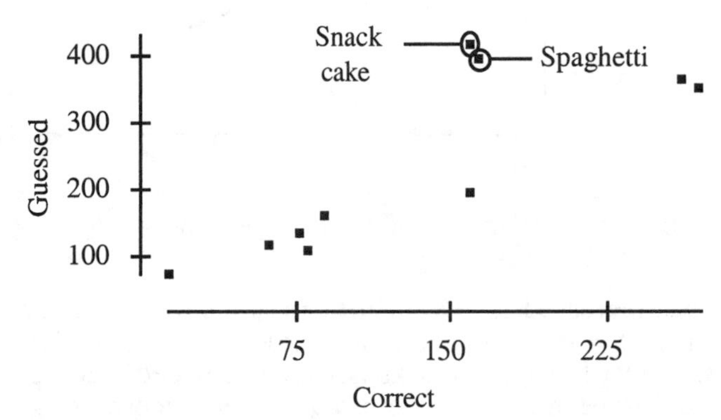

b) The least-squares regression line of guessed calories (y) on true calories (x) for all 10 points is = $58.5879 + 1.30356x$. The least-squares regression line of guessed calories on true calories after leaving out spaghetti and snack cake is $\hat{y} = 43.8814 + 1.14721x$.

c) The plot follows.

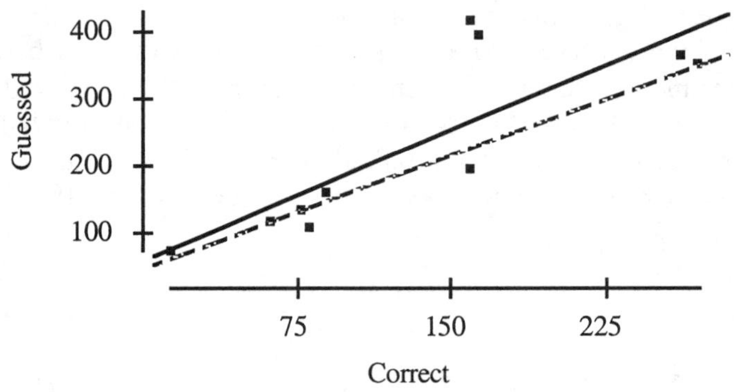

The points, taken together, are moderately influential. The regression line changes somewhat, but not dramatically.

2.55 a) If we let x denote pre-exam totals and y final exam scores, then we know $\bar{x} = 280$, $s_x = 30$, $\bar{y} = 75$, $s_y = 8$, and $r = 0.6$. Using this information, we compute the slope to be $b = r\frac{s_y}{s_x} = 0.6\frac{8}{30} = 0.16$ and the intercept to be $a = \bar{y} - b\bar{x} = 75 - 0.16 \times 280 = 30.2$.

b) Using the regression line, we would predict Julie's final exam score to be $30.2 + 0.16 \times 300 = 78.2$ (which we should probably round off to 78).

c) $r^2 = (0.6)^2 = 0.36$. Thus, only 36% of the variation in final exam scores is explained by the least-squares regression line. With 64% of the variation unexplained, the least-squares regression line would not be considered an accurate predictor of final exam score and Julie's actual score could have been much higher (or lower) than the predicted value of 78.

2.57 r^2 measures the fraction of the variance of one variable that is explained by least-squares regression on the other variable. The value 64% given in the problem refers to this fraction (expressed as a percent) and so in this case $r^2 = 0.64$. The correlation between number of days missed and salary increase is the square root of 0.64. It must be the negative square root, because smaller raises tended to be given to those who missed more days and so the variables are negatively associated. Thus, $r = -\sqrt{0.64} = -0.8$.

2.59 Octavio scored 10 points above the mean on his midterm, so his midterm exam score is $x = \bar{x} + 10$. If we substitute this into the equation of the least-squares regression line, we would predict his final exam score to be $46.6 + 0.41(\bar{x} + 10) = 46.6 + 0.41\bar{x} + 4.1$. We know that the least-squares regression line passes through the point $(\bar{x}, \bar{y})$ so $\bar{y} = 46.6 + 0.41\bar{x}$. Hence we would predict Octavio's final exam score to be $\bar{y} + 4.1$, i.e., we would predict Octavio to score 4.1 points above the class mean on the final exam.

2.61 a) Using statistical software, we find that the equation of the least-squares regression line for predicting the number of Target stores from the number of Wal-Mart stores is

Number of Target Stores = –0.525483 + 0.382345 × (Number of Wal-Mart Stores)

b) Because there are 254 Wal-Mart stores in Texas, the predicted number of Target stores in Texas is

Number of Target Stores = –0.525483 + 0.382345 × (254) = 96.590147

There are actually 90 Target stores in Texas so the residual is

residual = observed - predicted = 90 – 96.590147 = – 6.590147

c) Using statistical software, we find that the equation of the least-squares regression line for predicting the number of Wal-Mart stores from the number of Target stores is

Number of Wal-Mart Stores = 30.3101 + 1.13459 × (Number of Target Stores)

d) Because there are 90 Target stores in Texas, the predicted number of Wal-Mart stores in Texas is

Number of Wal-Mart Stores = 30.3101 + 1.13459 × (90) = 132.4232

There are actually 254 Wal-Mart stores in Texas, so the residual is

residual = observed – predicted = 254 – 132.4232 = 121.5768

SECTION 2.4

OVERVIEW

Plots of the **residuals,** which are the differences between the observed and predicted values of the response variable, are very useful for examining the fit of a regression line. Features to look for in a residual plot are unusually large values of the residuals (outliers), nonlinear patterns, and uneven variation about the horizontal line through zero (corresponding to uneven variation about the regression line).

The effects of **lurking variables,** variables other than the explanatory variable which may also affect the response, can often be seen by plotting the residuals versus such variables. Linear or nonlinear trends in such a plot are evidence of a lurking variable. If the time order of the observations is known, it is good practice to plot the residuals versus time order to see if time can be considered a lurking variable.

Influential observations are individual points whose removal would cause a substantial change in the regression line. Influential observations are often outliers in the horizontal direction.

Correlation and regression must be interpreted with caution. Plots of the data, including residual plots, help make sure the relationship is roughly linear and help to detect outliers and influential observations. The presence of lurking variables can make a correlation or regression misleading. Always remember that association, even strong association, does not imply a cause-and-effect relationship between two variables.

A correlation based on averages is usually higher than if we had data for individuals. A correlation based on data with a restricted range is often lower than would be the case if we had observed the full range of the variables.

APPLY YOUR KNOWLEDGE

2.63 We would expect the correlation for individual stocks to be lower in general. Correlations based on averages (such as the stock index) are usually too high when applied to individuals.

2.65 The reasoning assumes the correlation between the number of firefighters at a fire and the amount of damage done is due to causation, and that a greater number of firefighter causes more damage. Correlation does not imply causation. It is more plausible that a lurking variable, namely the size of the fire, is behind the correlation. Larger fires cause more damage and require more firefighters to combat them.

2.67 No. Correlation does not imply causation, i.e., the existence of a correlation between the size of a hospital and the median number of days that patients remain in the hospital does not imply that the size of the hospital causes the length of stay. It is possible that a lurking variable may explain the correlation. For example, larger hospitals often have better facilities and equipment that enables them to care for patients with very serious illnesses or injuries. The most seriously ill patients may be sent to larger hospitals rather than smaller hospitals, and this could account for the observed correlation.

SECTION 2.4 EXERCISES

2.69 There are several possible lurking variables. Intelligence may be a lurking variable. More intelligent students may have greater success in math and hence may be more inclined to take math courses in high school. More intelligent students are also more likely to be successful in college. Hard work may be a lurking variable. Students who work harder may be more successful in the classroom, including math, and thus may be more inclined to take math courses. Students who work harder are also more likely to be successful in college. Support from parents may be a lurking variable. Students who come from home environments that value success in school may be more successful in the classroom, including math, and thus may be more inclined to take math courses. These same students may also be more successful in college for the same reason. More generally, factors that may lead to students taking more math in high school may also lead to more success in college. If lurking variables are present, then requiring students to take algebra and geometry may have little effect on success in college. Remember, correlation does not imply causation.

2.71 a) A scatterplot of beef consumption y against beef price x follows.

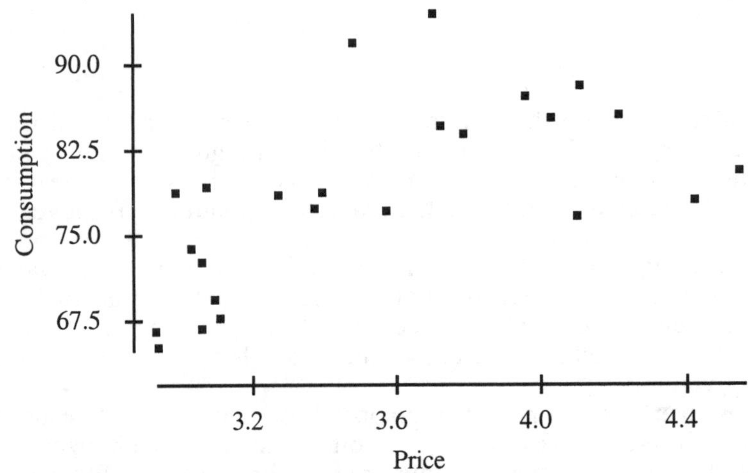

If the consumption of an item falls when price rises, we should see a negative association. However, the plot shows a modest positive association.

b) Using statistical software, we find the equation of the least-squares regression line to be $\hat{y} = 44.6954 + 9.59163x$. This line is displayed on the plot that follows. The proportion of the variation in beef consumption that is explained by the regression on beef price is just the square of the correlation, namely $r^2 = (0.604)^2 = 0.365$.

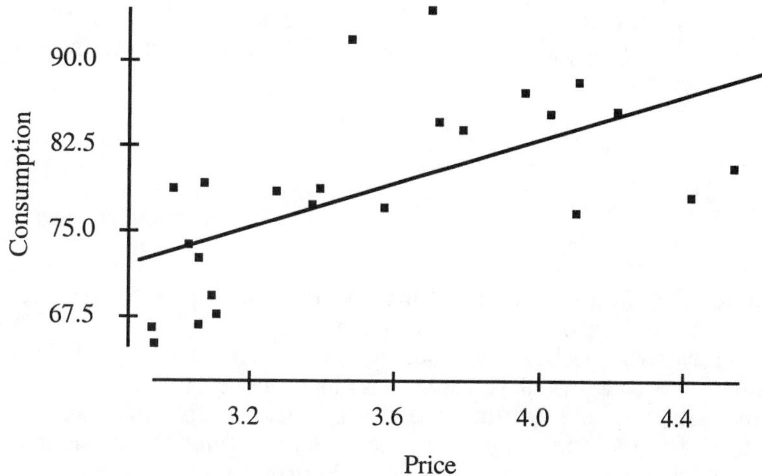

c) A plot of the residuals against time follows.

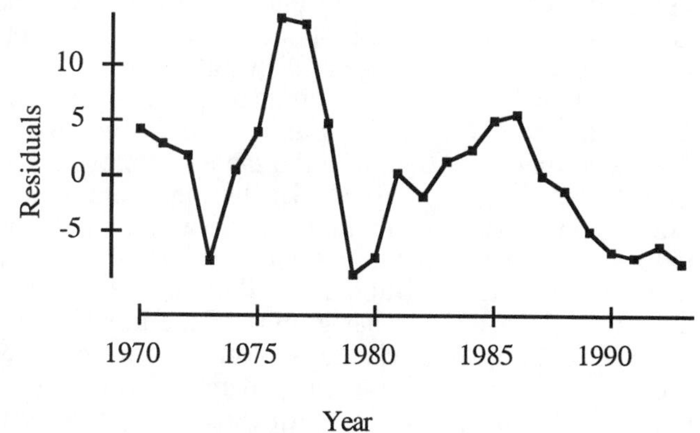

There are periodic fluctuations with two peaks around 1977 and 1986. There also appears to be an overall downward trend over time.

2.73 a) $r^2 = (0.9540)^2 = 0.9101$. This is the fraction of the variation in daily sales data that can be attributed to a linear relationship between daily sales on total item count.

b) We would expect the correlation to be smaller. Correlations based on averages are usually higher than the correlation one would compute for the individual values from which the averages were computed. Because totals are proportional to averages, the same would be true for totals.

2.75 One lurking variable might be intelligence. More intelligent individuals are more likely to seek more education. More intelligent individuals are also more likely to be successful in a job and advance to higher salary levels. Another lurking variable might be hard work or a desire to be successful. Men who work hard or desire success may do well in school complete more years of schooling. Men who work hard or desire success may also be successful in their jobs and advance to higher salary levels. Family background may also be a lurking variable. Wealthy or successful families may encourage children to seek more education and to strive for success (higher pay) in jobs. Families with a tradition of more education and high-paying jobs may encourage children to follow a similar career path.

2.77 a) The residuals are more widely scattered about the horizontal line (have larger absolute values) as the predicted value increases. This suggests that the regression model will predict low salaries more precisely because lower predicted salaries have smaller residuals and hence are closer to the actual salary.

b) There is a curved pattern in the plot. For very low and very high numbers of years in the majors, the residuals tend to be negative. For intermediate numbers of years the residuals tend to be positive. This means that the model will overestimate the salaries of players whoare new to the majors (a negative residual means that the observed values is less than the predicted value). This also means that the model will underestimate the salaries of players who have been in the major leagues about 8 years, and it will overestimate the salaries of players who have been in the majors more than 15 years.

SECTION 2.5

OVERVIEW

This section discusses techniques for describing the relationship between two or more categorical variables. To analyze categorical variables, we use counts (frequencies) or percents (relative frequencies) of individuals that fall into various categories. **A two-way table** of such counts is used to organize data about two categorical variables. Values of the **row variable** label the rows that run across the table, and values of the **column variable** label the columns that run down the table. In each cell (intersection of a row and column) of the table, we enter the number of cases for which the row and column variables have the values (categories) corresponding to that cell.

The **row totals** and **column totals** in a **two-way table** give the marginal distributions of the two variables separately. It is usually clearest to present these distributions as percents of the table total. Marginal distributions do not give any information about the relationship between the variables. **Bar graphs** are a useful way of presenting these marginal distributions.

The conditional distributions in a two-way table help us to see relationships between two categorical variables. To find the conditional distribution of the row variable for a specific value of the column variable, look only at that one column in the table. Express each entry in the column as a percent of the column total. There is a conditional distribution of the row variable for each column in the table. Comparing these conditional distributions is one way to describe the association between the row and column variables, particularly if the column variable is the explanatory variable. When the row variable is explanatory, find the conditional distribution of the column variable for each row and compare these distributions. Side-by-side bar graphs of the conditional distributions of the row or column variable can be used to compare these distributions and describe any association.

Data on three categorical variables can be presented in a **three-way table,** printed as separate two-way tables for each value of the third variable. An association between two variables that holds for each level of this third variable can be changed, or even reversed, when the data are **aggregated** by summing over all values of the third variable. **Simpson's paradox** refers to a reversal of an association by aggregation.

APPLY YOUR KNOWLEDGE

2.79 a) The data describe 1168 + 1823 + 1380 + 188 + 416 + 400 = 5375 students.

b) The number who smoke is 188 + 416 + 400 = 1004. Thus, the percent of the 5375 students who smoke is (1004/5375) × 100% = 18.68%.

c) The marginal distribution is obtained by computing the three column totals and, if desired, converting each to a percent of 5375. The results are summarized below.

	Neither parent smokes	One parent smokes	Both parents smoke
Total	1356	2239	1780
Percent	25.23%	41.66%	33.12%

2.81 We convert the number of students who smoke for each of the three groups to a percentage of the column total. The results are as follows.

	Neither parent smokes	One parent smokes	Both parents smoke
% students who smoke	(188/1356)×100% = 13.86%	(416/2239)×100% = 18.58%	(400/1780)×100% = 22.47%

The percent of students who smoke increases as the number of parents who smoke increases. Thus, the data support the belief that parents' smoking increases smoking in their children.

2.83 a) Adding the entries in the column labeled female, we find that 225 females responded. Adding the entries in the column labeled male, we see that 161 males responded. The conditional distributions for men and for women are computed by converting the entries in the table to percentages of these column totals. The results are given as percentages in the table below.

	Female	Male
Accounting	30.22%	34.78%
Administration	40.44%	24.84%
Economics	2.22%	3.73%
Finance	27.11%	36.65%

To facilitate comparisons, we plot the conditional distributions for females and males in a side-by-side bar graph.

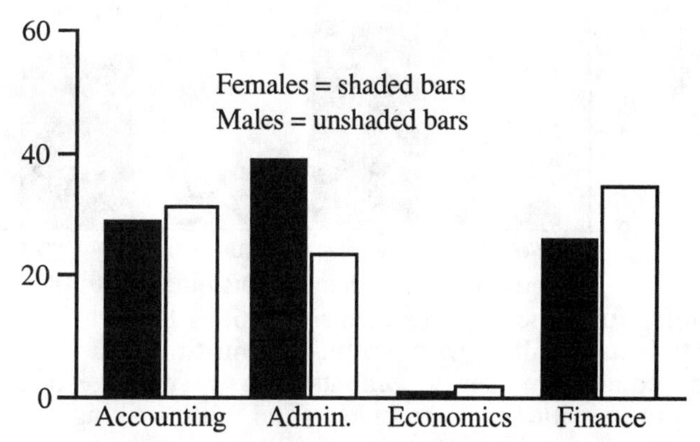

The most popular choice for women is administration, while for men accounting and finance are the most popular.

b) The total number of students who responded is 386 (the sum of all the entries in the table). The questionnaires were sent to all 722 members of the senior class. Thus, 722 – 386 = 336 students did not respond, or $(336/722)\times 100\% = 46.54\%$ did not respond.

2.85 a) Percent of Hospital A patients who died = $(63/2100)\times 100\% = 3\%$.
Percent of Hospital B patients who died = $(16/800)\times 100\% = 2\%$.

b) Percent of Hospital A patients who died who were classified as "poor" before surgery = $(57/1500)\times 100\% = 3.8\%$.
Percent of Hospital B patients who died who were classified as "poor" before surgery = $(8/200)\times 100\% = 4\%$.

c) Percent of Hospital A patients who died who were classified as "good" before surgery = $(6/600)\times 100\% = 1\%$.
Percent of Hospital B patients who died who were classified as "good" before surgery = $(8/600)\times 100\% = 1.3\%$.

d) If you are in "good" condition before surgery or if you are in "bad" condition before surgery, your chance of dying is lower in Hospital A, so choose Hospital A.

e) The majority of patients in Hospital A are in poor condition before surgery, while the majority of patients in Hospital B are in good condition before surgery. Patients in poor condition are more likely to die and this makes the overall number of deaths in Hospital A higher than in Hospital B, even though both types of patients fare better in Hospital A.

SECTION 2.5 EXERCISES

2.87 a) The percentage of all students enrolled part-time at two-year colleges who are 25 to 39 years old is $(421/2064) \times 100\% = 20.40\%$.

b) The percentage of all 25- to 39-year-old students (there are 421 + 1344 + 1234 + 1273 = 4272 thousand such students) who are enrolled part-time at two-year colleges is $(421/4272) \times 100\% = 9.85\%$.

2.89 There are 121 + 748 + 236 + 611 = 1716 thousands of older students. The distribution of these older students among the four types of colleges and universities is as follows.

	two-year part-time	two-year full-time	four-year part-time	four-year full-time
% older students	$\frac{121}{1716} \times 100\% = 7.05\%$	$\frac{748}{1716} \times 100\% = 43.59\%$	$\frac{236}{1716} \times 100\% = 13.75\%$	$\frac{611}{1716} \times 100\% = 35.61\%$

A bar graph that presents this distribution is given below.

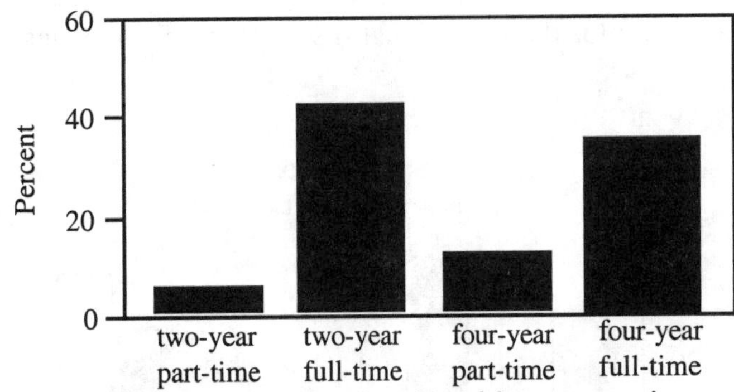

Older students tend to prefer full-time colleges and universities to part-time colleges and universities, with 89.2% of all older students enrolled in some full-time institution.

Another comparison could be made by computing the percentage of all students in each of the four types of colleges and universities that are older. The most striking feature of this comparison is that older students are the second most common group in four-year full-time colleges and universities. They are even more common than 18 to 24 year olds!

2.91 a) The two conditional distributions are as follows.

	Hired	Not hired
Applicants < 40	$(79/1227) \times 100\% = 6.44\%$	$(1148/1227) \times 100\% = 93.56\%$

	Hired	Not hired
Applicants ≥ 40	$(1/164) \times 100\% = 0.61\%$	$(163/164) \times 100\% = 99.39\%$

b) A graph showing the differences in distribution for the two age groups is given below.

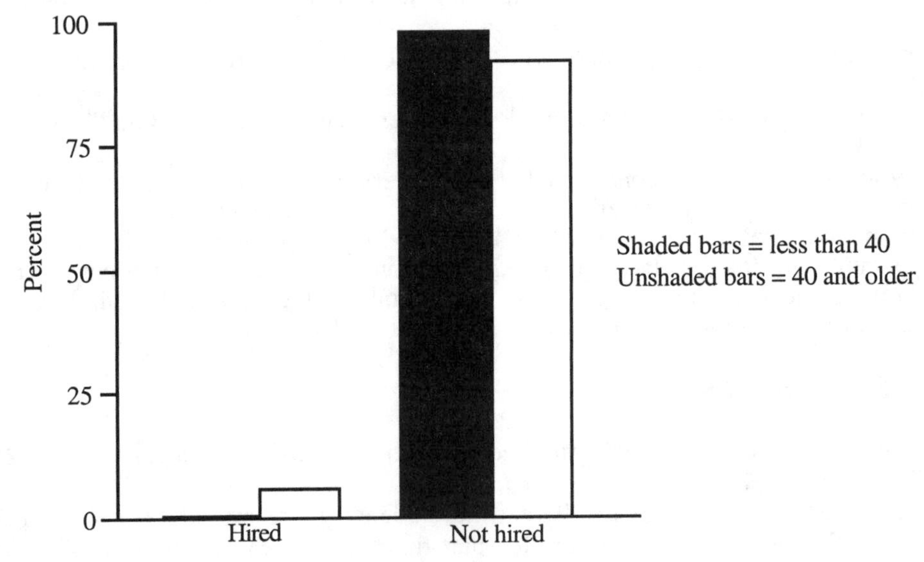

c) Although only a small percentage of all applicants are hired, the percentage (6.44%) of applicants who are less than 40 who are hired is more than 10 times the percentage (0.61%) of applicants who are 40 or older that are hired. This certainly makes it appear as though the company discriminates on the basis of age.

d) Lurking variables that might be involved are past employment history (why are older applicants without a job and looking for work?) or health.

2.93 a) The conditional distributions of relapse/no relapse for the three drugs is given below.

	Desipramine	Lithium	Placebo
Relapse	$(10/24) \times 100\% = 41.67\%$	$(18/24) \times 100\% = 75\%$	$(20/24) \times 100\% = 83.33\%$
No relapse	$(14/24) \times 100\% = 58.33\%$	$(6/24) \times 100\% = 25\%$	$(4/24) \times 100\% = 16.67\%$

A bar graph that displays this information follows.

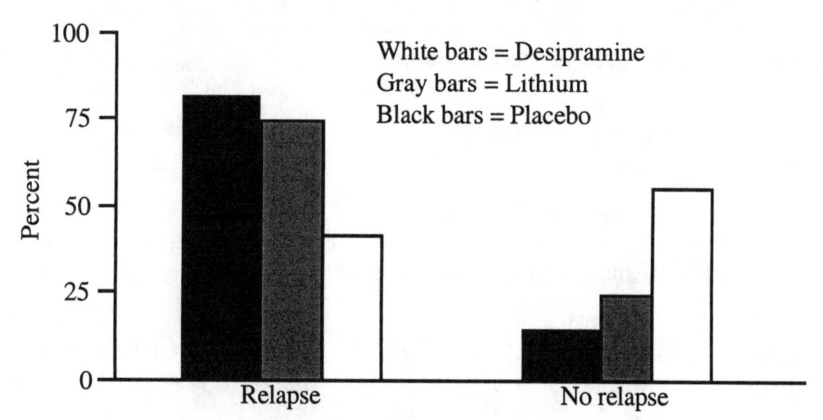

b) The data show that those taking desipramine had fewer relapses than those taking either lithium or a placebo. These results are interesting but association does not imply causation. Some concerns are the following. First, how strong is the association? Is the number of people in the study large enough that we can conclude the observed association is strong? Second, can we rule out lurking variables? For example, who determined which subjects received which treatments? Were the groups of subjects on each of the treatments similar or did they differ in some systematic way that might have affected the outcome? Did the subjects know anything about the treatments they received and would this knowledge affect their response to the treatment? Third, have similar results been observed in other studies with other groups in other contexts? Without additional information, one should be cautious about concluding that the study demonstrates that desipramine causes a reduction in relapses.

2.95 a) The sum of the entries in the married column is 2,598 + 20,129 + 28,923 + 8,270 = 59,920 thousand women. The entry in the total column in 59,918 thousand. The difference may be due to rounding off of the values in the married column (remember, these entries are thousands of women, so they have been rounded off to the nearest thousand).

b) The marginal distribution of marital status for all adult women is obtained by converting the entries in the row labeled Total to percents of the table total of 103,867. The results are as follows.

Never married	Married	Widowed	Divorced
21.05%	57.69%	10.54%	10.73%

A bar graph that displays this distribution follows.

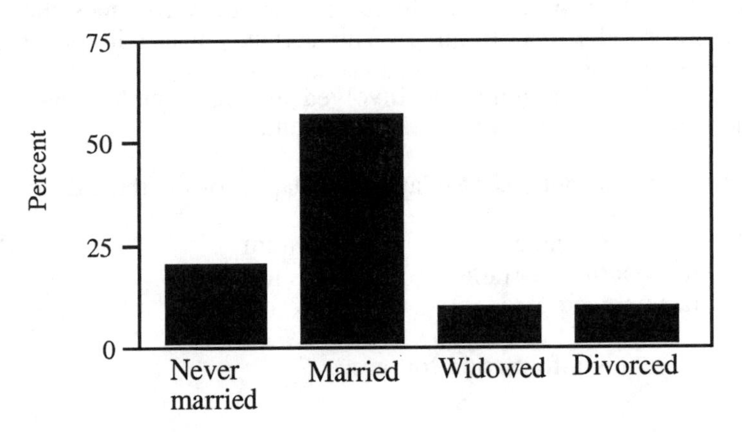

2.97 a) A combined two-way table is the following.

	Admit	Deny
Male	480 + 10 = 490	120 + 90 = 210
Female	180 + 100 = 280	20 + 200 = 220

b) The total number of male applicants is 490 + 210 = 700 and the total number of female applicants is 280 + 220 = 500. Using these values to convert the entries for males and females to percents of the total number of male and female applicants, respectively, gives the following.

	Admit	Deny
Male	$(490/700)\times 100\% = 70\%$	$(210/700)\times 100\% = 30\%$
Female	$(280/500)\times 100\% = 56\%$	$(220/500)\times 100\% = 44\%$

c) The percents for each school are as follows

	Business		Law	
	Admit	Deny	Admit	Deny
Male	$(480/600)\times 100\% = 80\%$	$(120/600)\times 100\% = 20\%$	$(10/100)\times 100\% = 10\%$	$(90/100)\times 100\% = 90\%$
Female	$(180/200)\times 100\% = 90\%$	$(20/200)\times 100\% = 10\%$	$(100/300)\times 100\% = 33\%$	$(200/300)\times 100\% = 67\%$

d) If we look at the tables, we see that it is easier to get into the Business School than the Law School. We also see that more men apply to the Business School than the Law School while more women apply to the Law School than the Business School. Because it is easier to get into the Business School, the overall admission rate for men appears relatively high (the large number of men applying to Business brings up the overall rate). It is hard to get into the Law School and this makes the overall admission rate for women appear low (the larger number of women applying to Law brings down the overall rate).

Chapter 2 Review Exercises

2.99 There appears to be a very slight negative association. High interest rates are weakly associated with lower stock returns. Because many other factors (lurking variables) affect stock returns, one should not conclude from these data that the observed association is evidence that high interest rates cause low stock returns (i.e., are bad for stocks).

2.101 Removing this point would make the correlation closer to 0. The point is an outlier in the horizontal direction and its location (high treasury bill return and below average stock return) is such that it accentuates the impression of a negative association and hence could strongly influence the regression line.

2.103 Suppose the return on Fund A is always twice that of Fund B. Then if Fund B increases by 10%, Fund B increases by 20%, and if Fund B increases by 20%, Fund A increases by 40%. This is a case where there is a perfect straight line relation between Fund A and B, as illustrated in the plot below.

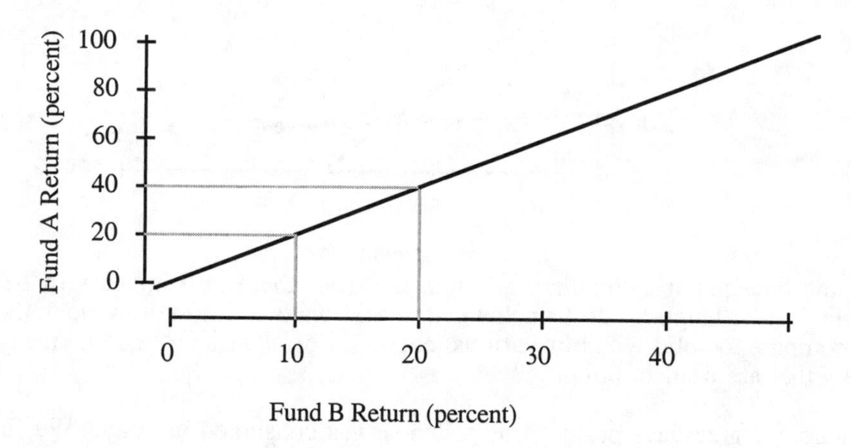

As a result, the two funds are perfectly correlated.

2.105 a) A scatterplot is given below with length on the horizontal axis as the predictor variable.

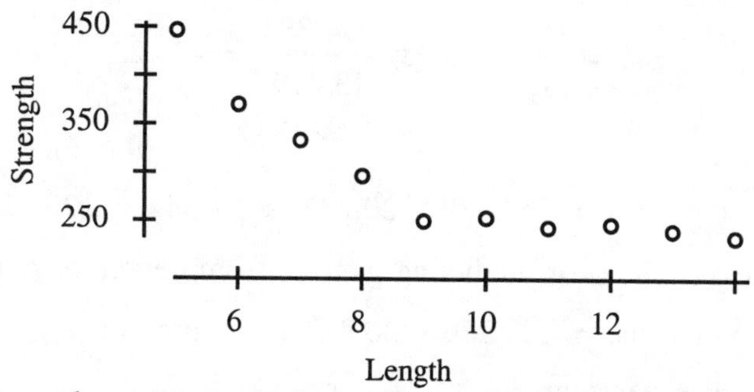

b) The overall pattern shows strength decreasing as length increases until length is 9, then the pattern is relatively flat with strength decreasing only very slightly for lengths greater than 9.

c) The equation of the least-squares line is Strength = 488.38 – 20.75 × Length. The line is displayed in the plot below.

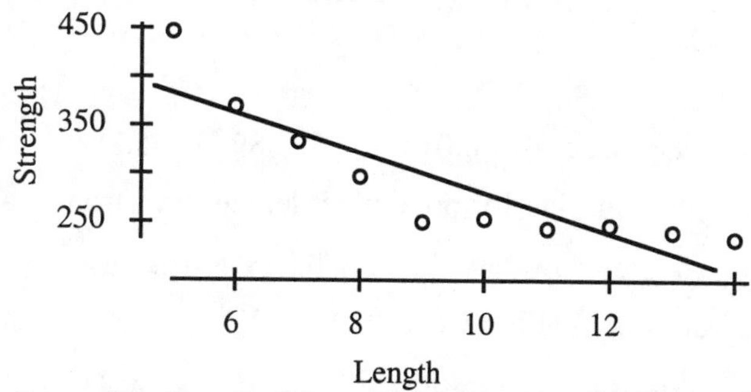

A straight line does not adequately describe these data because it fails to capture the "bend" in the pattern at Length = 9.

d) The equation for the lengths of 5 to 9 inches is 283.1 – 3.4 × Length. The equation for the lengths of 9 to 14 inches is 667.5 – 46.9 × Length. The two lines are displayed in the plot below.

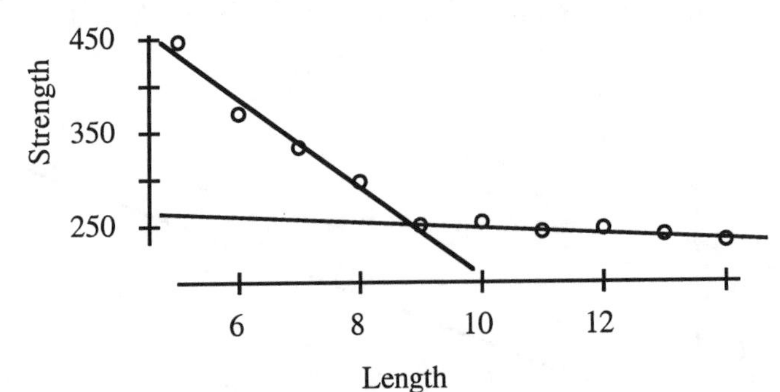

They describe the data much better than a single line (but there still appears to be a slight curved trend in the data for the lengths 5 to 9 inches). I would want to know how does the strength of the wood product compare to solid wood of various lengths. For what lengths is it stronger and are these common lengths that are used in building?

2.107 Let x denote degree-days per day and y denote gas consumed per day. We find

$$\bar{x} = 21.544,\ s_x = 13.419,\ \bar{y} = 558.889,\ s_y = 274.383, \text{ and } r = 0.989.$$

Thus, we compute the slope to be

$$b = r\frac{s_y}{s_x} = 0.989\frac{274.383}{13.419} = 20.222$$

and the intercept to be

$$a = \bar{y} - b\bar{x} = 558.889 - 20.222 \times 21.544 = 123.226.$$

The equation of the regression line for predicting gas use from degree-days is then

$$\text{gas use} = 123.226 + 20.222 \times (\text{degree-days}).$$

The slope has units cubic feet per degree-day.

If we want to find the equation of the regression line for predicting degree-days from gas use, we interchange the roles of x and y and compute the slope to be

$$b = 0.989\frac{13.419}{274.383} = 0.048$$

and the intercept to be

$$a = 21.544 - 0.048 \times 558.889 = -5.283.$$

Thus, the equation of the regression line for predicting degree-days from gas use is

$$\text{degree-days} = -5.283 + 0.048 \times (\text{gas use}).$$

The slope has units degree-days per cubic feet.

2.109 a) Using statistical software, we find that the equation of the least-squares regression line for predicting selling price from age at time of sale is

selling price = 189,226 – 1334.49 × age

b) For a house built in 2000 the age is 0, and so from the least-squares regression line we would expect the selling price to be

selling price = 189,226 – 1334.49 × 0 = \$189,226.

For a house built in 1999 the age is 1 and so from the least-squares regression line we would expect the selling price to be

selling price = 189,226 – 1334.49 × 1 = \$187,891.51.

For a house built in 1998 the age is 2, and so from the least-squares regression line we would expect the selling price to be

selling price = 189,226 – 1334.49 × 2 = \$186,557.02.

For a house built in 1997 the age is 3, and so from the least-squares regression line we would expect the selling price to be

selling price = 189,226 – 1334.49 × 3 = \$185,222.53.

We see that for each one-year increase in age the selling price drops by \$1334.49.

c) 1900 (age = 100) is within the range of the data used to calculate the least-squares regression line and 1899 (age = 101) is almost within the range, so we would probably trust the regression line to predict the selling price of a house in these years. 1850 (age = 150) is well outside the range of the data, and we would not trust the regression line to predict the selling price of such a house. In fact, the regression line predicts a house that is 150 years old to have a selling price of

selling price = 189,226 – 1334.49 × 150 = –\$10,947.50,

a negative value that makes no sense! It is dangerous to extrapolate to values outside the range of the data used to produce the regression line.

d) Using statistical software, we compute the correlation between selling price and age to be –0.682. The association is negative indicating that older houses are associated with lower selling prices and newer houses with higher selling prices.

2.111 We convert the table entries into percents of the column totals to get the conditional distribution of heart attacks and strokes for the aspirin group and for the placebo group.

	Aspirin group	Placebo group
Fatal heart attacks	(10/11,037)×100% = 0.09%	(26/11,034)×100% = 0.24%
Other heart attacks	(129/11,037)×100% = 1.17%	(213/11,034)×100% = 1.93%
Strokes	(119/11,037)×100% = 1.08%	(98/11,034)×100% = 0.89%

The data show that aspirin is associated with reduced rates of heart attacks but a slightly increased rate of strokes. In particular, aspirin (as compared to a placebo) reduces the rate of fatal heart attacks by a factor of 9/24 = 3/8 = 0.375 and reduces the rate of other heart attacks by a factor of about 1.17/1.93 = 0.6. Aspirin increases the rate of strokes by a factor of about 1.08/0.89 = 1.2.

Once again, association does not imply causation. However, the design of the study is such that it is difficult to identify lurking variables. Doctors were assigned at random to the treatments (aspirin or placebo), so there should not be any systematic differences in the groups that received the two treatments. The two treatments appeared and tasted similar, so doctors could not tell what treatment they were receiving and could not be influenced by that knowledge. The study is large, so the effects are strong. Because only doctors took part in the study, if doctors in some way differ from the population at large in a way that is associated with the occurrence of heart attacks and strokes, the results may not extend to the population at large. Overall, though, the results of the study are intriguing.

CHAPTER 3

PRODUCING DATA

SECTION 3.1

OVERVIEW

Chapters 1 and 2 describe methods for exploring data. Such **exploratory data analysis** is used to determine what the data tell us about the variables measured and their relations to each other. Conclusions apply to the data observed and may not generalize beyond these data.

Statistical inference produces answers to specific questions, along with a statement of how confident we are that the answer is correct. Answers are usually intended to apply beyond the data observed. This requires careful **production of data** appropriate for answering the specific questions asked.

Data can be produced in many ways. Done properly, sampling can yield reliable information about a population. **Observational studies** are investigations in which one simply observes the state of some population, usually with data collected by sampling. Even with proper sampling, data from observational studies are generally not appropriate for investigating cause-and-effect relations between variables. **Experiments** are investigations in which data are generated by active imposition of some treatment on the subjects of the experiment. Properly designed experiments are the best way to investigate cause-and-effect relations between variables.

The **population** is the entire group of individuals or objects about which we want information. The information collected is contained in a **sample** that is the part of the population we actually get to observe. How the sample is chosen – that is, the **design** – has a large impact on the usefulness of the data. A useful sample will be representative of the population and will help answer our questions. "Good" methods of collecting a sample include:

- **probability samples**
- **simple random samples,** also called **SRS**
- **stratified random samples**
- **multistage samples.**

All of these sampling methods involve some aspect of randomness through the use of a formal chance mechanism. Random selection is just one precaution that a person can take to reduce **bias,** the systematic favoring of a certain outcome. Samples we select using our own judgment, because they are convenient or "without forethought" (mistaking this for randomness) are usually biased in some way. This is why we use computers or a tool such as a **table of random digits** to help us select a sample.

A **voluntary response sample** includes people who choose to be in the sample by responding to a general appeal. They tend to be biased, as the sample is overrepresented by individuals with strong opinions, which are often negative.

Other kinds of bias to be on the lookout for include

• **nonresponse bias,** which occurs when individuals who are selected do not participate or cannot be contacted

• **undercoverage,** which occurs when some group in the population is given either no chance or a much smaller chance than other groups to be in the sample

•**response bias,** which occurs when individuals do participate but are not responding truthfully or accurately due to the way the question is worded, the presence of an observer, fear of a negative reaction from the interviewer, or any other such source.

These types of bias can occur even in a randomly chosen sample and we need to try to reduce their impact as much as possible.

Apply Your Knowledge

3.1 It is an observational study as information is gathered without imposing a treatment. The explanatory variable in this case is the consumer's gender and the response variables are whether or not the individual considers the particular features essential in a health plan. Although men and women may differ in their opinions, this is still observational data.

3.3 The state institutes a job-training program for manufacturing workers and five years later the unemployment rate for manufacturing workers has increased from 6% to 10%. Does this indicate that the program was ineffective? Many variables (lurking variables), such as a recession or factories leaving the area, could have changed during the five years to increase the unemployment rate . Perhaps, if there was no job training the unemployment rate would have increased even more. Observational studies are generally a poor way to gauge the effect of an intervention.

3.5 a) The population is all adult U.S. residents.
b) The population is U.S. households.
c) All regulators from the supplier, *or* just the regulators in the last shipment. The manufacturer is probably interested in all regulators from the supplier, but the sample is only from the last shipment.

3.7 To choose an SRS of six students to be interviewed, first label the members of the population by associating a two-digit number with each.

01 - Agarwal
02 - Anderson
03 - Baxter
04 - Bowman
05 - Brown
06 - Castillo
07 - Cross
08 - Dewald
09 - Fernandez
10 - Fleming
11 - Garcia
12 - Gates
13 - Goel
14 - Gomez
15 - Huang
16 - Kim
17 - Liao
18 - Mourning
19 - Naber
20 - Peters
21 - Pliego
22 - Puri
23 - Richards
24 - Rodriguez
25 - Santiago
26 - Shen
27 - Vega
28 - Wang

Now enter Table B and read two-digit groups until six students are chosen. Starting at line 139,

55|58|8 9|94|04| 70|70|8 4|10|98| 43|56|3 5|69|34| 48|39|4 5|17|19
12|97|5 1|32|58| 13|048

The selected sample is 04 - Bowman, 10 - Fleming, 17 - Liao, 19 - Naber, 12 - Gates, and 13 - Goel.

3.9 To choose an SRS of 10 retail outlets, first label the retailers from 001 to 440, a single three-digit number for each retailer (you could label from 000 to 439 or use two labels for each retailer, which would result in a different answer). Using the labels 001 to 440, enter Table B and read three-digit groups until 10 retailers are chosen. Starting at line 105,

955|92 9|400|7 69|971| 914|81 6|077|9 53|791| 172|97 5|933|5
68|417| 350|13 1|552|9 72|765| 850|89 5|706|7 50|211| 474|87
8|273|9 57|890| 208|07 4|7511

The selected sample are the retailers numbered 400, 077, 172, 417, 350, 131, 211, 273, 208, and 074.

3.11 There are 500 midsize accounts. We are going to sample 5% of them, which is 25. Label the accounts 001, 002, ..., 500 and select an SRS of 25 of the midsize accounts. There are 4400 small accounts. We are going to sample 1% of them, which is 44. Label the accounts 0001, 0002, ..., 4400 and select an SRS of 44 of the small accounts.

Starting at line 115, we first select five midsize accounts – that is, an SRS of size five – using the labels 001 through 500. Then continuing from where we left off in the table, we select five small accounts – that is, an SRS of size five – using the labels 0001 through 4400. Note that for the midsize accounts, we read from Table B using three-digit numbers, and for the small accounts we read from the table using four-digit numbers.

610|41 7|768|4 94|322| 247|09 7|3698| 1452|6 318|93 32|592
1|4459| 2605|6 314|24 80|371 6|

The first five midsize accounts are those with labels 417, 494, 322, 247, and 097. Continuing in the table, using four digits instead of three, the first five small accounts are those with labels 3698, 1452, 2605, 2480, and 3716.

3.13 You would expect that the higher rate of no-answer was probably during the second period as more families are likely to be gone for vacation. Nonresponse can always bias the results. In this case, those who are more affluent may be more likely to travel during the summer months and they would be underrepresented in the sample. Their views could be different, which would bias the results.

SECTION 3.1 EXERCISES

3.15 a) The individuals are the small businesses. The population of interest is "eating and drinking establishments" in the large city. (Undercoverage will occur for those establishments not listed in the Yellow Pages telephone directory.)

b) The individuals are the adults. The population of interest is the Congressman's constituents. (This is a voluntary sample and contains only those that felt strongly enough to write a letter. It is most likely a biased sample of the Congressman's constituents.)

c) The individuals are the auto insurance claims. The population is all claims filed in a given month.

3.17 The 1128 letters received are a voluntary response sample. These letters do not necessarily reflect the opinions of her constituents because persons with strong opinions on the subject are more likely to take the time to write.

3.19 To choose an SRS of three containers from the lot of 25, you must first label the 25 containers. We will use the two-digit numbers from 01 to 25. The assignment of the two-digit numbers to the containers is arbitrary and we have chosen to number across the rows.

A1096(01)	A1097(02)	A1098(03)	A1101(04)	A1108(05)
A1112(06)	A1113(07)	A1117(08)	A2109(09)	A2211(10)
A2220(11)	B0986(12)	B1011(13)	B1096(14)	B1101(15)
B1102(16)	B1103(17)	B1110(18)	B1119(19)	B1137(20)
B1189(21)	B1223(22)	B1277(23)	B1286(24)	B1299(25)

Using the labels 01 to 25, enter Table B and read two-digit groups until three containers are chosen. Starting at line 111

81|48|6 6|94|87| 60|51|3 0|92|97| 00|41|2 7|12|38 27|64|9 3|99|50
59|63|6 8|88|04| 04|63|4 7|11|97

The selected sample is 12 - B0986, 04 - A1101, and 11 - A2220. The number 04 appeared a second time and it was skipped as container A1101 was already in the sample. If multiple labels were used for each container, the process of selecting the sample would have been quicker, but in practice computer software is usually used to select the sample rather than the table of random numbers.

3.21 The numbering system for the 44 blocks is 1000, 1001, ..., 1005, 2000, 2001, ..., 2011, 3000, 3001, ..., 3025. You could represent these 44 blocks by their four-digit numbers and go to Table B, but because only 44 of the 10,000 four-digit numbers correspond to blocks, most four-digit numbers in the table would not correspond to blocks. As an alternative, the tracts could be renumbered from

01 to 44 according to the following scheme. The two-digit numbers assigned to each tract are in parentheses.

1000(1), 1001(2), ..., 1005(6),
2000(7), 2001(8), ..., 2011(18),
3000(19), 3001(20), ..., 3025(44)

Starting at line 125 in Table B and ignoring repeats

96746 12149 37823 71868 18442 35119 62103 39244

the five two-digit numbers selects are 21 (block 3002), 18 (block 2011), 23 (block 3004), 19 (block 3000), and 10 (block 2003).

3.23 a) We want to select five clusters out of 200, so we think of the 200 as five lists of 40 clusters. We choose one cluster from the first 40, and then every 40th cluster after that. The first step is to go to Table B, line 120, and choose the first two-digit random number you encounter that is one of the numbers 01, ..., 40.

35476

The selected number is 35, so the sample includes clusters numbered 35, 75, 115, 155, and 195.

b) Each individual is in exactly one systematic sample, and the systematic samples are equally likely to be chosen. In our previous example, there were 40 systematic samples, each containing five clusters. The chance of selecting any cluster is the chance of picking the systematic sample it is in, which is 1 in 40.

A simple random sample of size n would allow every set of n individuals an equal chance of being selected. Thus, in this exercise, using an SRS the sample consisting of the clusters numbered 1, 2, 3, 4, and 5 would have the same probability of being selected as any other set of five clusters. For a systematically selected sample, all samples of size n do not have the same probability of being selected. In our example, the sample consisting of the clusters numbered 1, 2, 3, 4, and 5 would have zero chance of being selected in a systematic sample because the numbers of the clusters do not all differ by 40. The sample we selected in (a), 35, 75, 115, 155, and 195 had a 1-in-40 chance of being selected, so all samples of five clusters are not equally likely, as required by simple random sampling.

3.25 Each name on the alphabetized list needs to be given a number — 001 to 500 for the females and 0001 to 2000 for the males. Starting at line 122, we first select five females, that is, an SRS of size five, using the labels 001 through 500. Then continuing from where we left off in the table, we select five males, that is, an SRS of size five, using the labels 0001 through 2000. Note that for the females we read from Table B using three-digit numbers, and for the males we read from the table using four-digit numbers.

138|73 8|159|8 95|052| 909|08 7|359|2 751|86 87|136 9|5761|
5458|0 815|07 27|102 5|6027| 5589|2 330|63 41|842 8|1868|

The first five females are those with labels 138, 159, 052, 087, and 359. Continuing in the table, using four digits instead of three, the first five males are those with labels 1369, 0815, 0727, 1025, and 1868.

3.27 a) "A national system of health insurance should be favored because it would provide health insurance for everyone and would reduce administrative costs." This question gives information only on one side of the issue and then asks for an opinion, which is a sure way to bias the response.

b) "The elimination of the tenure system at public institutions should be considered as a means of increasing the accountability of faculty to the public, while at the same time the question of whether such a move would have a deleterious effect on the academic freedom so important to such institutions cannot be ignored." The question is confusing as more than one idea is presented and it is unclear exactly what you are agreeing with.

SECTION 3.2

OVERVIEW

Experiments are studies in which one or more **treatments** are imposed on experimental **units** or **subjects.** A treatment is a combination of levels of the explanatory variables, called **factors.** The **design** of an experiment is a specification of the treatments to be used and the manner in which units or subjects are assigned to these treatments. The basic features of well-designed experiments are **control, randomization,** and **replication.**

Control is used to avoid confounding (mixing up) the effects of treatments with other influences such as lurking variables. One such lurking variable is the **placebo effect,** which is the response of a subject to the fact of receiving any treatment. The simplest form of control is **comparative experimentation,** which involve comparisons between two or more treatments. One of these treatments may be a **placebo** (fake treatment), and those subjects receiving the placebo are referred to as a **control group.**

Randomization uses a well-defined chance mechanism to assign subjects to treatments. It is used to create treatment groups that are similar, except for chance variation, prior to application of treatments. Randomized, comparative experiments are used to prevent **bias,** or systematic favoritism of certain outcomes. **Tables of random digits** or computer programs that generate random numbers are well-defined chance mechanisms that are used to carry out randomization. In either case, numerical labels are assigned to experimental units and random numbers from the table or computer software determine which labels (units) are assigned to which treatments.

Replication is the use of many units in an experiment and is used to reduce the effect of any chance variation between treatment groups arising from randomization. Replication increases the sensitivity of an experiment to differences in treatments.

Additional control in an experiment can be achieved by forming experimental units into **blocks** that are similar in some way, which is thought to affect the response. In a **block design,** units are first formed into blocks and then randomization is carried out separately in each block. **Matched pairs** are a simple form of blocking used to compare two treatments. In a matched pairs experiment, either the same unit (the block) receives both treatments in a random order or very similar units are matched in pairs (the blocks). In the latter case, one member of the pair receives one of the treatments and the other member the remaining treatment. Members of a matched pair are assigned to treatments using randomization.

Good experiments require attention to details. **Double-blind** experiments are ones in which neither the subject nor the person measuring the response is aware of what treatment is being used. **Lack of realism** in an experiment can prevent us from generalizing the results.

APPLY YOUR KNOWLEDGE

3.29 The subjects are the 300 sickle cell patients. There is only one factor and it is the type of medication given. The treatments are the two levels of medication — the drug hydroxyurea and the placebo. The response variable being measured is the number of pain episodes reported by each subject.

3.31 a) The individuals are the different batches, and the response variable is the yield of the chemical reaction.

b) There are two factors in the experiment — temperature and stirring rate. The treatments are the different temperature and stirring rate combinations. Because there are two levels of temperature and three levels of stirring rate, there are a total of six treatments. The diagram on the next page lays out the different treatment combinations in the design.

Factor B
Stirring Rate

Factor B Temperature	60 rpm	90 rpm	120 rpm
50 deg	1	2	3
60 deg	4	5	6

c) There are six treatments with two batches of the product at each treatment, so 12 batches or "individuals" are needed for the experiment.

3.33 The diagram below describes the experimental design. To assign the 20 pairs to the four treatments, give each of the pairs a two-digit number from 01 to 20. The first five selected are assigned to Group 1, the next five to Group 2 and the next five to Group 3. The remaining five are assigned to Group 4. Make sure to skip repeat numbers when going through the table as these pairs have already been assigned to a treatment. Starting at line 120,

35476	55972	39421	65850	04266	35435	43742	11937
71487	09984	29077	14863	61683	47052	62224	51025
13873	81598	95052	90908	73592	75186	87136	95761
54580	81507	27102	56027	55892	33063	41842	81868
71035	09001	43367	49497	72719	96758	27611	91596
96746	12149	37823	71868	18442	35119	62103	39244
96927	19931	36089	74192	77567	88741	48409	41903
43909	99477	25330	64359	40085	16925	85117	36071

Group 1 is pairs numbered 16, 04, 19, 07, and 10; Group 2 is 13, 15, 05, 09, and 08; Group 3 is 18, 03, 01, 06, and 11. The remaining five pairs are assigned to Group 4.

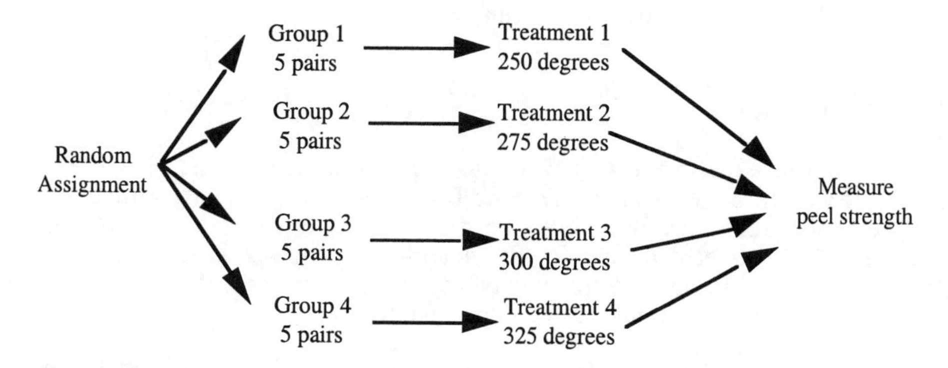

3.35 The electric company wants the only systematic differences in groups to be the treatments. Electric use varies from year to year depending on the weather. If charts or indicators are introduced in the second year, and the electric consumption in the first year is compared with the second year, you won't know if the observed differences are due to the introduction of the chart or lurking variables. For example, if the comparison is being made in the summer months, it is possible that the second year had a cooler summer, which reduced the need for air conditioning and reduced electric consumption, rather than the introduction of charts or indicators. A control group ensures that influences other than the introduction of the indicators or charts operate equally on all groups.

3.37 A statistically significant result means that it is unlikely the salary differences we are observing are just due to chance. No significant difference means the differences may just be due to chance variation.

3.39 This experiment suffers from lack of realism and limits the ability to apply the conclusions of the experiment to the situation of interest. The students know they are taking part in an experiment and there are no real consequences as to whether they win or lose the game. This is much different than a work setting where months have been spent on a project and the consequences of failure can be quite serious, both financially and professionally.

3.41 a) The subjects and their excess weights, rearranged in increasing order of excess weight, are listed below. The columns are the five blocks. We have labeled the subjects in each block from 1 to 4.

Block 1	Block 2	Block 3	Block 4	Block 5
1 Williams 22	1 Santiago 27	1 Brunk 30	1 Jackson 33	1 Birnbaum 35
2 Deng 24	2 Kendall 28	2 Obrach 30	2 Stall 33	2 Tran 35
3 Hernandez 25	3 Mann 28	3 Rodriguez 30	3 Brown 34	3 Nevesky 39
4 Moses 25	4 Smith 29	4 Loren 32	4 Cruz 34	4 Wilansky 42

b) We used lines 130 and 131 in Table B, which is given below.

69051 64817 87174 09517 8453|4 06489 87201| 97245

05007 1|6632 81|194 14873| 04197 85576 45195 96565

For Block 1, we read these from left to right, one digit at a time. The first label we encounter is assigned to Regimen A, the next to Regimen B, the next to Regimen C, and the remaining label is then automatically assigned to Regimen D. We have underlined digits corresponding to one of the labels, skipping repeats. The vertical lines indicates when we have completed a block. We summarize the results below.

Regimen A = Williams, Smith, Obrach, Brown, Birnbaum

Regimen B = Moses, Kendall, Loren, Stall, Wilansky

Regimen C = Hernandez, Santiago, Brunk, Jackson, Nevesky

Regimen D = Deng, Mann, Rodriguez, Cruz, Tran

SECTION 3.2 EXERCISES

3.43 This is a comparative experiment with two treatments — some students view the price history as a steady price and the other students view the price history with regular promotions that temporarily cut the price. The explanatory variable is how the price history is viewed. The response variable is the price the student would expect to pay for the detergent. Nothing is said in the problem about how the students are assigned to the two treatments, but it would be best if they had been assigned at random making this a randomized comparative experiment.

3.45 a) The subjects are the 210 children ages 4 to 12 years who participated in the study. The problem tells us nothing about how these children were selected to be in the study.

b) The factor is the set of choices that are presented to each subject. The levels correspond to the three sets of choices – Set 1, Set 2, and Set 3 – so there are three levels in this problem. (You might have thought the levels were the number of choices presented to the subject, but Sets 2 and 3 each has six choices and correspond to different levels of the factor.) The response variable is whether they chose a milk drink or a fruit drink. Because they had to choose one or the other, the response could just be whether or not they chose a milk drink (yes or no).

c)

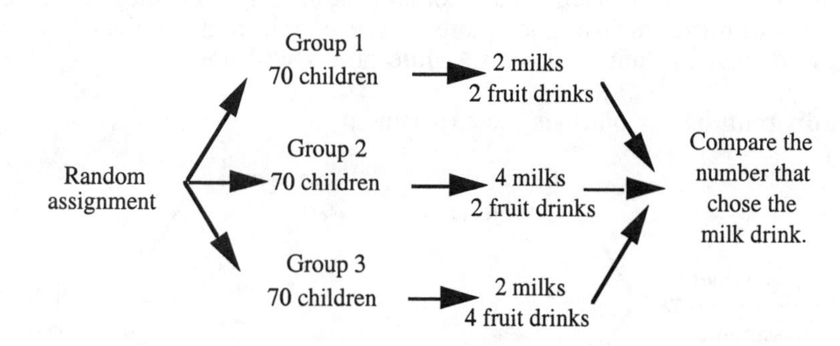

d) Table B starting at line 125 is reproduced here. Read across the row in groups of digits equal to the number of digits you used for your labels (for example, because you used three digits for labels, read line 125 in triples of digits). The five underlined numbers 119, 033, 199, 192, and 148 represent the first five children to be assigned to receive Set 1.

96746 12149 37823 71868 18442 35119 62103 39244
96927 19931 36089 74192 77567 88741 48409 41903

3.47 a) The more seriously ill patients may be assigned to the new method by their doctors. This might be done because experience tells the doctors that there is little chance of surgery being successful in the more serious cases and there is little harm in trying the new treatment on them. The seriousness of the illness would be a lurking variable that would make the new treatment look worse than it really is.

b)

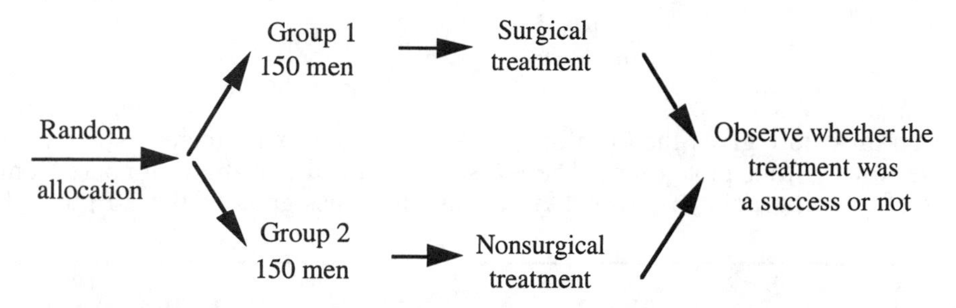

3.49 In a controlled scientific study, the effects of factors other than the treatment can be eliminated or accounted for. This ensures that the differences in improvement that are observed between the subjects can be attributed to the differences in treatments.

3.51 a) The diagram below outlines the experiment.

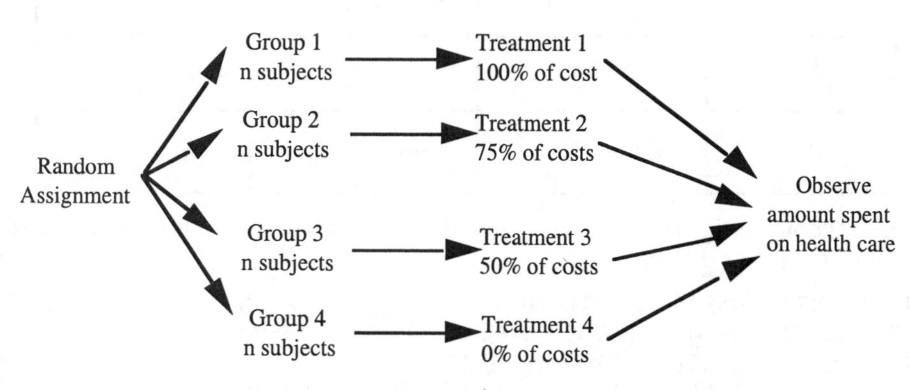

b) As a practical issue, it may take a long time to carry out the study. The experimenters will need to wait until enough claims have been filed to obtain the information they need. Ethically, we are going to assign subjects to different insurance plans; some people might object to this because some will be required to pay for part of their health care while others will not.

3.53 a) The diagram below outlines the experiment.

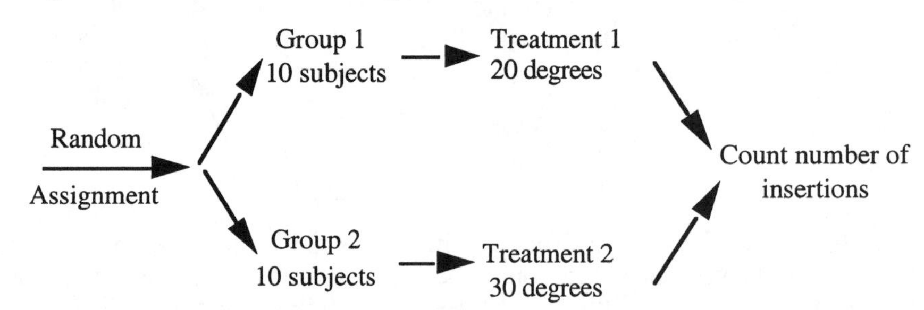

b) Each subject will do the task twice, once under each temperature condition. The experimenter randomly determines the order of the two treatments for each subject. The difference in the number of correct insertions at the two temperatures is the observed response.

3.55 The diagram below outlines a completely randomized design.

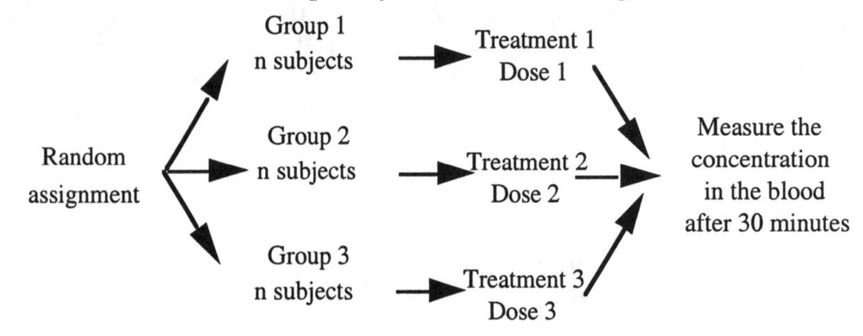

3.57 The diagram below gives the 20 plots available for the experiment. They are in a rectangular field with five rows of four plots each. The plots are assigned two-digit numbers from 01 to 20. The two methods of growing potatoes, A and B are randomly assigned to the 20 plots, 10 plots to each treatment.

01-A	02-A	03-B	04-B
05-B	06-A	07-A	08-A
09-A	10-A	11-B	12-B
13-B	14-B	15-B	16-A
17-B	18-B	19-A	20-A

To carry out the random assignment, enter Table B and read two-digit groups until 10 plots have been selected. Starting at line 145,

19687 12633 57857 95806 09931 02150 43163 58636
37609 59057 66967 83401 60705 02384 90597 93600
54973 86278 88737 74351 47500 84552 19909 67181
00694 05977 19664 65441 20903 62371 22725 53340

71546 05223 53946 68743 72460 27601 45403 88692
07511 88915

The first 10 plots selected are 19, 06, 09, 10, 16, 01, 08, 20, 02, and 07. These are assigned to method A and the remaining 10 are assigned to method B as in the diagram. When harvested, the number of tubers per plant and fresh weight of vegetable growth per plant will be measured.

SECTION 3.3

OVERVIEW

Statistical inference is the technique that allows us to use the information in a sample to draw conclusions about the population. To understand the idea of statistical inference, it is important to understand the distinction between **parameters** and **statistics.** A **statistic** is a number we calculate based on a sample from the population — its value can be computed once we have taken the sample, but its value varies from sample to sample. A statistic is generally used to estimate a population **parameter** that is a fixed but unknown number that describes the population.

The variation in a statistic from sample to sample is called **sampling variability.** It can be described through the **sampling distribution** of the statistic that is the distribution of values taken by the statistic in all possible samples of the same size from the population. The sampling distribution can be described in the same way as the distributions we encountered in Chapter 1. Three important features are

- a measure of center
- a measure of spread
- a description of the shape of the distribution.

The properties and usefulness of a statistic can be determined by considering its sampling distribution. If the sampling distribution of a statistic is centered (has its mean) at the value of the population parameter, then the statistic is **unbiased** for this parameter. This means that the statistic tends to neither overestimate nor underestimate the parameter.

Another important feature of the sampling distribution is its spread. If the statistic is unbiased and the sampling distribution has little spread or variability, then the statistic will tend to be close to the parameter it is estimating for most samples. The variability of a statistic is related to both the sampling design and the sample size n. Larger sample sizes give smaller spread (better estimates) for any sampling design. An important feature of the spread is that as long as the population is much larger than the sample (at least 10 times), the spread of the sampling distribution will depend primarily on the sample size, not the population size.

If the parameter p is the proportion of the population with a particular characteristic, then the statistic $\hat{p}$, the proportion in the sample with this characteristic, is an unbiased estimator. Provided the samples are selected at random, **probability** theory can be used to tell us about the distribution of a statistic.

APPLY YOUR KNOWLEDGE

3.59 The number 73% is a statistic as it is calculated from the sample of voters. The number 68% is a parameter as it describes the population of all registered voters.

3.61 a) The mean is $25.234 million which is somewhat smaller than the first two means.

b) The standard deviation is $30.788 million, which is also smaller than standard deviations calculated from the first two samples.

c) The median is $13.3 million which is close to the medians of the first two samples. It is possible that outliers were present in the first two samples, which explains the larger differences between the means and standard deviations than between the medians.

3.63 a) The range of the means of samples of 25 goes from approximately $10 million to close to $70 million.

b) The range of the means of samples of 10 goes from approximately $5 million to close to $80 million. The spread is larger than in part (a).

c) The range of the means of samples of 100 goes from approximately $18 million to slightly over $40 million. The spread is much smaller than in either (a) or (b).

d) The fact illustrated is that the sampling variability of a statistic decreases as the sample size increases.

3.65 a) The population is all people who live in Ontario. Because everyone uses health care, it is not restricted to adults, etc. The sample is the 61,239 residents of Ontario who were interviewed.

b) Yes. This is a very large sample and it is a probability sample, so we expect that the sample proportions are quite close to the population proportions.

SECTION 3.3 EXERCISES

3.67 The number 2.503 is a parameter because it describes the population, which is the carload of ball bearings. The number 2.515 is a statistic because it is calculated from the sample of 100 bearings that are inspected.

3.69 The management student is interested in obtaining information on the attitude of all students toward part-time work while attending college. The larger sample size suggested by the faculty advisor will decrease the sampling variability of the estimates and allow the student to obtain more accurate information about the population of interest. For very small samples, the sampling variability can be so large that the sample contains very little useful information.

3.71 The margin or error decreases as the sample size increases. Because the number of adults in the sample is 1025 + 472 = 1497 is larger than the number of men, the margin of error for men alone is larger.

3.73 a) The invoices use single-digit numbers. Starting at line 160 Table B,

98163 45944 34210 64158 76971 27689 82926 75957
43400 25831

the sample of size four are invoice numbers 9, 8, 1, and 6 corresponding to days past due of 6, 7, 12 and 15 with an average of $\bar{x} = 10$.

b) This process can be repeated nine more times, continuing where we left off in Table B to give

Sample	Days past due	$\bar{x}$
3,4,5,9	5,7,3,6	5.25
4,3,2,1	7,5,10,12	8.50
0,6,4,1	8,15,7,12	10.50
5,8,7,6	3,7,9,15	8,50
9,7,1,2	6,9,12,10	9.25
7,6,8,9	9,15,7,6	9.25
8,2,9,6	7,10,6,15	9.50
7,5,9,4	9,3,6,7	6.25
3,4,0,2	5,7,8,10	7.50

The histogram is given on the next page. The center of the histogram looks slightly below 9. If the number of repetitions were increased, the center of the histogram would be fairly close to the population mean of 8.2.

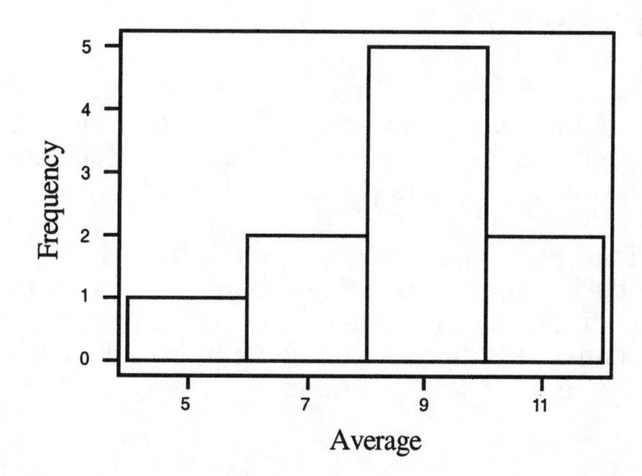

3.75 a) Going to line 101 in Table B and using the two-digit numbers 00 to 99 to represent the circles, the first sample is

19, 22, 39, 50, 34 with $\hat{p} = 0.6$

b) Continuing in lines 102 to 110, the next nine samples and their values of $\hat{p}$ are

73, 67, 64, 71, 50 with $\hat{p} = 0.4$
45, 46, 77, 17, 09 with $\hat{p} = 0.6$
52, 71, 13, 88, 89 with $\hat{p} = 0.4$
95, 59, 29, 40, 07 with $\hat{p} = 0.0$
68, 41, 73, 50, 13 with $\hat{p} = 0.6$
82, 73, 95, 78, 90 with $\hat{p} = 0.2$
60, 94, 07, 20, 24 with $\hat{p} = 0.8$
36, 00, 91, 93, 65 with $\hat{p} = 0.8$
38, 44, 84, 87, 89 with $\hat{p} = 0.6$

c) The histogram is given below.

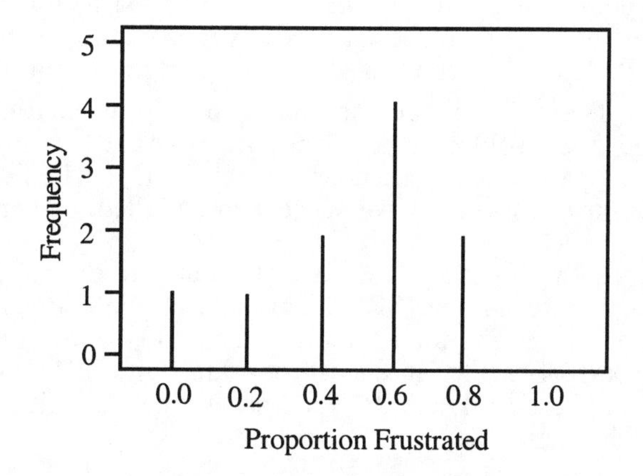

d) Four of the samples estimated the population proportion $p = 0.6$ exactly. The center of the 10 sample values looks reasonably close to 0.6, but if we took a larger number of samples 0.6 should be the center because $\hat{p}$ is an unbiased estimate of p. Recall that the definition of unbiased is that the center of the sampling distribution of $\hat{p}$ should be $p = 0.6$.

CHAPTER 3 REVIEW EXERCISES

3.77 This is a matched pairs experiments. The two types of muffin are the treatments and the subjects taste both treatments and indicate which they prefer. Both muffins could be placed in front of the subject, one on the left and one on the right, with the side randomized or if they are presented in order, the order in which the two muffins are presented should be randomized for each subject.

3.79 The wording of questions has the most important influence on the answers to a survey. Leading questions can introduce strong bias. The first question is presenting a rationale for differential pay and emphasizes fairness to the individual. The wording is leading subjects to agree. The second question is presenting a rationale for being fair and distributing wealth equally to improve society in general. It is also leading subjects to agree. But agreeing with these two statements is conflicting.

3.81 a) There may be systematic differences between the recitations attached to the two lecturers. There may be different lecturers and possibly different sets of recitation instructors associated with each lecture. Also, students often choose their lecture and possibly more motivated students have self-selected to be in one of the lectures. Any differences between the treatments may be due to these or other lurking variables rather than the treatments themselves, so this study design is a bad idea.

b) Randomly assign the 20 recitations to the two groups, 10 in each group. First number the 20 recitations from 01 to 20. To carry out the random assignment, enter Table B and read two-digit groups until 10 recitations have been selected. These are assigned to the online games. The remaining 10 recitations discuss markets. Starting at line 145,

19687 12633 57857 95806 09931 02150 43163 58636
37609 59057 66967 83401 60705 02384 90597 93600
54973 86278 88737 74351 47500 84552 19909 67181
00694 05977 19664 65441 20903 62371 22725 53340
71546 05223 53946 68743 72460 27601 45403 88692
07511

The first 10 recitations selected are 19, 06, 09, 10, 16, 01, 08, 20, 02, and 07. These are assigned to the online games and the remaining 10 are assigned to discussions. At the end of the experiment, students understanding of how markets set prices is determined and the two treatments compared.

3.83 The nonresponse rate can produce serious bias. Of the 13,000 questionnaires, 4814 responded for a 4814/13000 = 37% response rate. Businesses that responded to the initial questionnaire and those that did not could be different with respect to variables affecting their survival. Of the 4814 respondents, 2294 were selected for the study as they had started within 17 months and could be considered new businesses. These 2294 became the study "population". Of these 963 firms survived, 611 failed, 171 had been sold and no information could be obtained on the remaining 549/2294 = 24%. The final comparison was based on only the 963 + 611 firms, which may not be a representative sample of small businesses that have succeeded or failed, so generalization is difficult.

3.85 The four types of colleges stratify the population of faculty in the state into four groups. The basic stratified sampling design is to take an SRS from each of the four groups and measure their attitudes toward collective bargaining. The total sample size is 200, so you might select an SRS of 50 from each group. If the numbers of faculty in each group are quite different and the attitudes were known to be different, this might be taken into account in the stratified sampling design when choosing the number from each stratum.

3.87 Unfortunately, this is a sensitive question and many people are embarrassed to admit that they do not vote. When asking the question, the tendency is for more people to indicate that they have voted then actually did. A careful survey should ask related questions, such as where do people usually vote, etc., to try to adjust for nontruthful answers to the questions. The Gallup poll includes such questions when trying to estimate the proportions that will vote for each candidate in an election.

3.89 a) The response variable is whether or not the subject gets colon cancer. The explanatory variable are the four different supplement combinations—daily beta carotene, daily vitamins C and E, all three vitamins, and a daily placebo.

b) The design of the experiment is outlined below. The 864 subjects were divided into four groups of equal size.

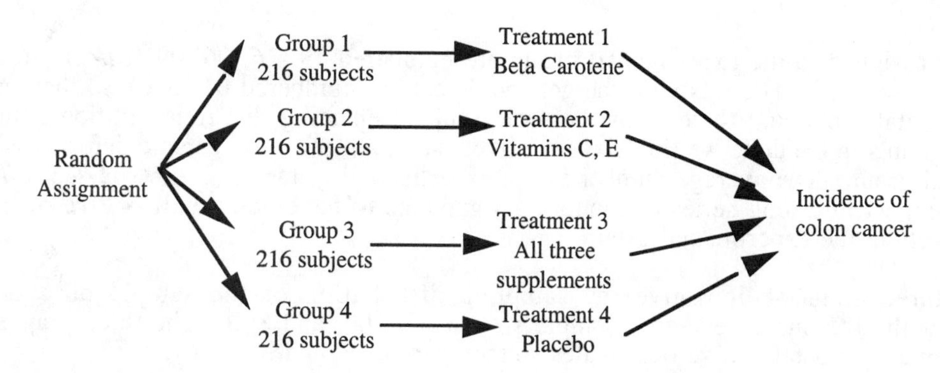

c) The subjects are given three-digit labels 001 to 864. Starting at line 118 and reading in groups of three,

731|90   3|253|3   04|470| 296|69   84407

the first five subjects assigned to the beta carotene group are subjects 731, 253, 304, 470, and 296.

d) The study was double-blind means that neither the subjects nor those evaluating the subjects' responses knew which treatments were applied to individual subjects.

e) "No significant difference" means that any observed differences are due to chance. There is no evidence of any differences between the four treatments in reducing the incidence of colon cancer.

f) People who eat lots of fruits and vegetables tend to eat a diet that is high in fiber. There is also evidence that a diet high in fiber reduces the incidence of colon cancer, so it could be the fiber rather than the antioxidants that is responsible for the observed benefits of fruits and vegetables. Also, those who watch their diets may exercise more, not smoke, and have other lifestyle differences that could relate to getting colon cancer.

3.91 a) There are two factors. The first is the "storage method," which has three levels — freshly harvested, stored for a month at room temperature and stored for a month refrigerated. The second factor is "when cooked," which has two levels — immediately or after an hour at room temperature. The treatments are the six combinations of the levels of the two factors. The response is the rating of the color and flavor by the judges.

b) One possible design is to take the group of judges and divide them at random into six groups. One group is assigned to each treatment. This is a randomized comparative experiment but has the disadvantage that the response variable is very subjective, and there must be training of the subjects to get meaningful responses. Part (c) offers an alternative design that is often used for this type of sensory evaluation.

c) Having each subject taste fries from each of the six treatments in a random order is a block design and eliminates the variability between subjects. Each subject could try and rank the fries from best to worst, compare them in pairs, or try and assign a score for color and flavor.

3.93 They should not be told which burger comes from Wendy's or McDonald's and in fact shouldn't be told which two burger chains are being compared as identification of the burger might be easy. It would be best to make the hamburgers in such a way that they look alike in terms of size, bun, and condiments. Each subject would be presented with the two burgers in a random order. The randomization of the order should be done by a flip of a coin for each subject.

3.95 The rats are numbered 01 to 30. Now enter Table B and read two-digit groups until 15 rats are selected for the experimental group. Starting at line 145,

19687 12633 57857 95806 09931 02150 43163 58636
37609 59057 66967 83401 60705 02384 90597 93600
54973 86278 88737 74351 47500 84552 19909 67181
00694 05977 19664 65441 20903 62371 22725 53340
71546 05223 53946 68743 72460 27601 45403 88692
07511 88915

The 15 rats assigned to the experimental group are numbered 19, 26, 06, 09, 10, 21, 16, 01, 23, 08, 20, 22, 02, 07, and 18. The rats with the genetic defect are numbered 01 to 10, so there are seven in the experimental group and three in the control group. Repeating this four additional times starting at different points in the table we find six, two, five, and seven rats with genetic defects assigned to the experimental group. The average number in the experimental group is $(7 + 6 + 2 + 5 + 7) / 5 = 5.4$. We would expect in a long series of random assignments to have about half or five of the rats with genetic defects in the experimental group.

3.97 The three graphs below give the sampling distribution of the sample proportion from a population with 60% successes and a sample size of 50, 200, and 800. The three graphs are drawn using the same horizontal and vertical scales so they can be compared easily.

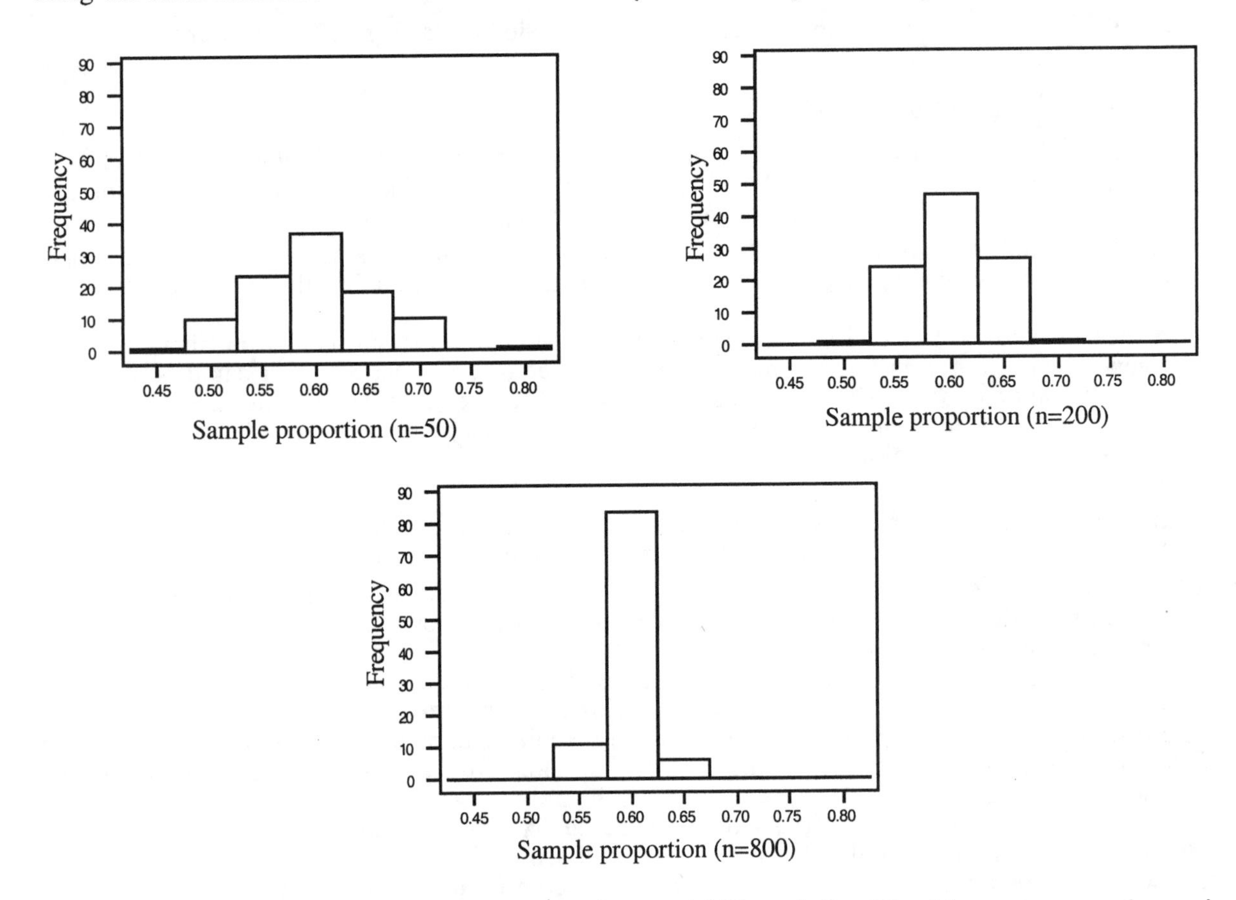

Increasing the sample size decreases the sampling variability of $\hat{p}$. The histograms are becoming more concentrated about their center, which is $p = 0.6$, so the chance of getting a value of $\hat{p}$ far from 0.6 becomes smaller as the sample size increases.

CHAPTER 4

PROBABILITY AND SAMPLING DISTRIBUTIONS

SECTION 4.1

OVERVIEW

A process or phenomenon is called **random** if its outcome is uncertain. Although individual outcomes are uncertain, when the process is repeated a large number of times the underlying distribution for the possible outcomes begins to emerge. For any outcome, its **probability** is the proportion of times, or the relative frequency, with which the outcome would occur in a long series of repetitions of the process. It is important that these repetitions or trials be **independent** for this property to hold.

You can study random behavior by carrying out physical experiments such as coin tossing or rolling of a die, or you can simulate a random phenomenon on the computer. Using the computer is particularly helpful when we want to consider a large number of trials.

APPLY YOUR KNOWLEDGE

4.1 We spun a nickel 50 times and got 22 heads. From this, we estimate the probability of heads to be 22/50 = 0.44. Your results will probably differ somewhat, but your estimate of the probability of heads will be the number of heads you got in 50 spins divided by 50.

SECTION 4.1 EXERCISES

4.3 The first 200 digits contain 21 0's (three in the first line, five in the second, six in the third, four in the fourth, and three on the fifth line of the table). The proportion of 0's is thus 21/200 = 0.105.

4.5 We tossed a thumbtack 100 times and it landed with the point up 39 times. Thus, the approximate probability of landing point up is 39/100 = 0.39. You may get different results depending on the type of thumbtack you use. Thumbtacks come in different sizes and slightly different shapes and these characteristics may affect your results.

4.7 a) 0 = "This event is impossible. It can never occur."

b) 1 = "This event is certain. It will occur on every trial."

c) 0.01 = "This event is very unlikely, but it will occur once in a while in a long sequence of trials."

d) 0.6 = "This event occurs more often than not."

4.9 a) We used Minitab to simulate the 100 customers. We used the random data command under the Calc menu in Minitab to generate a sequence of 100 Bernoulli trials with probability of success 0.5. The result of each trial is a 0 or 1. We let 1 represent "buy" and 0 "not buy." The sequence of 100 0's and 1's we obtained is given on the next page.

1 1 1 1 0 1 1 1 0 0 0 0 0 1 0 1 1 0 0 1 1 1 1 0 1

0 1 1 0 0 0 1 1 1 0 0 0 0 1 0 1 1 0 1 0 0 1 0 0 0

1 0 0 1 0 0 1 0 0 1 1 1 1 0 0 1 1 0 0 1 1 0 0 1 1

0 0 1 0 1 1 0 0 0 1 0 0 1 0 1 1 0 1 1 1 1 0 0 1 0

The number of "buys" (1's) in the sequence is 50. Thus, the percent of simulated customers that bought a new computer is 50/100 × 100% = 50%. You may get somewhat different results, but your percent will probably be fairly close to 50%.

b) The longest run of "buys" (1's) is 4. This occurred four times and these are underlined in the sequence displayed in part (a). The longest run of "not buys" is five. This occurred once and is underlined in the sequence displayed in part (a).

4.11 a) We used Minitab to simulate the 100 draws of 20 people from the population in Exercise 4.10. We used the random data command under the Calc menu in Minitab to generate 100 draws of size 20 from a binomial with probability of success 0.65. The results of the 100 draws we obtained are given below.

12 14 13 14 12 15 10 13 15 12 10 11 11 12 11 11 11 15 18 14

14 16 11 16 13 08 17 11 13 14 10 15 11 14 09 09 13 14 14 11

16 17 13 12 12 14 12 12 11 05 16 16 12 13 12 08 11 11 09 09

14 14 14 10 11 13 12 12 13 08 10 09 09 12 13 10 12 11 14 16

15 16 13 16 12 18 14 09 16 11 15 15 15 16 15 09 12 13 10 11

We have underlined all the cases in these 100 draws where at least 14 people had a favorable opinion. There are 37 such cases, so we the approximate probability that out of 20 people at least 14 have a favorable opinion of Microsoft is 37/100 = 0.37. Different simulations will give different answers, but generally the results should not be too different from about 0.42.

b) We converted the 100 counts above into percents by dividing each count by 20 and multiplying by 100%. A histogram of these percents is

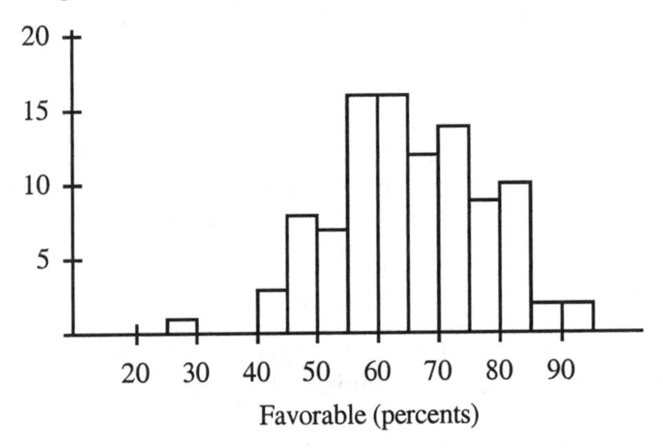

The shape is roughly symmetric and bell-shaped, the center appears to be about 65% and (other than the outlier at 25%) the values range from 40% to 90% with most of the distribution between 45% and 85%.

c) We simulated 100 draws of 320 people and converted the counts to percents. The results are presented below.

65.00 63.75 66.88 73.75 64.06 58.44 72.50 69.38 66.56 63.75

63.75 63.75 66.56 64.69 63.44 65.00 65.63 64.06 65.94 61.88

64.38 67.50 66.56 68.44 68.44 64.69 66.88 66.25 60.94 63.75

69.69 63.13 64.68 63.44 65.31 64.69 69.38 67.19 66.25 71.88

62.50 67.50 63.75 63.75 66.56 65.31 62.50 64.38 65.63 68.13

65.63 65.31 66.56 61.88 63.75 67.19 63.13 66.56 69.06 65.63

68.75 60.63 69.69 66.88 59.38 66.25 67.19 65.94 62.81 65.94

64.69 60.31 62.50 70.00 62.19 63.75 67.50 63.44 64.69 62.50

66.25 62.81 69.69 65.94 65.63 62.50 68.75 70.31 61.56 62.19

65.94 62.81 62.81 62.50 63.44 67.19 61.88 67.19 65.31 66.25

A histogram of these 100 percents is

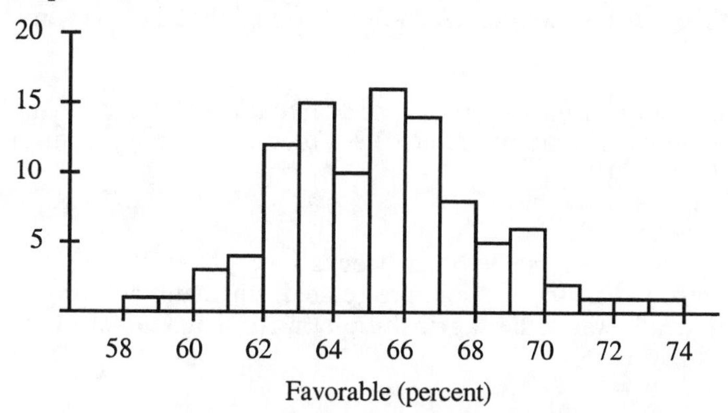

The shape is roughly symmetric and bell-shaped, the center is about 65% or 66%, and the values range from 58% to 74% with most of the distribution between 62% and 70%.

d) Both distributions are roughly symmetric and bell-shaped, and both have center near 65%. The histogram in (b) is more spread out than the histogram in (c).

SECTION 4.2

OVERVIEW

The description of a random phenomenon begins with the **sample space,** which is the list of all possible outcomes. A set of outcomes is called an **event.** Once we have determined the sample space, a **probability model** tells us how to assign probabilities to the various events that can occur. There are four basic rules that probabilities must satisfy.

- Any probability is a number between 0 and 1.
- All possible outcomes together must have probability 1.
- The probability that an event does not occur is 1 minus the probability that the event occurs.
- If two events have no outcomes in common, the probability that one or the other occurs is the sum of their individual probabilities.

In a sample space with a finite number of outcomes, probabilities are assigned to the individual outcomes and the probability of any event is the sum of the probabilities of the outcomes that it contains. In some special cases, the outcomes are all **equally likely** and the probability of any event A is just computed as

$$P(A) = (\text{number of outcomes in } A)/(\text{number of outcomes in } S).$$

Events are **disjoint** if they have no outcomes in common. In this special case, the probability that one or the other event occurs is the sum of their individual probabilities. This is the addition rule for disjoint events, namely

$$P(A \text{ or } B) = P(A) + P(B).$$

APPLY YOUR KNOWLEDGE

4.13 a) S = {any number (including fractional values) between 0 and 24 hours}.
b) S = {any integer value between 0 and 11,000}.
c) $S = \{0, 1, 2, 3, 4, 5, 6, 7, 8, 9, 10, 11, 12\}$.
d) There are several possible answers. Because it is difficult to say precisely what the lowest possible and highest possible annual salaries for CEOs are, we might take S = {any number 0 or larger}. If we know the minimum and maximum salaries for CEOs, we might take S = {any number between the known minimum and maximum salaries of CEOs}.

e) It is difficult to say precisely what the minimum and maximum possible weight gains are. In fact, it is possible that the rats lose weight. As a result, we might take S = {any possible number, either positive or negative}.

4.15 The events are disjoint so P(death was either agriculture-related or manufacturing-related)
= P(death was agriculture-related) + P(death was manufacturing-related)
= 0.134 + 0.119
= 0.253.

We also find P(death was related to some other occupation)
= P(death was neither agriculture-related nor manufacturing-related)
= 1 – P(death was either agriculture-related or manufacturing-related)
= 1 – 0.253
= 0.747.

4.17 For a legitimate model, each probability must be between 0 and 1 (inclusive) and the probabilities of the outcomes must sum to 1. We conclude the following.

Model 1: Not legitimate. The probabilities of the six outcomes do not sum to 1 (they sum to 6/7).
Model 2: Legitimate.
Model 3: Not legitimate. The probabilities of the six outcomes do not sum to 1 (they sum to 7/6).
Model 4: Not legitimate. Some of the probabilities are greater than 1 (and the probabilities of the six outcomes sum to 8 rather than 1).

4.19 a) Because the probabilities of the outcomes must sum to 1, we find P(completely satisfied) = 1 – P(somewhat satisfied) – P(somewhat dissatisfied) – P(completely dissatisfied)
= 1 – 0.47 – 0.12 – 0.02 = 1 – 0.61 = 0.39

b) P(dissatisfied) = P(somewhat dissatisfied) + P (completely dissatisfied) = 0.12 + 0.02 = 0.14

4.21 a) Recall from geometry that the area of a triangle is (1/2) × height × base. In Figure 4.6, we see that the height is 1 and the length of the base is 2. Thus, the area is (1/2) × 1 × 2 = 1.

b) The desired area is shaded in the plot below.

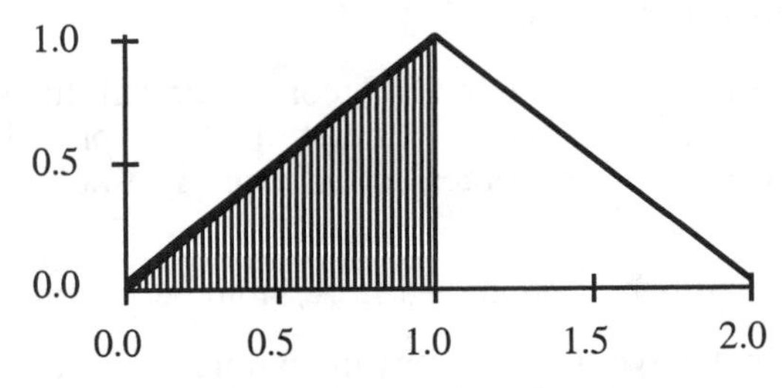

The shaded region is a triangle with a height of 1 and a base of length 1 so the area = (1/2) × height × base = (1/2)×1×1 = 0.5. Thus, the probability that T is less than 1 is 0.5.

c) The desired area is shaded in the plot below.

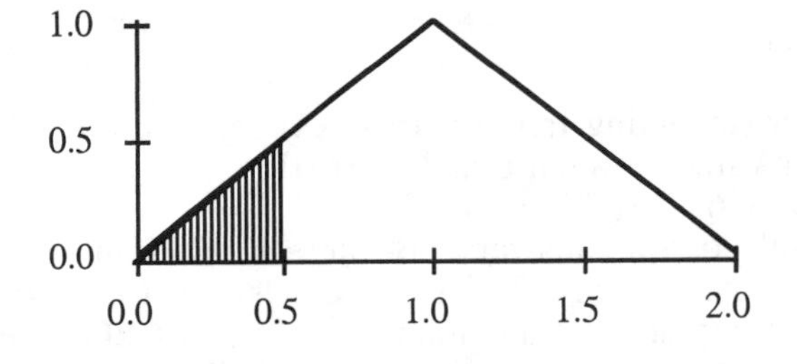

The shaded region is a triangle with a height of 0.5 and a base of length 0.5 so the area = (1/2) × height × base = (1/2) × 0.5 × 0.5 = 0.1125. Thus, the probability that T is less than 0.5 is 0.1125.

4.23 a) In symbols, we seek $P(Y > 300)$. 300 is the mean and we know that half of the area under the normal curve is above the mean so $P(Y > 300) = 0.5$.

b) In symbols, we seek $P(Y > 370)$. The mean is 300 and the standard deviation is 35, so 370 is two standard deviations above the mean. We know that the area within two standard deviations of the mean is 0.95, thus the area more than two standard deviations away from the mean is 1 − 0.95 = 0.05. The normal density curve is symmetric, so half of this area is corresponds to scores that are more than two standard deviations above the mean, namely scores above 370. Thus $P(Y > 370)$ = (1/2) × 0.05 = 0.025.

SECTION 4.2 EXERCISES

4.25 a) P(not farmland) = 1 − P(farmland) = 1 − 0.92 = 0.08.

b) Because farmland and forest are disjoint categories of land,

P(either farmland or forest) = P(farmland) + P(forest) = 0.92 + 0.01 = 0.93.

c) P(something other than farmland or forest) = P(not farmland nor forest)
= 1 − P(either farmland or forest)
= 1 − 0.93 = 0.07.

4.27 a) P(blue) = 1 − P(not blue) = 1 − P(brown) − P(red) − P(yellow) − P(green) − P(orange)
= 1 − 0.3 − 0.2 − 0.2 − 0.1 − 0.1 = 1 − 0.9 = 0.1.

b) P(blue) = 1 − P(not blue) = 1 − P(brown) − P(red) − P(yellow) − P(green) − P(orange)
= 1 − 0.2 − 0.2 − 0.2 − 0.1 − 0.1 = 1 − 0.8 = 0.2.

c) The events M&M is red, M&M is yellow, and M&M is orange are disjoint so

P(plain M&M is red, yellow, or orange) = P(plain M&M is red) + P(plain M&M is yellow)
+ P(plain M&M is orange)
= 0.2 + 0.2 + 0.1 = 0.5.

P(peanut M&M is red, yellow, or orange) = P(peanut M&M is red) + P(peanut M&M is yellow)
+ P(peanut M&M is orange)
= 0.2 + 0.2 + 0.1 = 0.5.

4.29 For a legitimate model, each probability must be between 0 and 1 (inclusive) and the probabilities of the outcomes must sum to 1. We conclude the following.

a) Legitimate.

b) Not legitimate. The events are disjoint but the probabilities sum to 1.6 which is greater than 1.

c) Not legitimate. The probabilities of the outcomes do not sum to 1 (they sum only to 0.8).

4.31 a) $P(A) = P(<10) + P(10–49)$ = 0.09 + 0.20 = 0.29.

$P(B) = P(500–999) + P(1000–1999) + P(\geq 20{,}000)$ = 0.09 + 0.05 + 0.04 = 0.18.

b) In words, "A does not occur" means the farm is 50 acres or more in size. Using rule 3 we find

$P(A$ does not occur$) = 1 − P(A)$ = 1 − 0.29 = 0.71.

c) In word, "A or B" means the farm is less than 50 acres or 500 or more acres in size. By the addition rule and using the results of part (a), we find $P(A \text{ or } B) = P(A) + P(B)$ = 0.29 + 0.18 = 0.47.

4.33 a) The eight arrangements of preferences are NNN, NNO, NON, ONN, NOO, ONO, OON, and OOO. Because all eight arrangements are equally likely, and there probabilities must sum to 1, each must have probability 1/8 = 0.125.

b) Because the arrangements are disjoint, $P(X = 2)$ = P(NNO, NON, or ONN) = P(NNO) + P(NON) + P(ONN) = 1/8 + 1/8 + 1/8 = 3/8 = 0.375.

c) The event $X = 0$ corresponds to the outcome OOO, which has probability 1/8. The event $X = 1$ corresponds to outcomes NOO, ONO, or OON, which sum to probability 3/8. The event $X = 3$

corresponds to the outcome NNN, which has probability 1/8. A table listing the values of X and the corresponding probabilities is thus

X	0	1	2	3
$P(X)$	1/8 = 0.125	3/8 = 0.375	3/8 = 0.375	1/8 = 0.125

4.35 a) All the probabilities given are between 0 and 1 and they sum to 1 so this is a legitimate probability distribution.

b) $P(X \geq 5) = P(X = 5) + P(X = 6) + P(X = 7) = 0.07 + 0.03 + 0.01 = 0.11.$

c) $P(X > 5) = P(X = 6) + P(X = 7) = 0.03 + 0.01 = 0.04.$

d) $P(2 < X \leq 4) = P(X = 3) + P(X = 4) = 0.17 + 0.15 = 0.32.$

e) $P(X \neq 1) = 1 - P(X = 1) = 1 - 0.25 = 0.75.$

f) P(a randomly chosen household contains more than two persons)
$= P(X > 2) = P(X = 3) + P(X = 4) + P(X = 5) + P(X = 6) + P(X = 7)$
$= 0.17 + 0.15 + 0.07 + 0.03 + 0.01 = 0.43.$

SECTION 4.3

OVERVIEW

A **random variable** is a variable whose value is a numerical outcome of a random phenomenon. The restriction to numerical outcomes makes the description of the probability model simpler and allows us to begin to look at some further properties of probability models in a unified way. If we toss a coin three times and record the sequence of heads and tails, then an example of an outcome would be HTH, which would not correspond directly to a random variable. On the other hand, if we were keeping track only of the number of heads on the three tosses, then the outcome of the experiment would be 0, 1, 2, or 3 and would correspond to the value of the random variable X = number of heads.

The two types of random variables we will encounter are **discrete** and **continuous** random variables. The **probability distribution** of a random variable tells us about the possible values of X and how to assign probabilities to these values. A discrete random variable has a finite number of values, and the probability distribution is a list of the possible values of X and the probabilities assigned to these values. The probability distribution can be given in a table or using a **probability histogram.** For any event described in terms of X, the probability of the event is just the sum of the probabilities of the values of X included in the event.

A continuous random variable takes all values in some interval of numbers. Probabilities of events are determined using a **density curve.** The probability of any event is the area under the curve corresponding to the values that make up the event. For density curves that involve regular shapes, such as rectangles or triangles, we can compute probabilities of events using simple geometrical arguments. The **normal distribution** is another example of a continuous probability distribution, and probabilities of events for normal random variables are computed by standardizing and referring to Table A as was done in Section 1.3.

In Chapter 1, we introduced the concept of the distribution of a set of numbers or data. The distribution describes the different values in the set and the frequency or relative frequency with which those values occur. The mean of the numbers is a measure of the center of the distribution and the standard deviation is a measure of the variability or spread. These concepts are also used to describe features of a random variable X. The probability distribution of a random variable indicates the possible values of the random variable and the probability (relative frequency in repeated observations) with which they occur. The **mean** μ_X of a random variable X describes the center or balance point of the probability distribution or density curve of X. If X is a discrete random variable having possible values $x_1, x_2, \ldots, x_k$ with corresponding probabilities $p_1, p_2, \ldots, p_k$, the mean μ_X is the average of the possible values weighted by the corresponding probabilities, i.e.,

$$\mu_X = x_1p_1 + x_2p_2 + \ldots + x_kp_k$$

The mean of a continuous random variable is computed from the density curve but computations require more advanced mathematics.

The **variance** σ_X^2 of a random variable X is the average squared deviation of the values of X from the mean. For a discrete random variable

$$\sigma_X^2 = (x_1 - \mu_X)^2 p_1 + (x_2 - \mu_X)^2 p_2 + \ldots + (x_k - \mu_X)^2 p_k$$

The **standard deviation** $\boldsymbol{\sigma_X}$ is the positive square root of the variance. The standard deviation measures the variability of the distribution of the random variable X about its mean. The variance of a continuous random variable, such as the mean, is computed from the density curve. Again, computations require more advanced mathematics.

The mean and variances of random variables obey the following rules. If a and b are fixed numbers, then

$$\mu_{a+bX} = a + b\mu_X$$

$$\sigma^2_{a+bX} = b^2\sigma^2_X$$

If X and Y are any two random variables having correlation ρ, then

$$\mu_{X+Y} = \mu_X + \mu_Y$$

$$\sigma^2_{X+Y} = \sigma^2_X + \sigma^2_Y + 2\rho\sigma_X\sigma_Y$$

$$\sigma^2_{X-Y} = \sigma^2_X + \sigma^2_Y - 2\rho\sigma_X\sigma_Y$$

If X and Y are independent random variables, then

$$\sigma^2_{X+Y} = \sigma^2_X + \sigma^2_Y$$

$$\sigma^2_{X-Y} = \sigma^2_X + \sigma^2_Y$$

APPLY YOUR KNOWLEDGE

4.37 a) The random variable is continuous because all times (all numbers) greater than 0 are possible without any separation between values.

b) The random variable is discrete because it is a count and takes only values 0, 1, 2, 3, or 4.

c) The random variable is continuous because any number 0 or larger is possible without any separation between values.

d) The random variable is discrete. Household size can only take values 1, 2, 3, 4, 5, 6, or 7.

4.39 a) All the probabilities given are between 0 and 1 and they sum to 1 so this is a legitimate probability distribution. A probability histogram that displays this distribution is

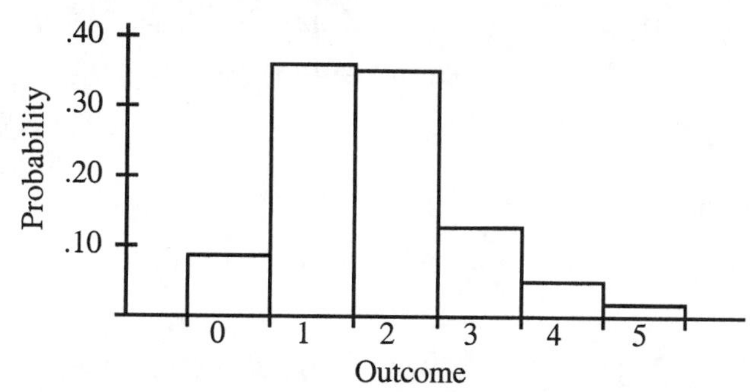

b) The event $\{X \geq 1\}$ means the household chosen at random owns at least 1 car. $P(X \geq 1) = 1 - P(X = 0) = 1 - 0.09 = 0.91$.

c) The percent households that have more cars than the garage can hold is the percent of households that have three or more cars. This percent is the probability (expressed as a percent) that a randomly chosen household owns three or more cars. This probability is

$$P(X \geq 3) = P(X = 3) + P(X = 4) + P(X = 5) = 0.13 + 0.05 + 0.02 = 0.20.$$

Expressed as a percent, we conclude that 20% of all households have more cars than the garage can hold.

4.41 a) The probability that a tire lasts more than 50,000 miles is $P(X > 50{,}000)$. Because 50,000 is the mean and the area to the right of the mean under a normal curve is 0.5, $P(X > 50{,}000) = 0.5$.

b) To calculate $P(X > 60{,}000)$, we will use the techniques for calculating normal probabilities that we learned in Chapter 1.

$$P(X > 60{,}000) = P\left(\frac{X - 50{,}000}{5{,}500} > \frac{60{,}000 - 50{,}000}{5{,}500}\right) = P(Z > 1.82)$$
$$= 1 - P(Z \leq 1.82) = 1 - 0.9656 = 0.0344.$$

c) The normal distribution is continuous, so $P(X = 60{,}000) = 0$ (the area under a single point is 0). Thus, using the result of part (b) $P(X \geq 60{,}000) = P(X > 60{,}000) + P(X = 60{,}000) = 0.0344 + 0 = 0.0344$.

4.43 We calculate $\mu = (10)(0.50) + (20)(0.25) + (30)(0.15) + (40)(0.10) = 5 + 5 + 4.5 + 4 = 18.5$. What this tells us is that if we record the size of the hard drive chosen by many customers in the 60 day period and compute the average of these sizes, the average will be close to 18.5. Unfortunately, knowing μ is not very helpful to the computer maker because it is not one of the possible choices and it does not indicate which choice is most popular. More helpful would be the actual distribution of choices. This would indicate which is the most popular choice, the second most popular choice, etc.

4.45 a) $\mu_X = (200)(0.4) + (300)(0.4) + (400)(0.2) = 80 + 120 + 80 = 280$.

$\mu_Y = (100)(0.3) + (200)(0.50) + (300)(0.15) + (400)(0.05) = 30 + 100 + 45 + 20 = 195$.

b) Profit at the mall is 25 times the number X of phones sold, or $25X$. Thus, the mean profit for the mall is

$$\mu_{25X} = 25\mu_X = 25(280) = 7000.$$

Profit downtown is 35 times the number Y of phones sold, or $35Y$. Thus, the mean profit for downtown is

$$\mu_{35Y} = 35\mu_Y = 35(195) = 6825.$$

c) The combined profit is $25X + 35Y$, so the combined mean profit is

$$\mu_{25X + 35Y} = 25\mu_X + 35\mu_Y = 7000 + 6825 = 13{,}825.$$

4.47 Arranging the calculation in a table, we find

y_i	p_i	$y_i p_i$	$(y_i - p_i)^2 p_i$	
300	0.4	120	$(300 - 445)^2(0.4) =$	8410.00
500	0.5	250	$(500 - 445)^2(0.5) =$	1512.50
750	0.1	75	$(750 - 445)^2(0.1) =$	9302.50
		$\mu_Y = 445$	$\sigma_Y^2 =$	19225.00

and $\sigma_Y = \sqrt{19225} = 138.65$.

4.49 a) Arranging our calculations in a table, we find for the mall location

x_i	p_i	$x_i p_i$	$(x_i - p_i)^2 p_i$	
200	0.4	80	$(200 - 280)^2(0.4) =$	2560
300	0.4	120	$(300 - 280)^2(0.4) =$	160
400	0.2	80	$(400 - 280)^2(0.2) =$	2880
		$\mu_X = 280$	$\sigma_Y^2 =$	5600

and $\sigma_X = \sqrt{5600} = 74.83$.

b) Arranging our calculations in a table, we find for the downtown location

y_i	p_i	y_ip_i	$(y_i - p_i)^2p_i$
100	0.30	30	$(100 - 195)^2(0.30) = 2707.50$
200	0.50	100	$(200 - 195)^2(0.50) = 12.50$
300	0.15	45	$(300 - 195)^2(0.15) = 1653.75$
400	0.05	20	$(400 - 195)^2(0.05) = 2101.25$
		$\mu_Y = 195$	$\sigma_Y^2 = 6475.00$

and $\sigma_Y = \sqrt{6475} = 80.47$.

4.51 We compute

$$\mu_{X-Y} = \mu_X - \mu_Y = 1100 - 1000 = 100.$$

To compute σ_{X-Y} we calculate

$$\begin{aligned}\sigma_{X-Y}^2 = \sigma_X^2 + \sigma_Y^2 - 2\rho\sigma_X\sigma_Y &= (100)^2 + (80)^2 - 2(0.4)(100)(80)\\ &= 10{,}000 + 6400 - 6400 = 10{,}000\end{aligned}$$

Thus, $\sigma_{X-Y} = \sqrt{10{,}000} = 100$. If the correlation between two variables is positive, then large values of one tend to be associated with large values of the other, resulting in a relatively small difference. Also, small values of one tend to be associated with small values of the other, again resulting in a relatively small difference. This suggests that when two variables are positively associated, they vary together and the difference tends to stay relatively small and vary little. Hence, the variance of the difference will tend to be relatively small.

SECTION 4.3 EXERCISES

4.53 a) Probability histograms of the two distributions are given below.

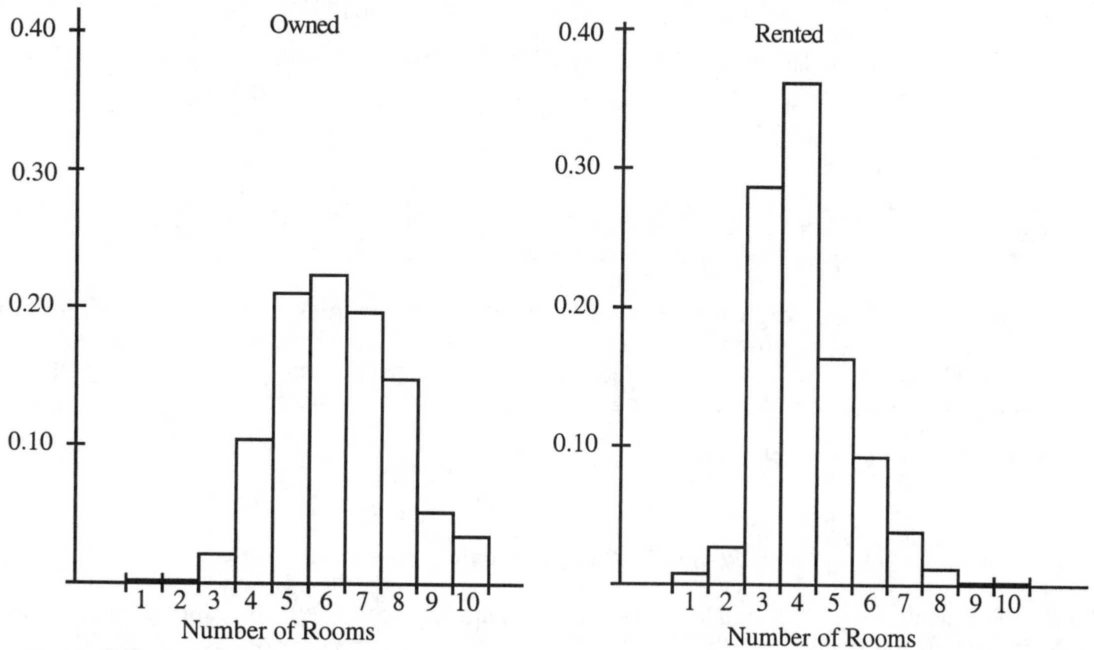

Two important differences between the histograms are (i) the center of the distribution of the number of rooms of owner-occupied units is larger than (to the right of) the center of the distribution of the number of rooms of renter-occupied units and (ii) the spread of the distribution of the number of rooms of owner-occupied units is slightly larger than the spread of the distribution of the number of rooms of renter-occupied units.

b) μ_{owned} = (1)(0.003) + (2)(0.002) + (3)(0.023) + (4)(0.104) + (5)(0.210) + (6)(0.224) + (7)(0.197) + (8)(0.149) + (9)(0.053) + (10)(0.035)
= 0.003 + 0.004 + 0.069 + 0.416 + 1.050 + 1.344 + 1.379 + 1.192 + 0.477 + 0.350 = 6.284.

μ_{rented} = (1)(0.008) + (2)(0.027) + (3)(0.287) + (4)(0.363) + (5)(0.164) + (6)(0.093) + (7)(0.039) + (8)(0.013) + (9)(0.003) + (10))0.003)
= 0.008 + 0.054 + 0.861 + 1.452 + 0.820 + 0.558 + 0.273 + 0.104 + 0.027 + 0.030 = 4.187.

We see that the mean number of rooms for owner-occupied units is larger than the mean number of rooms for renter-occupied units. This reflects the fact that the center of the distribution of the number of rooms of owner-occupied units is larger than the center of the distribution of the number of rooms of renter-occupied units.

4.55 Examine the histograms given in Exercise 4.53. The histogram for the number of rooms for renter-occupied units has four predominant bars (those corresponding to units with 3, 4, 5, and 6 rooms). The histogram for the number of rooms for owner-occupied units has five predominant bars (those corresponding to units with 4, 5, 6, 7, and 8 rooms). This makes the histogram for renter-occupied units appear more peaked and less spread out than the histogram for owner-occupied units.

To find the standard deviation for these distributions, we compute the variances, using the results from Exercise 4.53 that $\mu_{\text{owned}} = 6.284$ and $\mu_{\text{rented}} = 4.187$. We find

$$\sigma^2_{\text{owned}} = (1 - 6.284)^2(0.003) + (2 - 6.284)^2(0.002) + (3 - 6.284)^2(0.023) + (4 - 6.284)^2(0.104) + (5 - 6.284)^2(0.210) + (6 - 6.284)^2(0.224) + (7 - 6.284)^2(0.197) + (8 - 6.284)^2(0.149) + (9 - 6.284)^2(0.053) + (10 - 6.284)^2(0.035) = 2.69204.$$

$$\sigma^2_{\text{rented}} = (1 - 4.187)^2(0.008) + (2 - 4.187)^2(0.027) + (3 - 4.187)^2(0.287) + (4 - 4.187)^2(0.363) + (5 - 4.187)^2(0.164) + (6 - 4.187)^2(0.093) + (7 - 4.187)^2(0.039) + (8 - 4.187)^2(0.013) + (9 - 4.187)^2(0.003) + (10 - 4.187)^2(0.003) = 1.71174.$$

Thus $\sigma_{\text{owned}} = \sqrt{2.69204} = 1.64074$ and $\sigma_{\text{rented}} = \sqrt{1.71174} = 1.30833$. This is consistent with our observations about the spread of the distributions based on the probability histograms.

4.57 There are 1000 three-digit numbers (000 to 999). If you pick a number with three different digits, there are six possible orders (abc, acb, bac, bca, cab, cba). Thus, with such a number there are six out of 1000 numbers that can be selected that will match yours and your probability of winning is 6/1000 and your probability of not winning is 994/1000. Your expected payoff is therefore

(payout if you win)P(winning) + (0 payout if you do not win)P(not winning)
= ($83.33)(6/1000) + $0
= ($499.98/1000) = $0.49998.

At a cost of $1 to play, your long-run "winnings" would be $0.49998 – $1 = –$0.50002.

4.59 a) We would expect X and Y to be independent because they correspond to events that are widely separated in time.

b) Experience suggests that the amount of rainfall on one day is not closely related to the amount of rainfall on the next. Thus, we might expect X and Y to be independent or, because they are not so widely separated in time, perhaps slightly dependent.

c) Orlando and Disney World are close to each other. Rainfall usually covers more than just a very small geographic area, we would expect that if it is raining in Orlando, there is a good chance it is raining in Disney World also. Thus, we would not expect X and Y to be independent.

4.61 a) If X is the time to bring the part from the bin to its position on the automobile chassis and Y is the time required to attach the part to the chassis, then the total time for the entire operation is $X + Y$. Thus, $\mu_{X+Y} = \mu_X + \mu_Y = 11 + 20 = 31$ seconds.

b) The decrease in the standard deviation will not affect the mean. $\mu_{X+Y} = \mu_X + \mu_Y$ and does not depend on the standard deviation of X or Y.

c) Whether the correlation is 0.8 or 0.3, the answer in (a) will remain the same. $\mu_{X+Y} = \mu_X + \mu_Y$ and does not depend on the correlation between X and Y.

4.63 If X and Y are independent,

$$\sigma^2_{X+Y} = \sigma^2_X + \sigma^2_Y = 2^2 + 4^2 = 4 + 16 = 20$$

and thus

$$\sigma_{X+Y} = \sqrt{20} = 4.47.$$

If X and Y are dependent with correlation 0.3, then

$$\sigma^2_{X+Y} = \sigma^2_X + \sigma^2_Y + 2\rho\sigma_X\sigma_Y = 2^2 + 4^2 + (2)(0.3)(2)(4) = 4 + 16 + 4.8 = 24.8$$

and thus

$$\sigma_{X+Y} = \sqrt{24.8} = 4.98.$$

If X and Y are positively correlated, then large values of X and Y tend to occur together resulting in a very large value of $X + Y$. Likewise, small values of X and Y tend to occur together resulting in a small value of $X + Y$. Thus, $X + Y$ exhibits larger variation (a greater spread in values) when X and Y are positively correlated than if they are not.

4.65 $\mu_X = (\mu + \sigma)(0.5) + (\mu - \sigma)(0.5) = \mu$.

$$\begin{aligned}\sigma^2_X &= (\mu + \sigma - \mu_X)^2(0.5) + (\mu - \sigma - \mu_X)^2(0.5)\\ &= (\mu + \sigma - \mu)^2(0.5) + (\mu - \sigma - \mu)^2(0.5)\\ &= \sigma^2(0.5) + \sigma^2(0.5)\\ &= \sigma^2.\end{aligned}$$

Thus $\sigma_X = \sqrt{\sigma^2} = \sigma$.

4.67 a) If we assume that military sales X and civilian sales Y are independent, then using the results of Exercise 4.47,

$$\sigma^2_{X+Y} = \sigma^2_X + \sigma^2_Y = 7{,}800{,}000 + 12{,}763.75 = 7{,}812{,}763.75.$$

Thus, $\sigma_{X+Y} = \sqrt{7{,}812{,}763.75} = 2795.13$.

b)
$$\begin{aligned}\sigma^2_Z &= \sigma^2_{2000X+3500Y}\\ &= 2000^2\,\sigma^2_X + 3500^2\,\sigma^2_Y\\ &= 2000^2(7{,}800{,}000) + 3500^2(12{,}763.75)\\ &= 3.135635594\times10^{13}.\end{aligned}$$

Thus, $\sigma_Z = \sqrt{3.135635594 \times 10^{13}} = 5{,}599{,}674$.

4.69 a) The two students are selected at random and thus we expect their scores to be unrelated or independent.

b) $\mu_{\text{female - male}} = \mu_{\text{female}} - \mu_{\text{male}} = 120 - 105 - 15.$

$$\sigma^2_{\text{female - male}} = \sigma^2_{\text{female}} + \sigma^2_{\text{male}} = 28^2 + 35^2 = 284 + 1225 = 2009.$$

Thus, $\sigma_{\text{female - male}} = \sqrt{2009} = 44.82$.

c) We cannot find the probability that the woman selected scores higher than the man selected because we do not know the probability distribution for the scores of women or men. All we know are the means and standard deviations of these distributions.

4.71 $\sigma^2_{0.8W+0.2Y} = (0.8)^2\,\sigma^2_W + (0.2)^2\,\sigma^2_Y = (0.64)(4.64)^2 + (0.04)(6.75)^2 = 15.60.$

Thus, $\sigma_{0.8W+0.2Y} = \sqrt{15.60} = 3.95$. This is smaller than the result in Exercise 4.70, as we would expect because we no longer include the positive term $2\rho\sigma_{0.8W}\sigma_{0.2Y}$. The mean return remains the same. $\mu_{0.8W+0.2Y} = \mu_{0.8W} + \mu_{0.2Y}$, and this does not depend on the correlation.

4.73 The general rule tells us that if $\rho_{XY} = 1$,

$$\sigma^2_{X+Y} = \sigma^2_X + \sigma^2_Y + 2\rho_{XY}\sigma_X\sigma_Y = \sigma^2_X + \sigma^2_Y + 2\sigma_X\sigma_Y = (\sigma_X + \sigma_Y)^2.$$

Taking square roots, we have $\sigma_{X+Y} = \sigma_X + \sigma_Y$.

4.75 a) $\mu_X = (540)(0.1) + (545)(0.25) + (550)(0.3) + (555)(0.25) + (560)(0.1)$
$= 54 + 136.25 + 165 + 138.75 + 56 = 550.$

$$\sigma^2_X = (540 - 550)^2(0.1) + (545 - 550)^2(0.25) + (550 - 550)^2(0.3) + (555 - 550)^2(0.25) + (560 - 550)^2(0.1) = 10 + 6.25 + 0 + 6.25 + 10 = 32.5.$$

Thus $\sigma_X = \sqrt{32.5} = 5.7$.

b) Using the rules for means and variances, we find

$$\mu_{X - 550} = \mu_X - 550 = 550 - 550 = 0.$$

$$\sigma^2_{X-550} = \sigma^2_X = 32.5.$$

Thus $\sigma_{X - 550} = \sqrt{32.5} = 5.7$.

c) Using the rules for means and variances we find

$$\mu_Y = \mu_{(9X/5)+32} = (9/5)\mu_X + 32 = (9/5)550 + 32 = 990 + 32 = 1022.$$

$$\sigma^2_Y = \sigma^2_{(9X/5)+32} = (9/5)^2\,\sigma^2_X = (9/5)^2(32.5) = 105.3.$$

Thus $\sigma_Y = \sqrt{105.3} = 10.26$

SECTION 4.4

OVERVIEW

This section examines properties of the **sample mean** $\bar{x}$. When we want information about a population mean μ for some variable, we usually select an SRS and use the **sample mean** $\bar{x}$ to estimate the unknown parameter μ. The law of large numbers relates the mean of a set of data to the mean of a random variable and says that the average of the values of X observed in many trials approaches μ_X.

If we select an SRS of size n from a large population with mean μ and standard deviation σ, the sample mean $\bar{x}$ has a sampling distribution with

$$\text{mean} = \mu_{\bar{x}} = \mu$$

and

$$\text{standard deviation} = \sigma_{\bar{x}} = \frac{\sigma}{\sqrt{n}}$$

This implies that the sample mean is an unbiased estimator of the population mean and is less variable than a single observation.

Linear combinations (such as sums or means) of independent normal random variables have normal distributions. In particular, if the population has a normal distribution, the sampling distribution of $\bar{x}$ is normal. Even if the population does not have a normal distribution, for large sample sizes the sampling distribution of $\bar{x}$ computed from an SRS is approximately normal. In particular, the **central limit theorem** states that for large n, the sampling distribution of $\bar{x}$ computed from an SRS is approximately $N(\mu, \frac{\sigma}{\sqrt{n}})$ for any population with mean μ and finite standard deviation σ.

APPLY YOUR KNOWLEDGE

4.77 These are statistics. Pfeiffer undoubtedly tested only a sample of all the models produced by Apple and these means are computed from these samples.

4.79 The law of large numbers says that in the long run, the average payout to Joe will be 60 cents. However, Joe pays $1.00 to play each time, so in the long run his average winnings are $0.60 – $1.00 = –$0.40. Thus, in the long run, if Joe keeps track of his net winnings and computes the average per bet, he will find that he loses an average of 40 cents per bet.

4.81 That is not right. The law of large numbers tells us that the long-run average will be close to 34%. Six of seven at bats is hardly the "long-run." Furthermore, the law of large numbers says nothing about the next event. It only tells us what will happen if we keep track of the long-run average. It is not correct to say that Tony Gwynn is due for a hit after six hitless at-bats.

4.83 a) Using statistical software, we compute the mean of the 10 sizes to be $\mu = 69.4$.

b) Using line 120 in Table B, beginning at the left, our SRS is companies 3, 5, 4, and 7. The mean size of these companies is $\bar{x} = (58 + 73 + 72 + 66)/4 = 269/4 = 67.25$.

c) If we repeat this process nine more times, continuing in line 120 of Table B where we left off (and when we reach the end of a line, continuing on to the next) we get

Sample (Companies selected)	$\bar{x}$
6, 5, 9, 7	(65 + 73 + 62 + 66)/4 = 66.50
2, 3, 9, 4	(80 + 58 + 62 + 72)/4 = 68.00
2, 1, 6, 5	(80 + 62 + 65 + 73)/4 = 70.00
8, 5, 0, 4	(74 + 73 + 82 + 72)/4 = 75.25
2, 6, 3, 5	(80 + 65 + 58 + 73)/4 = 69.00
4, 3, 5, 7	(72 + 58 + 73 + 66)/4 = 67.25
4, 2, 1, 9	(72 + 80 + 62 + 62)/4 = 69.00
3, 7, 1, 4	(58 + 66 + 62 + 72)/4 = 64.50
8, 7, 0, 9	(74 + 66 + 82 + 62)/4 = 71.00

A histogram of these 10 values is given on the next page. The center appears to be at about 69, which is close to the value of μ = 69.4 computed in (a).

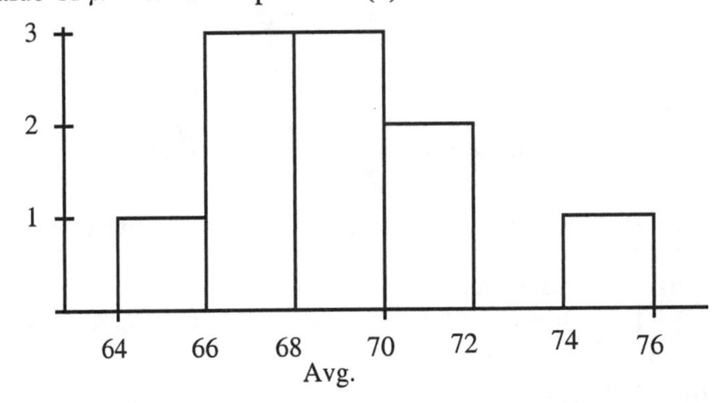

4.85 a) To say that $\bar{x}$ is an unbiased estimator of μ means that $\bar{x}$ will neither systematically overestimate or underestimate μ in repeated use. This also means that if we take many, many samples, calculate $\bar{x}$ for each, and compute the average of these $\bar{x}$ values, this average would be close to μ.

b) Statistics computed from large samples are less variable (in repeated use) than those from small samples. If we draw a large sample from a population, compute the value of some statistic (such as $\bar{x}$), repeat this many times, and keep track of our results, these results will vary less from sample to sample then the results we would obtain if our samples were small.

4.87 The sample size of 200 is reasonably large, so by the central limit theorem we might expect the sampling distribution of $\bar{x}$ to be approximately $N(1.6, 1.2/\sqrt{200}) = N(1.6, 0.085)$. Thus,

$$P(\bar{x} > 2) = P(\frac{\bar{x} - 1.6}{0.085} > \frac{2 - 1.6}{0.085}) = P(z > 4.76)$$

which is approximately 0. (4.76 is well outside the range of values in our normal tables and so we know that this probability is much less than 0.0002; the probability that a standard normal random variable is greater than the largest value in the table, namely 3.49.)

SECTION 4.4 EXERCISES

4.89 19 is a parameter because it is the mean of the population of all businesses in the United States. 14 is a statistic because it is the mean of a sample of businesses in North Dakota.

4.91 The gambler pays \$1.00 for an expected payout of \$0.947. His expected winnings are therefore \$0.947 – \$1.00 = –\$0.053 per bet. In other words, the expected losses of the gambler are \$0.053 per bet. The law of large numbers tells us that if the gambler makes a large number of bets on red, keeps track of his net winnings, and computes the average of these, this average will be close to –\$0.053. In other words, he will find that he loses about 5.3 cents per bet on average.

4.93 a) The individual weighings have a $N(123, 0.08)$ distribution. Thus, the mean $\bar{x}$ of a sample of $n = 3$ weighings will follow a $N(123, 0.08/\sqrt{3}) = N(123, 0.0462)$ distribution.

b) The probability that the lab reports a mean weight of 124 or higher is

$$P(\bar{x} \geq 124) = P(\frac{\bar{x} - 123}{0.0462} \geq \frac{124 - 123}{0.0462}) = P(z \geq 21.65)$$

which is approximately 0. (21.65 is well outside the range of values in our normal tables and so we know that this probability is much less than 0.0002, the probability that a standard normal random variable is greater than the largest value in the table, namely 3.49.)

4.95 a) The weight X of an individual bottle varies according to a $N(298, 3)$ distribution. Thus,

$$P(X < 295) = P(\frac{X - 298}{3} < \frac{295 - 298}{3}) = P(z < -1) = 0.1587.$$

b) The mean contents, $\bar{x}$, of the bottles in a six-pack (assuming these bottles can be viewed as a random sample from the population of all bottles) will vary according to a $N(298, 3/\sqrt{6}) = N(298, 1.225)$ distribution. Thus, the probability that the mean contents of the bottles in a six-pack will be less than 295 ml is

$$P(\bar{x} < 295) = P(\frac{\bar{x} - 298}{1.225} < \frac{295 - 298}{1.225}) = P(z < -2.45) = 0.0071.$$

4.97 If we assume these 400 radiators are a random sample from the population of radiators made by the supplier, then we might reasonably apply the central limit theorem because 40 is a large sample size. We would expect $\bar{x}$ to vary approximately according to an $N(0.15, 0.4/\sqrt{400}) = N(0.15, 0.02)$ distribution and thus the middle 95% will be within two standard deviations ($2 \times 0.02 = 0.04$) of the mean 0.15. In other words, the range $0.15 - 0.04 = 0.11$ to $0.15 + 0.04 = 0.19$ will contain approximately 95% of the many $\bar{x}$'s.

4.99 a) If we assume the 52 weeks represent a random sample of weekly expenses, we might apply the central limit theorem (52 would be a moderate sample size) and argue that the average weekly postal expense, $\bar{x}$, for a period of 52 weeks will vary approximately according to an $N(312, 58/\sqrt{52}) = N(312, 8.043)$ distribution. Thus,

$P(\bar{x} > 400) = P(\frac{\bar{x} - 312}{8.043} > \frac{400 - 312}{8.043}) = P(z > 10.94)$, which is approximately 0. (10.94 is well outside the range of values in our normal tables and so we know that this probability is much less than 0.0002; the probability that a standard normal random variable is greater than the largest value in the table, namely 3.49.)

b) We would need to know the distribution of weekly postal expenses to compute that probability that postage for a particular week will exceed $400. In part (a), we applied the central limit theorem (assuming 52 was a sufficiently large sample size and our sample can be considered a random sample), which does not require that we know the distribution of weekly expenses.

4.101 Sheila's mean glucose level, $\bar{x}$, will vary according to an $N(125, 10/\sqrt{4}) = N(125, 5)$ distribution. We wish to find L such that $P(\bar{x} > L) = 0.05$. We know that $P(\bar{x} > L) =$ $P(\frac{\bar{x} - 125}{5} > \frac{L - 125}{5}) = P(Z > \frac{L - 125}{5})$ and from our normal tables we know that $P(Z > 1.65) = 0.05$. Thus, we must have that $\frac{L - 125}{5} = 1.65$. Solving for L, we find $L = 5 \times 1.65 + 125 = 8.25 + 125 = 133.25$.

CHAPTER 4 REVIEW EXERCISES

4.103 Because exactly one of Brown, Chavez, and Williams will be promoted, the rules of probability require that
1 = *P*(Brown promoted) + *P*(Chavez promoted) + *P*(Williams promoted) = 0.25 + 0.2 + *P*(Williams promoted) so *P*(Williams promoted) = 1 – 0.25 – 0.2 = 0.55.

4.105a) This is a legitimate assignment of probabilities because all the probabilities listed are between 0 and 1 and they sum to 1.

b) *P*(worker female) = *P*(worker female and occupation A) + *P*(worker female and occupation B) + *P*(worker female and occupation C) + *P*(worker female and occupation D) + *P*(worker female and occupation E) + *P*(worker female and occupation F) = 0.09 + 0.20 + 0.08 + 0.01 + 0.04 + 0.01 = 0.43.

c) *P*(not in occupation F) = 1 – *P*(occupation F) = 1 – *P*(male and occupation F or female and occupation F) = 1 – {*P*(male and occupation F) + *P*(female and occupation F)} = 1 – {0.03 + 0.01} = 1 – 0.04 = 0.96.

d) *P*(occupation D or E) = *P*(male and occupation D) + *P*(female and occupation D) + *P*(male and occupation E) + *P*(female and occupation E) = 0.11 + 0.01 + 0.12 + 0.04 = 0.28.

e) *P*(not in occupation D or E) = 1 – *P*(occupation D or E) = 1 – 0.28 = 0.72.

4.107 Because the die are balanced, all outcomes for each are equally likely. If we assume the rolls are independent, then all possible combinations of outcomes when both are rolled are equally likely. There are 12 possible outcomes, listed below, and each must have a probability of 1/12.

	Die 1					
Die 2	1	2	3	4	5	6
0	1/12	1/12	1/12	1/12	1/12	1/12
6	1/12	1/12	1/12	1/12	1/12	1/12

For each outcome, we can compute the total of the faces that are showing to get the probability distribution

Y	1	2	3	4	5	6	7	8	9	10	11	12
$P(Y)$	1/12	1/12	1/12	1/12	1/12	1/12	1/12	1/12	1/12	1/12	1/12	1/12

4.109 Let $X_1, X_2, X_3, X_4, X_5, X_6, X_7, X_8, X_9, X_{10}, X_{11}$, and X_{12} denote the weights of the 12 eggs in a carton. Each X_i varies according to an $N(65, 5)$ distribution. If these are a random sample, their weights are independent and

$$\mu_{X_1+X_2+X_3+X_4+X_5+X_6+X_7+X_8+X_9+X_{10}+X_{11}+X_{12}} = \mu_{X_1}+\mu_{X_2}+\mu_{X_3}+\mu_{X_4}+\ldots+\mu_{X_{12}} = 12\times 65 = 780$$

$$\sigma^2_{X_1+X_2+X_3+X_4+X_5+X_6+X_7+X_8+X_9+X_{10}+X_{11}+X_{12}} = \sigma^2_{X_1}+\sigma^2_{X_2}+\sigma^2_{X_3}+\sigma^2_{X_4}+\ldots+\sigma^2_{X_{12}} = 12\times 5^2 = 300$$

$$\sigma_{X_1+X_2+X_3+X_4+X_5+X_6+X_7+X_8+X_9+X_{10}+X_{11}+X_{12}} = \sqrt{300} = 17.32.$$

Thus, the distribution of the sum is also normal (the sum is just the mean time 12 and we know that the sampling distribution of the mean is normal) and so the weight of a carton varies according to a $N(780, 17.32)$ distribution. Hence, letting Y denote the weight of a carton,

$$\begin{aligned} P(750 < Y < 825) &= P\left(\frac{750-780}{17.32} < \frac{Y-780}{17.32} < \frac{825-780}{17.32}\right) \\ &= P(-1.73 < z < 2.60) \\ &= P(z < 2.60) - P(z < -1.73) \\ &= 0.9953 - 0.0418 \\ &= 0.9535. \end{aligned}$$

4.111 a) This is a legitimate assignment of probabilities because all the probabilities listed are between 0 and 1 and they sum to 1.

b) The mean grade is μ = (4)(0.18) + (3)(0.32) + (2)(0.34) + (1)(0.09) + (0)(0.07) = 0.72 + 0.96 + 0.68 + 0.09 + 0 = 2.45.

4.113 If one sells only 12 policies, the total sales would be 12 × ($250 + extra charges for costs and profits) = $3000 + 12 × (extra charges for costs and profits). If a single home is destroyed by fire, replacement costs could be several hundred thousands of dollars. The money received from 12 policies (unless extra charges for costs and profits are huge) would not cover these replacement costs. Although the chance that a home will be destroyed by fire is small, the risk to the company is too great. In addition (unless extra charges for costs and profits are huge), sales totaling $3000 + 12 × (extra charges for costs and profits) is not very lucrative.

If one sells thousands of policies, one can appeal to the law of large numbers and feel confident that the mean loss per policy will be close to $250. Thus, the company can be reasonably sure that the charges for extra costs and profits will be available for these costs and profits, and that the company will make money. The more policies the company sells, the better off the company will be.

4.115 The sum of all the probabilities listed in the table must equal 1 if this is a legitimate probability model. Thus, the missing probability must have the property that it makes the sum equal to 1 and so

$$P(\text{Age at death} \geq 26) = 1 - 0.00183 - 0.00186 - 0.00189 - 0.00191 - 0.00193 = 0.99058.$$

Using this result, we calculate

$$\begin{aligned}\mu_X &= (-99{,}750)(0.00183) + (-99{,}500)(0.00186) + (-99{,}250)(0.00189) + (-99{,}000)(0.00191) + \\ &\quad (-98{,}750)(0.00193) + (1250)(0.99058) \\ &= 303.35.\end{aligned}$$

4.117 $\sigma_X^2 = (-99{,}750 - 303.35)^2(0.00183) + (-99{,}500 - 303.35)^2(0.00186) + (-99{,}250 - 303.35)^2(0.00189) + (-99{,}000 - 303.35)^2(0.00191) + (-98{,}750 - 303.35)^2(0.00193) + (1250 - 303.35)^2(0.99058) = 94236826.6$. Thus, $\sigma_X = \sqrt{94236826.6} = 9707.57$.

4.119 a) Because we round to the nearest year, only integer values are possible. The smallest number of years that an employee can have been with the company is 0 (if they are new). The maximum number is not completely clear. It is unlikely that someone will have been with the company as long as 75 years, but we might set the upper bound to be higher just to be safe. The choice is a bit arbitrary. We have chosen 100 because is a nice, round number, but other choices are possible. In any case, our sample space becomes $S = \{0, 1, 2, 3, ..., 100\}$

b) Because we are allowing only a finite number of possible values, X is discrete.

c) We included 101 possible values (0 to 100). Different choices are possible depending on the value you chose as representing the maximum number of years someone could be with the company.

4.121 a) Presumably, W can range from 0 ml (no ink in the cartridge) to 35 ml (the maximum amount of ink possible) with any value in between possible. Thus, a reasonable sample space S for W is S = {all numbers between 0 and 35ml with no gaps}.

b) S is a continuous sample space. All values between 0 and 35 ml are possible with no gaps.

c) We included an infinite number of possible values.

CHAPTER 5

PROBABILITY THEORY

SECTION 5.1

OVERVIEW

In Chapter 4, the four basic rules of probability and the notion of a probability model were introduced. In this section, some additional laws of probability are described. Recall that two events A and B are **disjoint** if they have no outcomes in common. If events $A, B, C, \ldots$ are all disjoint in pairs, then the more general addition rule is given by

$$P(\text{at least one of these events occurs}) = P(A) + P(B) + P(C) + \cdots$$

Events are **independent** if knowledge that one event has occurred does not alter the probability that we would assign to the second event. The mathematical definition of independence leads to the **multiplication rule** for independent events. If A and B are independent, then

$$P(A \text{ and } B) = P(A)P(B).$$

In any particular problem, we can use this definition to check if two events are independent by seeing if the probabilities multiply according to the definition. However, most of the time, independence is assumed as part of the probability model.

Many students confuse independent and disjoint events once they have seen both definitions. Remember, disjoint events have no outcomes in common, and when two events are disjoint you can compute $P(A \text{ or } B) = P(A) + P(B)$ in this special case. The probability being computed is that one *or* the other event occurs. Disjoint events cannot be independent because once we know that A has occurred, then the probability of B occurring becomes 0 (B cannot have occurred as well — this is the meaning of disjoint). The multiplication rule can be used to compute the probability that two events occur *simultaneously*, $P(A \text{ and } B) = P(A)P(B)$, in the special case of independence.

The general addition rule for two events allows the computation of $P(A \text{ or } B)$ even when the two events are not disjoint, namely

$$P(A \text{ or } B) = P(A) + P(B) - P(A \text{ and } B)$$

The four basic rules, the multiplication rule and the general addition rule allow us to compute the probabilities of events in many random phenomena.

APPLY YOUR KNOWLEDGE

5.1 a) Recall that events are independent if knowledge that one event has occurred does not alter the probability that we would assign to the second event. This exercise should seem similar to Example 5.1 of the text in which two cards are selected at random from a deck, with the key difference being that the number of first-year college students is much larger than the number of cards in a deck. Suppose we have selected a first-year college student at random and recorded their academic rank in high school. Regardless of the rank of the first person selected, when we select the second person the percentages of first-year college students remaining in each of the five categories is virtually unchanged. So the probability that the rank of the second student falls into the five categories is (almost) unaffected by the rank of the first student and we consider their high school ranks as independent. This idea is discussed in more detail in Section 5.2 on page 331 of the text.

b) Let A be the event that the first student is in the top 20% and B be the event that the second student is in the top 20% of their high school class. From the table, $P(A) = P(B) = 0.41$. Both students being in the top 20% corresponds to the event "A and B." Because the two events are independent, by the multiplication rule for independent events

$$P(A \text{ and } B) = P(A)P(B) = (0.41)(0.41) = 0.1681.$$

c) Let A be the event that the first student is in the top 20% and B be the event that the second student is in the lowest 20% of their high school class. From the table, $P(A) = 0.41$ and $P(B) = 0.01$. The first student being in the top 20% and the second student being in the lowest 20% corresponds to the event "A and B". Because the two events are independent, by the multiplication rule for independent events

$$P(A \text{ and } B) = P(A)P(B) = (0.41)(0.01) = 0.0041.$$

5.3 a) Let L be the event that a call reaches a live person, and N be the event that the call doesn't reach a live person. Because these two events are complements, we could have written N as L^c For any call, $P(L) = 0.2$ is the probability that a call reaches a live person. Using the complement rule, $P(N) = 0.8$ is the probability that the call doesn't reach a live person. As part of the probability model, we are told that calls are independent. Let N_i be the event that the ith call does not reach a live person, where $P(N_i) = 0.8$. Then

$$\begin{aligned} P(\text{none of the calls reaches a live person}) &= P(N_1 \text{ and } N_2 \text{ and } N_3 \text{ and } N_4 \text{ and } N_5) \\ &= P(N_1)P(N_2)P(N_3)P(N_4)P(N_5) \\ &= (0.8)^5 = 0.32768. \end{aligned}$$

b) In this part the probability a call reaches a live person is smaller, so the probability that none of the calls reaches a live person is larger. Using the same notation as in part (a), $P(L) = 0.08$ and $P(N) = 1 - P(L) = 1 - 0.08 = 0.92$. Because $P(N_i) = 0.92$ and the calls are still independent,

$$\begin{aligned} P(\text{none of the calls reaches a live person}) &= P(N_1 \text{ and } N_2 \text{ and } N_3 \text{ and } N_4 \text{ and } N_5) \\ &= P(N_1)P(N_2)P(N_3)P(N_4)P(N_5) \\ &= (0.92)^5 = 0.65908. \end{aligned}$$

5.5 The string of lights will remain bright for three years provided all 20 lights do not fail. The probability of any light failing during the a three-year period is 0.02, so the probability of B, the event that a light remains bright for three years is $P(B) = 1 - 0.02 = 0.98$ by the complement rule. Let B_i be the event that the ith light remains bright, where $P(B_i) = 0.98$ and the B_i are assumed independent.

$$\begin{aligned} P(\text{string remains bright}) &= P(\text{all lights remain bright}) \\ &= P(B_1 \text{ and } B_2 \text{ and } B_3 \text{ and } \cdots \text{ and } B_{20}) \\ &= P(B_1)P(B_2)P(B_3) \cdots P(B_{20}) \\ &= (0.98)^{20} = 0.66761. \end{aligned}$$

5.7 In the Venn diagram below, the events coffee drinker, tea drinker, and cola drinker are represented by square regions that overlap. The resulting Venn diagram consists of disjoint regions for which the percentages can all be evaluated from the information given. We will proceed in several stages to get all the information provided onto the Venn diagram. First, we are told that 15% drink both coffee and tea, which is the region shaded in Figure A. Because we are also given that 5% drink all three beverages, this allows us to determine that 10% must drink coffee and tea but not cola. We are also told that 5% drink only tea, which is the region that is shaded in Figure B. Because 25% drink tea, this allows to fill in the remaining region of tea drinkers, namely 5% drink tea and cola but not coffee. In figure C, we use the information that 25% drink both coffee and cola, which is the shaded region. Because 5% drink all three, the additional piece of information is that 20% must drink coffee and cola but not tea. For the final regions, we know that 55% drink coffee so that 20% must drink *only* coffee and 45% drink cola so that 15% must drink *only* cola. This leads to figure D. Check that the percentages in Figure D satisfy all the information given.

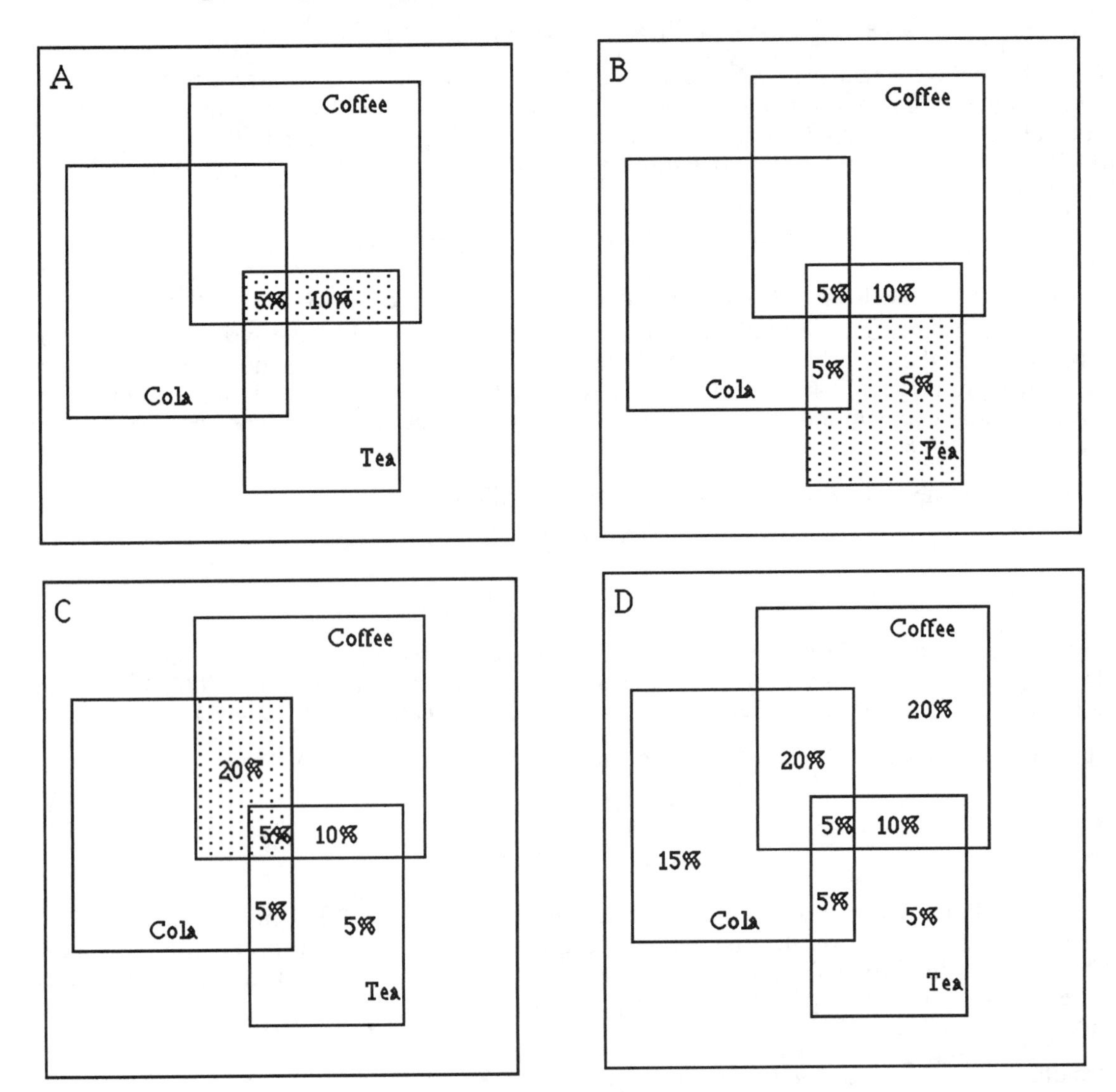

a) With the Venn diagram in Figure D, it is easy to read off that 15% drink only cola.

b) If we add all the percentages in Figure D, this corresponds to the percentage that drink either cola, coffee, or tea and totals to 80%. The remaining 20% drink none of these beverages.

SECTION 5.1 EXERCISES

5.9 The event that you win at least once is the complement to the event that you lose every time. In this case, it will be easier to calculate the probability of the complementary event as in Example 5.4 of the text on page 323. Let W be the event that you win on one play and L be the event that you lose on one play. You are told that $P(W) = 0.02$, so $P(L) = 0.98$ by the complement rule. Winning or losing on different plays are independent. Let L_i be the event that the ith play is a loss, where $P(L_i) = 0.98$.

$$\begin{aligned} P(\text{win at least once}) &= 1 - P(\text{lose on every play}) \\ &= 1 - P(L_1 \text{ and } L_2 \text{ and } L_3 \text{ and } L_4 \text{ and } L_5) \\ &= 1 - P(L_1)P(L_2)P(L_3)P(L_4)P(L_5) \\ &= 1 - (0.98)^5 = 0.09608. \end{aligned}$$

5.11 a) The three years are independent. If U indicates a year for the price being up and D indicates a year for the price being down, you need to compute

$$P(UUU) = (0.65)^3 = 0.2746$$

b) The probability of the price being down in any given year is $1 - 0.65 = 0.35$. Because the years are independent, the probability of the price being down in the third year is 0.35, regardless of what has happened in the first two years.

c) This problem must be set up carefully and done in steps.

$$P(\text{moves in the same direction in the next two years}) = P(UU \text{ or } DD) = P(UU) + P(DD).$$

The last equality follows because the events UU and DD are disjoint so that $P(UU \text{ or } DD) = P(UU) + P(DD)$. Now, using the independence of two successive years, $P(UU) = (0.65)^2 = 0.4225$, and $P(DD) = (0.35)^2 = 0.1225$. Putting this together,

$$P(\text{moves in the same direction in the next two years}) = 0.4225 + 0.1225 = 0.5450.$$

5.13 This is a straightforward application of the general addition rule. A is the event that Consolidated Builders wins the first contract and B is the event that they win the second contract. We are given $P(A) = 0.6$, $P(B) = 0.5$ and $P(A \text{ and } B) = 0.3$. To find $P(A \text{ or } B)$, the probability that Consolidated will win at least one of the jobs, use the general addition rule

$$P(A \text{ or } B)= P(A) + P(B) - P(A \text{ and } B) = 0.6 + 0.5 - 0.3 = 0.8.$$

5.15 Independence is often assumed as part of a probability model, and we use it to calculate probabilities. In this problem, we are given several probabilities and need to use the definition of independence to check if two events are independent by seeing if the probabilities multiply appropriately. We are told that $P(A) = 0.6$, $P(B) = 0.5$ and $P(A \text{ and } B) = 0.3$. From the definition of independence, we know that A and B are independent if $P(A \text{ and } B) = P(A)P(B)$. Checking the definition, we see that

$$0.3 = (0.6)(0.5),$$

so that A and B are independent.

5.17 As an example of our notation, O_H is the event that the husband has type O blood and O_W is the event that the wife has type O blood. The probability that both wife and husband share the same blood type is

$$\begin{aligned} P(\text{share the same blood type}) &= P(O_H O_W \text{ or } A_H A_W \text{ or } B_H B_W \text{ or } AB_H AB_W) \\ &= P(O_H O_W) + P(A_H A_W) + P(B_H B_W) + P(AB_H AB_W) \end{aligned}$$

because these four events are disjoint. Because the blood types of husband and wife are taken to be independent, the probability that both husband and wife have type O blood is $P(O_H O_W) = P(O_H)P(O_W) = (0.45)(0.45) = 0.2025$, and similarly for the other three probabilities. This gives

$$P(\text{share the same blood type}) = (0.45)(0.45) + (0.40)(0.40) + (0.11)(0.11) + (0.04)(0.04) = 0.3762.$$

5.19 The probabilities of the different combinations of "*ABO*-type" and "Rh-factor type" are given below where the independence of Rh-factor and "*ABO*-type" is used in the calculation. If you sum up the eight combinations, they must sum to 1 because these are the only possible outcomes.

$P(O$ and Rh-positive)	$= P(O)P$(Rh-positive)	$= (0.45)(0.84)$	$= 0.3780$
$P(O$ and Rh-negative)	$= P(O)P$(Rh-negative)	$= (0.45)(0.16)$	$= 0.0720$
$P(A$ and Rh-positive)	$= P(A)P$(Rh-positive)	$= (0.40)(0.84)$	$= 0.3360$
$P(A$ and Rh-negative)	$= P(A)P$(Rh-negative)	$= (0.40)(0.16)$	$= 0.0640$
$P(B$ and Rh-positive)	$= P(B)P$(Rh-positive)	$= (0.11)(0.84)$	$= 0.0924$
$P(B$ and Rh-negative)	$= P(B)P$(Rh-negative)	$= (0.11)(0.16)$	$= 0.0176$
$P(AB$ and Rh-positive)	$= P(AB)P$(Rh-positive)	$= (0.04)(0.84)$	$= 0.0336$
$P(AB$ and Rh-negative)	$= P(AB)P$(Rh-negative)	$= (0.04)(0.16)$	$= 0.0064$
			1.0000

5.21 a) There are 36 possible outcomes when throwing two balanced dice and they are equally likely. Of the 36 outcomes, there are only two that result in an 11, so the probability of an 11 is 2 / 36. The three throws of the pair of dice are independent, so if E corresponds to the event that an 11 results, the probability of three 11's in three independent throws is

$$P(EEE) = P(E)P(E)P(E) = \left(\frac{2}{36}\right)^3 = 0.000171$$

b) For the event "throw an 11," $P = 2 / 36$, so the odds against throwing an 11 are

$$\frac{1-P}{P} = \frac{1-(2/36)}{(2/36)} = 17$$

or 17 to 1 odds. For the odds against throwing three straight 11's, we need to use the value for P that was computed in part (a) in the formula for the odds. This gives the odds against three straight 11's as

$$\frac{1-P}{P} = \frac{1-(2/36)^3}{(2/36)^3} = 5831$$

or 5831 to 1. The writer's first statement is correct. When computing the odds for the three tosses, the writer multiplied the odds, which is not the correct way to compute the odds for the three throws.

SECTION 5.2

OVERVIEW

One of the most common situations giving rise to a **count** X is the **binomial setting.** It consists of four assumptions about how the count was produced. They are

- the number n of observations is fixed
- the n observations are all independent
- each observation falls into one of two categories called "success" and "failure"
- the probability of success p is the same for each observation

When these assumptions are satisfied, the number of successes X has a **binomial distribution** with n trials and success probability p, denoted by $B(n, p)$. For smaller values of n, the probabilities for X can be found easily using statistical software. Table C in the text gives the probabilities for certain combinations of n and p, and there is also an exact formula. For large n, the **normal approximation** can be used.

For a large population containing a proportion p of successes, the binomial distribution is a good approximation to the number of successes in an SRS of size n, provided the population is at least 10 times larger than the sample.

The mean and standard deviation for the binomial count X can be found using the formulas,

$$\mu_X = np$$

$$\sigma_X = \sqrt{np(1-p)}$$

When n is large, the count X is approximately $N(np, \sqrt{np(1-p)})$. This approximation should work well when $np \geq 10$ and $n(1-p) \geq 10$.

The exact **binomial probability formula** is given by

$$P(X = k) = \binom{n}{k} p^k (1-p)^{n-k}$$

where $k = 0, 1, 2, ..., n$ and $\binom{n}{k}$ is the **binomial coefficient.**

APPLY YOUR KNOWLEDGE

5.23 Although each birth is a boy or girl, we are not counting the number of successes in a fixed number of births. The number of observations (births) is random. The assumption of a fixed number of observations is violated.

5.25 a) For a binomial random variable, the possible values of X are 0, 1, ..., n. Because $n = 5$ in this example, the possible values of X are 0, 1, 2, 3, 4, 5.

b) To calculate the probability of each value of X, we can use the binomial formula, statistical software or Table C in the text. In this exercise, the use of the binomial formula is illustrated.

$$P(X = 0) = \binom{5}{0}(.25)^0(.75)^5 = \frac{5!}{0!5!}(0.2373) = 0.2373$$

$$P(X = 1) = \binom{5}{1}(.25)^1(.75)^4 = \frac{5!}{1!4!}(0.25)(.3164) = 0.3955$$

$$P(X = 2) = \binom{5}{2}(.25)^2(.75)^3 = \frac{5!}{2!3!}(.0625)(.4219) = 0.2637$$

$$P(X = 3) = \binom{5}{3}(.25)^3(.75)^2 = \frac{5!}{3!2!}(.0156)(.5625) = 0.0879$$

$$P(X = 4) = \binom{5}{4}(.25)^4(.75)^1 = \frac{5!}{4!1!}(.0039)(.75) = 0.0146$$

$$P(X = 5) = \binom{5}{5}(.25)^5(.75)^0 = \frac{5!}{5!0!}(.0010) = 0.0010$$

The probability histogram is given on the next page.

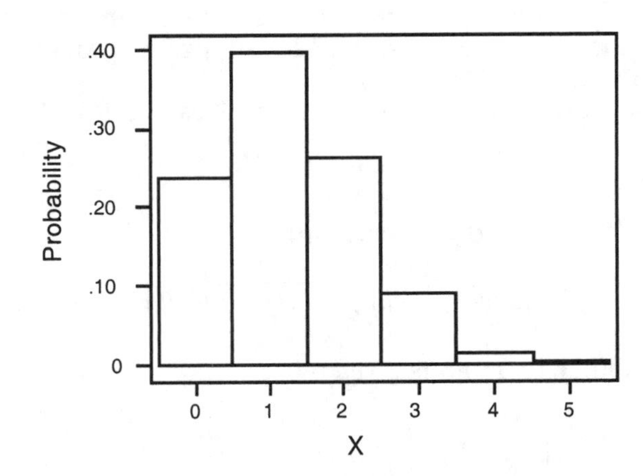

5.27 X is the number of players among the 20 who graduate. According to the university's claim, X should have the binomial distribution with $n = 20$ and $p = 0.8$. You need to find the probability that exactly 11 out of 20 players graduate, or evaluate $P(X = 11)$. The exact binomial probability formula is given by

$$P(X = k) = \binom{n}{k} p^k (1 - p)^{n-k}$$

where $k = 0, 1, 2, ..., n$ and $\binom{n}{k}$ is the binomial coefficient. Plugging the appropriate values of n, k, and p in the formula we get

$$P(X = 11) = \binom{20}{11}(0.8)^{11}(0.2)^9 = \frac{20!}{11!9!}(0.8)^{11}(0.2)^9 = 0.0074$$

5.29 X, the number of respondents in the survey who seek nutritious food when eating out, has a binomial distribution with $n = 20$ and $p = 0.4$. To find the mean and standard deviation of the number of "yes" responses, you need to enter these values of n and p into the general formula for the mean and standard deviation of the binomial distribution. The mean of X is $\mu = np = 20(0.4) = 8$, and the standard deviation of X is

$$\sigma = \sqrt{np(1-p)} = \sqrt{20(0.4)(0.6)} = 2.191$$

5.31 a) X is the number of players among the 20 who graduate. According to the university's claim, X should have the binomial distribution with $n = 20$ and $p = 0.8$. The formula for the mean is $\mu = np = 20(0.8) = 16$.

b) The formula for the standard deviation of X is

$$\sigma = \sqrt{np(1-p)} = \sqrt{20(0.8)(0.2)} = 1.789$$

c) When $p = 0.9$, the standard deviation of X is

$$\sigma = \sqrt{np(1-p)} = \sqrt{20(0.9)(0.1)} = 1.342$$

and when $p = 0.99$, the standard deviation of X is

$$\sigma = \sqrt{np(1-p)} = \sqrt{20(0.99)(0.01)} = 0.445$$

As the value of p gets closer to 1, there is less variability in the values of X. The same is true as the value of p gets closer to zero.

5.33 a) X, the number of respondents in the survey who seek nutritious food when eating out, has a binomial distribution with n = 200 and p = 0.4. The probability that 100 or more respondents answer "yes," $P(X \geq 100)$, can be found using statistical software or the normal approximation for the binomial distribution. To apply the normal approximation, we first check that np = 200(0.4) = 80 and $n(1 - p)$ = 200(0.6) = 120 are both greater than 10, which they are. The next step is to evaluate the mean and standard deviation of the binomial distribution. In this case, $\mu = np$ = 200(0.4) = 80, and the standard deviation of X is

$$\sigma = \sqrt{np(1-p)} = \sqrt{200(0.4)(0.6)} = 6.928.$$

Next, we act as though X had the N(80, 6.928) distribution.

$$\begin{aligned} P(X \geq 100) &= P\left(\frac{X-80}{6.928} \geq \frac{100-80}{6.928}\right) \\ &= P(Z \geq 2.89) \\ &= 1 - 0.9981 = 0.0019 \end{aligned}$$

This probability is extremely small suggesting that p may be greater than 0.4.

b) In Exercise 5.28, $P(X \geq 10)$ was evaluated for a sample of size 20 and found to be 0.2447. In both cases, you were evaluating the chance that a sample would have at least 50% successes when the population has 40% successes. This calculation depends on the sample size, with the probability decreasing as the sample size gets larger. The proportion in the sample will be closer to 40% for larger sample sizes, so the chance of the proportion in the sample being as large as 50% decreases.

SECTION 5.2 EXERCISES

5.35 a) Although the number of observations is fixed at 20, the meaning of p, the probability a machinist will perform satisfactorily on the test, is unclear. If the 20 machinists selected are an SRS from a large population of machinists, then the binomial would be appropriate.

b) The number of observations is fixed at 100, the number of persons to be chosen at random. Each observation results in a "yes", they would participate in the study if given the chance, or "no," they would not. The success probability p is the proportion of adult residents in the city who would agree to participate if given the chance. We know that the count of successes in an SRS from a large population containing a proportion p of successes is well approximated by the binomial distribution. Thus the binomial distribution is an appropriate model here.

5.37 a) X has approximately a binomial distribution with n = 10, the sample size and p = 0.25, the proportion of successes (those never having been married) in the population.

b) In this part, you can use statistical software, Table C in the text, or the binomial formula. Using the formula, the values that need to be substituted for evaluating binomial probabilities are n = 10, k = 2, and p = 0.25. This gives

$$P(X=2) = \binom{10}{2}(0.25)^2(0.75)^8 = \frac{10!}{2!8!}(0.25)^2(0.75)^8 = (45)(0.0625)(0.10012)$$

$$= 0.2816$$

c) You need to evaluate $P(X \leq 2) = P(X=0) + P(X=1) + P(X=2)$

$$= \binom{10}{0}(0.25)^0(0.75)^{10} + \binom{10}{1}(0.25)^1(0.75)^9 + \binom{10}{2}(0.25)^2(0.75)^8$$

$$= 0.0563 + 0.1877 + 0.2816 = 0.5256,$$

where you need to remember that 0! = 1 and $(0.25)^0 = 1$ when applying the formulas.

d) The mean of X is $\mu = np = 10(0.25) = 2.5$, and the standard deviation of X is

$$\sigma = \sqrt{np(1-p)} = \sqrt{10(0.25)(0.75)} = 1.37.$$

5.39 a) X has a binomial distribution with $n = 5$ (the number of years to be observed) and $p = 0.65$ (the probability the index will increase in any given year). The independence of years is assumed as part of the model.

b) Because $n = 5$, the possible values are X are 0, 1, 2, 3, 4, 5.

c) To calculate the probability of each value of X, we can use the binomial formula or statistical software This is very similar to Exercise 5.25 in which the use of the binomial formula was illustrated. The only difference is that $p = 0.65$ in this exercise and p was 0.25 in Exercise 5.25. The probabilities listed on the next page were obtained using the Minitab software.

```
Binomial with n = 5 and p = 0.650000
         x                P(X = x)
         0.00              0.0053
         1.00              0.0488
         2.00              0.1811
         3.00              0.3364
         4.00              0.3124
         5.00              0.1160
```

The probability histogram corresponding to this distribution is given below.

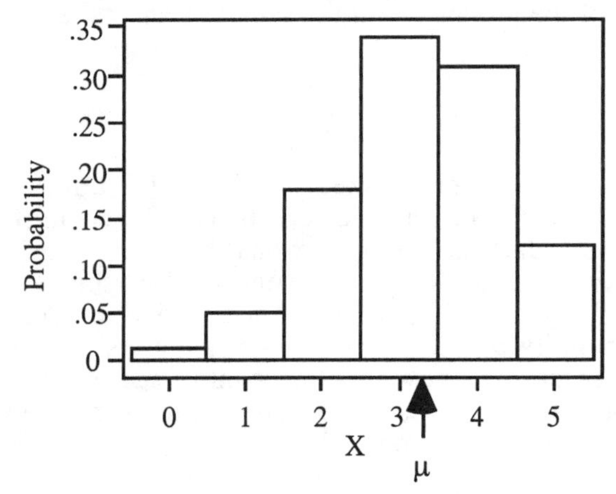

d) The mean of X is $\mu = np = 5(0.65) = 3.25$ and is indicated on the histogram in part (c). The standard deviation of X is

$$\sigma = \sqrt{np(1-p)} = \sqrt{5(0.65)(0.35)} = 1.067$$

5.41 a) X, the number of questions Jodi gets correct has a binomial distribution with $n = 100$ and $p = 0.75$. For Jodi to score 70% or lower, she must get 70 or fewer questions correct. $P(X \leq 70)$ can be found using statistical software or the normal approximation for the binomial distribution. To apply the normal approximation, we first check that $np = 100(0.75) = 75$ and $n(1 - p) = 100(0.25) = 25$ are both greater than 10, which they are. The next step is to evaluate the mean and standard deviation of the binomial distribution. In this case, $\mu = np = 100(0.75) = 75$, and the standard deviation of X is

$$\sigma = \sqrt{np(1-p)} = \sqrt{100(0.75)(0.25)} = 4.33.$$

Next, we act as though X had the $N(75, 4.33)$ distribution.

$$P(X \le 70) = P\left(\frac{X-75}{4.33} \le \frac{70-75}{4.33}\right)$$
$$= P(Z \le -1.15)$$
$$= 0.1251.$$

The chance that Jodi will score below 70% or lower is about 12.5%.

b) In this part, the exam consists of 250 questions, so X, the number of questions Jodi gets correct has a binomial distribution with n = 250 and p = 0.75. For Jodi to score 70% or lower, she must get (.70)(250) = 175 or fewer questions correct. $P(X \le 175)$ can be found using statistical software or the normal approximation for the binomial distribution. To apply the normal approximation, we first check that np = 250(0.75) = 187.5 and $n(1-p)$ = 250(0.25) = 62.5 are both greater than 10, which they are. The next step is to evaluate the mean and standard deviation of the binomial distribution. In this case, $\mu = np$ = 250(0.75) = 187.5, and the standard deviation of X is

$$\sigma = \sqrt{np(1-p)} = \sqrt{250(0.75)(0.25)} = 6.85.$$

Next, we act as though X had the $N(187.5, 6.85)$ distribution.

$$P(X \le 175) = P\left(\frac{X-187.5}{6.85} \le \frac{175-187.5}{6.85}\right)$$
$$= P(Z \le -1.82)$$
$$= 0.0344.$$

or slightly over a 3% chance that Jodi will score 70% or lower.

5.43 a) The Binomial distribution with n observations and probability p of success gives a good approximation to the sampling distribution of the count of successes in an SRS of size n from a large population containing proportion p of successes. In this case, we believe p in the population (businesses listed in the Yellow Pages) is about 0.5 and the number of observations (size of sample) is 150. The Binomial distribution with n = 150 and p = 0.5 is reasonable.

b) The mean of the binomial is $\mu = np$ = 150(0.50) = 75, so we expect 75 businesses to respond.

c) To apply the normal approximation we first check that np = 150(0.5) = 75 and $n(1-p)$ = 100(0.5) = 75 are both greater than 10, which they are. The next step is to evaluate the mean and standard deviation of the binomial distribution. From part (b), $\mu = np$ = 100(0.75) = 75, and the standard deviation of X is

$$\sigma = \sqrt{np(1-p)} = \sqrt{150(0.5)(0.5)} = 6.12.$$

Next, we act as though X had the $N(75, 6.12)$ distribution.

$$P(X \le 70) = P\left(\frac{X-75}{6.12} \le \frac{70-75}{6.12}\right)$$
$$= P(Z \le -82)$$
$$= 0.2061.$$

There is about a 20% chance that fewer than 70 businesses will respond.

d) The mean is $\mu = np = n(0.50)$. To increase the mean number of respondents to 100 requires that n be increased to 200.

5.45 a) X has a binomial distribution with n = 1500 and p = 0.12. The mean of X is $\mu = np$ = 1500(0.12) = 180, and the standard deviation of X is

$$\sigma = \sqrt{np(1-p)} = \sqrt{1500(0.12)(0.88)} = 12.586$$

b) The normal approximation can be used because both np = 180 and $n(1 - p)$ = 1320 are greater than 10. Using the mean and variance evaluated in (a) gives the approximation

$$P(X \leq 170) = P\left(\frac{X-180}{12.586} \leq \frac{170-180}{12.586}\right) = P(Z \leq -0.79) = 0.2148.$$

SECTION 5.3

OVERVIEW

The **Poisson distribution** is another distribution for count random variables. It is appropriate in the **Poisson setting:** the number of successes in any unit of measure is independent of the number of successes in any other nonoverlapping interval; the probability of a success in any unit of measure is proportional to the size of the unit and is the same for all units of equal size; the probability of two or more successes in a unit approaches zero as the size of the unit becomes smaller. If the Poisson setting is appropriate, then X, the number of successes or events in a fixed unit of time, has the Poisson distribution with mean μ. The standard deviation of X is $\sigma = \sqrt{\mu}$, and the probability that X takes any value 0, 1, 2, ... is given by the Poisson probability

$$P(X = k) = \frac{e^{-\mu}\mu^k}{k!} \qquad k = 0,1,2,3, \ldots$$

Although probabilities can be found using the previous formula, events of interest can often be expressed in terms of cumulative probabilities of the form $P(X \leq k)$ and these are most easily evaluated by software. In a Poisson model with mean μ per unit of space or time, the count of success in a units of space or time is a Poisson random variable with mean $a\mu$.

APPLY YOUR KNOWLEDGE

5.47 a) In a Poisson model with mean μ per unit of space or time, the count of success in a units is a Poisson random variable with mean $a\mu$. The accident counts per month follow a Poisson distribution with a mean of 7 per month, with a month being the basic unit of time. Because there are 12 months in a year, a = 12 and μ = 7, and the number of accidents in a year is a Poisson random variable with mean 12 × 7 = 84 accidents.

b) If X is the number of accidents per year, then we know from part (a) that X has a Poisson distribution with a mean of 84. You are asked to evaluate $P(X \leq 66)$ using statistical software. Using EXCEL as in Example 5.16 of the text, $P(X \leq 66)$ = 0.0248. The chance of having this few accidents is quite small suggesting that the initiative has reduced the accident rate.

SECTION 5.3 EXERCISES

5.49 a) X, the number of vehicles passing the mile marker in 15 minutes, has a Poisson distribution with a mean of 48.7. You are asked to evaluate $P(X \geq 50)$. Statistical software gives cumulative probabilities of the form $P(X \leq k)$, so we must write $P(X \geq 50)$ as an expression involving a cumulative probability.

$$P(X \geq 50) = 1 - P(X \leq 49).$$

You need to be careful when using the complement rule for discrete distributions. The complement of 50 or more (≥ 50) is 49 or fewer (≤ 49). The output from the Minitab software for the cumulative probability $P(X \leq 49)$ is

```
Poisson with mu = 48.7000

         x          P(X <= x)
     49.00            0.5550
```

which gives $P(X \geq 50) = 1 - P(X \leq 49)$ = 1 – 0.5550 = 0.4450.

b) The standard deviation for a Poisson random variable with mean μ is

$$\sigma = \sqrt{\mu} = \sqrt{48.7} = 6.98$$

In a Poisson model with mean μ per unit of space or time, the count of success in *a* units is a Poisson random variable with mean *a*μ. The number of vehicles passing the mile marker follows a Poisson distribution with a mean 48.7 vehicles per 15-minute period, so a 15-minute period is the basic unit of time. Because there are two 15-minute periods in a 30-minute period, *a* = 2 and μ = 48.7, so *X*, the number of vehicles passing the marker in a 30-minute interval, is a Poisson random variable with mean 2 × 48.7 = 97.4 accidents. The standard deviation for a 30-minute period is

$$\sigma = \sqrt{\mu} = \sqrt{97.4} = 9.87$$

c) From part (b), *X*, the number of vehicles passing the marker in a 30-minute interval, is a Poisson random variable with mean 2 × 48.7 = 97.4 accidents. The probability a hundred or more vehicles passes corresponds to $P(X \geq 100)$. Writing this as an expression involving a cumulative probability and using statistical software gives

$$P(X \geq 100) = 1 - P(X \leq 99) = 1 - .5905 = 0.4095$$

5.51 a) The number of work-related deaths in the United States has an approximate Poisson distribution with a mean of 17 per day. The standard deviation for a Poisson random variable with mean μ is

$$\sigma = \sqrt{\mu} = \sqrt{17} = 4.12 \text{ deaths.}$$

b) The probability of 10 or fewer work-related deaths in one day is $P(X \leq 10)$, where X has a Poisson distribution with a mean of 17. This is already a cumulative probability and using statistical software we have

$$P(X \leq 10) = 0.0491$$

c) The probability of more than 30 work-related deaths in one day is $P(X > 30)$, where *X* has a Poisson distribution with a mean of 17. Writing this as an expression involving a cumulative probability and using statistical software gives

$$P(X > 30) = 1 - P(X \leq 30) = 1 - .9986 = 0.0014$$

where you need to be careful using the complement rule. The complement of more than 30 (> 30) is 30 or fewer (≤ 30).

5.53 a) *X*, the number of calls received between 8 a.m. and 9 a.m., has a Poisson distribution with a mean of 14. The probability of at least five calls is $P(X \geq 5)$. Writing this as an expression involving a cumulative probability and using statistical software gives

$$P(X \geq 5) = 1 - P(X \leq 4) = 1 - .0018 = 0.9982$$

b) In a Poisson model with mean μ per unit of space or time, the count of success in *a* units is a Poisson random variable with mean *a*μ. The number of calls received between 8 a.m. and 9 a.m. has a Poisson distribution with a mean of 14 calls so that one hour, or 60 minutes, is the basic unit of time. The interval 8:15 a.m. to 8:45 a.m. corresponds to a 30-minute interval, so *a* = 1 / 2 and μ = 14. *X*, the number of calls received between 8:15 a.m. to 8:45 a.m. is a Poisson random variable with mean $1/2 \times 14 = 7$ calls. The probability of at least five calls is $P(X \geq 5)$. Writing this as an expression involving a cumulative probability and using statistical software gives

$$P(X \geq 5) = 1 - P(X \leq 4) = 1 - .1730 = 0.8270$$

where the answer differs from part (a) because we are using a mean of 7 rather than a mean of 14 in the calculation.

c) In a Poisson model with mean μ per unit of space or time, the count of success in a units is a Poisson random variable with mean $a\mu$. The number of calls received between 8 a.m. and 9 a.m. has a Poisson distribution with a mean of 14 calls so that one hour, or 60 minutes, is the basic unit of time. The interval 8:15 a.m. to 8:30 a.m. corresponds to a 15-minute interval, so $a = 1/4$ and $\mu = 14$. X, the number of calls received between 8:15 a.m. to 8:30 a.m. is a Poisson random variable with mean $1/4 \times 14 = 3.5$ calls. The probability of at least five calls is $P(X \geq 5)$. Writing this as an expression involving a cumulative probability and using statistical software gives

$$P(X \geq 5) = 1 - P(X \leq 4) = 1 - .7254 = 0.2746$$

where the answer differs from part (a) because we are using a mean of 3.5 rather than a mean of 14 in the calculation.

5.55 a) The number of defects in large sheets of plastic can be modeled as a Poisson distribution with a mean of 2.3 defects per square yard. The standard deviation for a Poisson random variable with mean μ is

$$\sigma = \sqrt{\mu} = \sqrt{2.3} = 1.52 \text{ defects.}$$

b) The probability the inspector finds more than five defects in a randomly chosen square yard is $P(X > 5)$, where X has a Poisson distribution with a mean of 2.3. Writing this as an expression involving a cumulative probability and using statistical software gives

$$P(X > 5) = 1 - P(X \leq 5) = 1 - .9700 = 0.0300$$

c) We know $P(X > 5) = 0.03$ from part (a). In order for $P(X > k) \geq 0.15$, k must be smaller than 5. Trying $k = 4$ gives $P(X > 5) = 0.0838$, so $k = 4$ is not small enough. Trying $k = 3$ gives $P(X > 5) = 0.2007$, so $k = 3$ is the largest value of k that will work.

SECTION 5.4

OVERVIEW

This section discusses the probability analog of the conditional distribution described in Section 2.5 of the text. The **conditional probability** of an event B given an event A is denoted $P(B|A)$ and is defined by

$$P(B|A) = \frac{P(A \text{ and } B)}{P(A)}$$

when $P(A) > 0$. In practice it can often be determined directly from the information given in a problem.

Any two events A and B satisfy the **general multiplication rule** $P(\text{A and B}) = P(\text{A})P(\text{B}|\text{A})$. In the special case that A and B are **independent,** $P(\text{B}|\text{A}) = P(\text{B})$, and the general multiplication rule reduces to the multiplication rule for independent events, namely $P(\text{A and B}) = P(\text{A})\text{P}(\text{B})$.

When $P(A)$, $P(B|A)$ and $P(B|A^C)$ are known, Baye's rule can be used to calculate $P(A|B)$ as

$$P(A|B) = \frac{P(B|A)P(A)}{P(B|A)P(A) + P(B|A^c)P(A^c)}$$

In problems with several stages, it is helpful to draw a tree diagram to guide you in the use of the multiplication and addition rules

APPLY YOUR KNOWLEDGE

5.57 This is an application of the general multiplication rule. We are given $P(A) = 0.46$ and $P(B|A) = 0.32$. The event that a randomly chosen employed person is a woman holding a managerial or professional position corresponds to the event "A and B." By the general multiplication rule,

$$\begin{aligned} P(\text{woman holding a managerial or professional position}) &= P(A \text{ and } B) \\ &= P(A)P(B|A) \\ &= (0.46)(0.32) \\ &= 0.1472. \end{aligned}$$

5.59 a) The calculation of the four probabilities in this exercise is a straightforward application of the definition of conditional probability. The probability that a person is employed, *given the information that the person is male,* is

$$P(\text{employed}|\text{male}) = \frac{3{,}927}{9{,}058} = 0.4335$$

b) $P(\text{male}|\text{employed}) = \dfrac{3{,}927}{8{,}240} = 0.4766$

c) $P(\text{female}|\text{unemployed}) = \dfrac{446}{966} = 0.4617$

d) $P(\text{unemployed}|\text{female}) = \dfrac{446}{9{,}116} = 0.0489$

5.61 a) This is an application of the definition of conditional probability. We are given that A is the event that a household is prosperous and B is the event that the household is educated. In addition, $P(A) = 0.134$, $P(B) = 0.254$ and $P(A \text{ and } B) = 0.080$. The conditional probability that a household is educated, given that it is prosperous, is

$$P(B|A) = \frac{P(A \text{ and } B)}{P(A)} = \frac{0.080}{0.134} = 0.5970$$

b) The conditional probability that a household is prosperous, given that it is educated is

$$P(A|B) = \frac{P(A \text{ and } B)}{P(B)} = \frac{0.080}{0.254} = 0.3150$$

c) If the events A and B were independent, then $P(B|A) = P(B)$, which is not the case. $P(B|A) > P(B)$, which means that a household is more likely to be educated given that you the household is prosperous than in the general population.

5.63 In many problems involving probability, the most difficult part is setting up the notation. In this problem we are told that A is the event that the screen came from Brightscreens and B is the event that a randomly chosen screen is unsatisfactory. We are told that the company receives 55% of its screens from Screensource and the remainder from Brightscreen. Thus, A^C is the event that the screen came from Screensource (there are only two suppliers), $P(A^C) = 0.55$ and $P(A) = 0.45$. The remaining two probabilities given are conditional probabilities. Screensource supplies 1% unsatisfactory screens corresponds to $P(B|A^C) = 0.01$ and 4% of the screens from Brightsource are unsatisfactory corresponds to $P(B|A) = 0.04$. We are asked to evaluate the probability a screen came from Brightscreen given that it is unsatisfactory, or $P(A|B)$. This follows from Baye's rule as

$$
\begin{aligned}
P(A|B) &= \frac{P(B|A)P(A)}{P(B|A)P(A) + P(B|A^c)P(A^c)} \\
&= \frac{(0.04)(0.45)}{(0.04)(0.45) + (0.01)(0.55)} \\
&= 0.7660
\end{aligned}
$$

Note that the notation has been set up to make the events correspond exactly to the definition of Baye's rule given in the text. The problems will become a little harder when you need to define the events on your own.

SECTION 5.4 EXERCISES

5.65 The first thing to do is to define the events of interest, determine which probabilities are given and then which probabilities need to be computed. Let A be the event that a college student likes country music and B be the event that a college student likes gospel music. We are given that $P(A) = 0.40$, $P(B) = 0.30$, and $P(A \text{ and } B) = 0.10$.

a) The problem asks you to find $P(B|A)$. The formula is

$$P(B|A) = \frac{P(A \text{ and } B)}{P(A)} = \frac{0.10}{0.40} = 0.25$$

Note that the roles of A and B have been switched in the definition of conditional probability. It is best if you first define the events and then adjust the formula, rather than starting with the formula and trying to determine which event must be A and which must be B to make the original formula work.

b) The problem asks you to find $P(B|A^C)$. The formula is

$$P(B|A^C) = \frac{P(A^C \text{ and } B)}{P(A^C)}$$

By the complement rule, $P(A^C) = 1 - P(A) = 1 - 0.40 = 0.60$. In addition, if you draw a Venn diagram, you can verify that

$$P(A \text{ and } B) + P(A^C \text{ and } B) = P(B), \text{ so that}$$

$$P(A^C \text{ and } B) = P(B) - P(A \text{ and } B) = 0.30 - 0.10 = 0.20.$$

This gives

$$P(B|A^C) = \frac{P(A^C \text{ and } B)}{P(A^C)} = \frac{0.20}{0.60} = 0.3333.$$

5.67 You want use geometric arguments to evaluate

$$P(Y < 1/2 | Y > X) = \frac{P(Y < 1/2 \text{ and } Y > X)}{P(Y > X)}$$

In the above, we have just used the definition of conditional probability where A corresponds to "$Y < 1/2$" and B corresponds to "$Y > X$." The three figures on the next page are of the square $0 \le x \le 1$ and $0 \le y \le 1$. The figure on the left has the region corresponding to "$Y > X$" shaded, and because probabilities correspond to areas we see that $P(Y > X) = 1/2$. The middle figure has the region corresponding to "$Y < 1/2$" shaded and the figure on the right has the region "$Y < 1/2$ and $Y > X$" shaded. The area of the shaded triangle in the figure on the right is $1/8$, so $P(Y < 1/2 \text{ and } Y > X) = 1/8$.

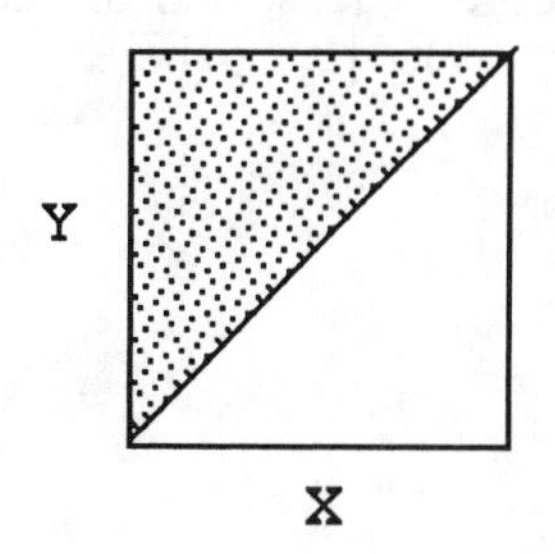

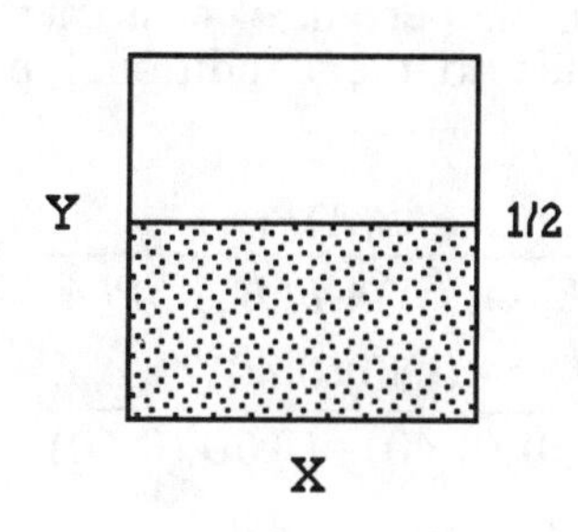

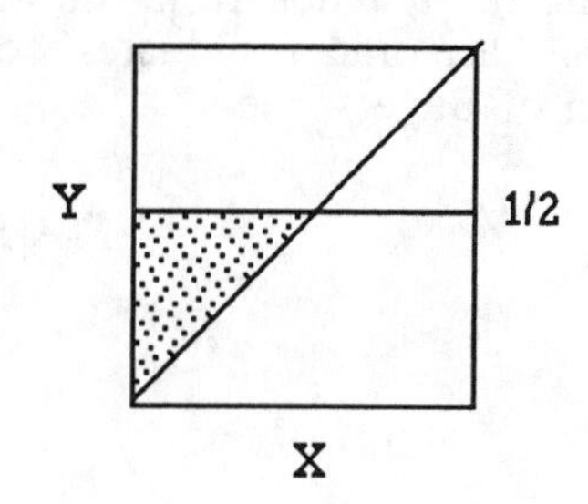

Putting this all together gives

$$P(Y<1/2|Y>X)=\frac{P(Y<1/2 \text{ and } Y>X)}{P(Y>X)}=\frac{1/8}{1/2}=0.25$$

5.69 Let A be the event that a would be MBA student is currently an undergraduate and A^C be the event that a would be MBA student is a college graduate. You are given that $P(A) = 0.4$ and $P(A^C) = 0.6$. Let B be the event that a customer scores at least 600. We are told that $P(B|A) = 0.50$ and $P(B|A^C) = 0.70$. This information is organized in the tree diagram below.

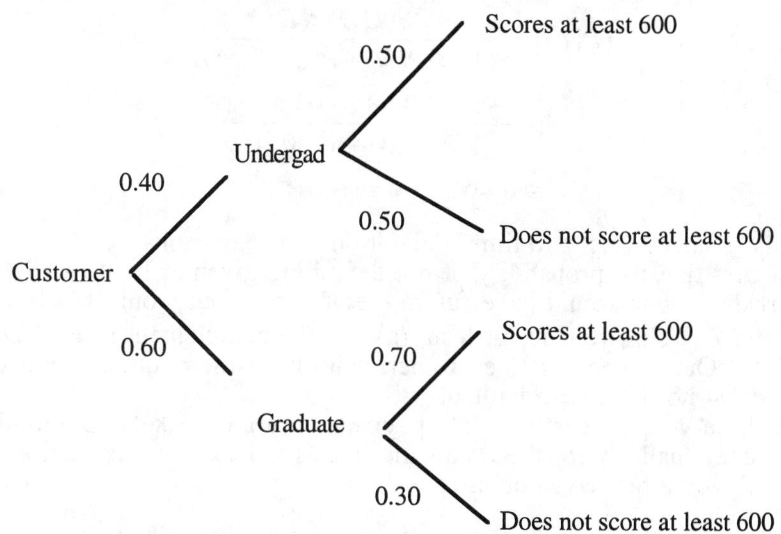

a) You can use the tree diagram or the general multiplication rule to do this part. Being an undergraduate *and* scoring at least 600 is the event A and B. To find the probability of A and B, you can use the general multiplication rule

$$P(A \text{ and } B) = P(A)P(B|A) = (0.4)(0.5) = 0.20.$$

or the appropriate path in the tree diagram. Being an graduate *and* scoring at least 600 is the event A^C and B. To find the probability of A^C and B, you can again use the general multiplication rule or follow the appropriate path in the tree diagram to obtain

$$P(A^C \text{and } B) = P(A^C)P(B|A^C) = (0.6)(0.7) = 0.42.$$

b) B is the event that a customer scores at least 600 and you need to evaluate $P(B)$.

$$P(B) = P(A \text{ and } B) + P(A^C \text{ and } B) = 0.20 + 0.42 = 0.62$$

We have used the fact that the event B corresponds to (A and B) or ($A^C$and B) and then added the two probabilities as in problem 5.65. You can use a Venn diagram or the tree diagram to convince yourself that this is correct.

5.71 Using the notation from problem 5.69, the percent of customers who score at least 600 on the GMAT that are undergraduates corresponds to the conditional probability $P(A|B)$. This is an application of Baye's rule

$$P(A|B) = \frac{P(B|A)P(A)}{P(B|A)P(A) + P(B|A^c)P(A^c)}$$
$$= \frac{(0.50)(0.40)}{(0.50)(0.40) + (0.70)(0.60)}$$
$$= 0.323$$

5.73 You need to first define the events of interest and determine which probabilities are given. Let D be the event that a credit customer defaults and L be the event that a credit customer is late on two or more payments. We are told that 88% of cardholders who defaulted on their payments were late with two or more payments. This is a conditional probability, $P(L|D) = 0.88$. We are also told that $P(D) = 0.03$ and $P(L|D^C) = 0.40$.

a) You are first asked to find $P(D|L)$. This is an example of Baye's rule where we are using different letters for the events intentionally so that you can concentrate on understanding the technique.

$$P(D|L) = \frac{P(L|D)P(D)}{P(L|D)P(D) + P(L|D^c)P(D^c)}$$
$$= \frac{(0.88)(0.03)}{(0.88)(0.03) + (0.40)(0.97)}$$
$$= 0.064$$

Only about 6% of those late at least two times default on their payments.

b) You must first find the probability of not defaulting given at least two late payments. This corresponds to individuals who would have future credit denied but would not have defaulted. This corresponds to $P(D^C|L)$. Using your result from (a) and the complement rule, $P(D^C|L) = 1 - P(D|L) = 1 - 0.064 = 0.936$. Out of every 100 customers who have their future credit denied, we would expect almost 94% or 94 not to have defaulted.

c) Although those with at least two late payments are more likely to default than others, the probability is still quite small. With the credit manager's policy, the vast majority of those whose future credit is denied would not have defaulted.

5.75 Independence is often assumed as part of a probability model, and we use it to calculate probabilities. In this problem, we are given several probabilities and need to use the definition of independence to check if two events are independent by seeing if the probabilities multiply appropriately. We are told that $P(A) = 0.6$, $P(B) = 0.5$ and $P(A \text{ and } B) = 0.3$. From the definition of independence, we know that A and B are independent if $P(A \text{ and } B) = P(A)P(B)$. Checking the definition, we see that

$$0.3 = (0.6)(0.5),$$

so that A and B are independent.

5.77 You need to first define the events of interest and determine which probabilities are given. Let A be the event that a product fails to conform to specifications and B be the event that a product is selected for complete inspection. We are told that 55% of nonconforming items are selected for inspection. This is the conditional probability, $P(B|A) = 0.55$. We are also told that $P(A) = 0.08$ and $P(B|A^C) = 0.20$. You are asked to find $P(A|B)$. This is an application of Baye's rule

$$P(A|B) = \frac{P(B|A)P(A)}{P(B|A)P(A) + P(B|A^c)P(A^c)}$$
$$= \frac{(0.55)(0.08)}{(0.55)(0.08) + (0.20)(0.92)}$$
$$= 0.1930$$

CHAPTER 5 REVIEW EXERCISES

5.79 The Binomial distribution with n observations and probability p of success gives a good approximation to the sampling distribution of the count of successes in an SRS of size n from a large population containing proportion p of successes. In this case, a success is a leaking tank, $p = 0.25$ and the number of observations (size of sample) is 15. The Binomial distribution with $n = 15$ and $p = 0.25$ is appropriate.

a) The mean of the binomial is $\mu = np = 150(0.50) = 3.75$ leaking tanks.

b) If X is the number of leaking tanks, you need to evaluate

$$P(X \geq 10) = P(X = 10) + P(X = 11) + \cdots + P(X = 15)$$

You can use the formula to calculate the individual probabilities on the right or you can look up these probabilities in Table C provided in the text. The alternative is to rewrite $P(X \geq 10)$ in terms of a cumulative probability and use statistical software to evaluate

$$P(X \geq 10) = 1 - P(X \leq 9) = 1 - 0.9992 = 0.0008.$$

c) In this case, X, the number of leaking tanks, has the Binomial distribution with $n = 1000$ and $p = 0.25$. To apply the normal approximation, we first check that $np = 1000(0.25) = 250$ and $n(1 - p) = 750$ are both greater than 10, which they are. The next step is to evaluate the mean and standard deviation of the binomial distribution. The mean is $\mu = np = 1000(0.25) = 250$, and the standard deviation of X is

$$\sigma = \sqrt{np(1-p)} = \sqrt{1000(0.25)(0.75)} = 13.69.$$

Next, we act as though X had the $N(250, 13.69)$ distribution. The probability at least 275 tanks are leaking is

$$P(X \geq 275) = P\left(\frac{X - 250}{13.69} \geq \frac{275 - 250}{13.69}\right)$$
$$= P(Z \geq 1.83)$$
$$= 1 - 0.9664 = 0.0336$$

5.81 The Binomial distribution with n observations and probability p of success gives a good approximation to the sampling distribution of the count of successes in an SRS of size n from a large population containing proportion p of successes. In this case, a success is contacting a Macintosh user, $p = 0.05$, and the number of observations (size of sample) is 25,000. The Binomial distribution with $n = 25{,}000$ and $p = 0.05$ is appropriate.

a) The mean of the binomial is $\mu = np = 25{,}000(0.05) = 1250$ Macintosh users.

b) In this case, X, the number of Macintosh users receiving the flyer, has the Binomial distribution with $n = 25{,}000$ and $p = 0.05$. To apply the normal approximation, we first check that $np = 25{,}000(0.05) = 1250$ and $n(1 - p) = 23750$ are both greater than 10, which they are. The next step is to evaluate the mean and standard deviation of the binomial distribution. The mean is $\mu = np = 25{,}000(0.05) = 1250$, and the standard deviation of X is

$$\sigma = \sqrt{np(1-p)} = \sqrt{25{,}000(0.05)(0.95)} = 34.46.$$

Next, we act as though X had the $N(1250, 34.46)$ distribution. The probability at least 1245 Macintosh users receive the flyer is

$$P(X \geq 1245) = P\left(\frac{X-1250}{34.46} \geq \frac{1245-1250}{34.46}\right)$$
$$= P(Z \geq -0.15)$$
$$= 1 - 0.4404 = 0.5596$$

5.83 a) An important assumption of the Binomial is that the probability of a success p is the same for each observation. To ensure that this is true, it is important to take the observations under similar conditions, such as the same location and time of day.

b) The probability of the driver being male may be different for observations made outside a church on Sunday morning than on campus after a dance. If half the observations are taken under each of these conditions, the assumption that the probability of success is the same for each observation will probably be violated.

c) X the number of male drivers has a binomial distribution with p = 0.85 and the number of observations is n = 10. To evaluate $P(X \leq 8)$, the probability that the man is driving eight or fewer cars, you can use statistical software to obtain the cumulative probability $P(X \leq 8)$ = 0.4557. Alternatively, you can use the binomial formula or the tables, but in this case it would be easier to evaluate $P(X \geq 9)$ and then use the fact that $P(X \leq 8) = 1 - P(X \geq 9)$.

d) X, the number of male drivers, has a binomial distribution with p = 0.85 and the number of observations is n = 100, the total number of observations taken by the class. To evaluate $P(X \leq 80)$, the probability that the man is driving eight or fewer cars, you can use statistical software to obtain the cumulative probability $P(X \leq 80) = 0.1065$. If software is not available, then you can use the normal approximation to evaluate $P(X \leq 80)$.

5.85 You need to first define the events of interest and determine which probabilities are given. Let I be the event that infection occurs and F be the event that the repair fails. We are told $P(I)$ = 0.03, $P(F)$ = 0.14, and $P(I \text{ and } F)$ = 0.01. The event we are interested in is that the operation succeeds and is free from infection. This is the complement of the event "failure or infection," and using our notation we need to evaluate the probability of the event $(F \text{ or } I)^C$. You should draw a Venn diagram to convince yourself of this. Summarizing and using the complement rule, we have

$$P(\text{success and free from infection}) = P[(F \text{ or } I)^C] = 1 - P(F \text{ or } I)$$

The addition rule can now be used to evaluate

$$P(F \text{ or } I) = P(I) + P(F) - P(I \text{ and } F) = 0.03 + 0.14 - 0.01 = 0.16.$$

Combining these results gives

$$P(\text{success and free from infection}) = 1 - P(F \text{ or } I) = 1 - 0.16 = 0.84,$$

or 84% of the surgeries succeed and are free from infection.

5.87 a) The probability of being in the labor force is the percentage of all those in the labor force relative to the total population, or

$$\frac{12{,}073+36{,}855+33{,}331+37{,}281}{27{,}325+57{,}221+45{,}471+47{,}371} = \frac{119{,}540}{177{,}388} = 0.674$$

b) The probability of being in the labor force given being a college graduate is a conditional probability and is evaluated as

$$\frac{37{,}281}{47{,}371} = 0.787$$

c) If the events "in labor force" and "college graduate" were independent, we should have

$$P(\text{in labor force}) = P(\text{in labor force} \mid \text{college graduate})$$

which you can see by comparing the answers in (a) and (b) is not true. College graduates are more likely to be in the labor force than those in the general population.

5.89 a) The information in the problem can be summarized in the tree diagram below.

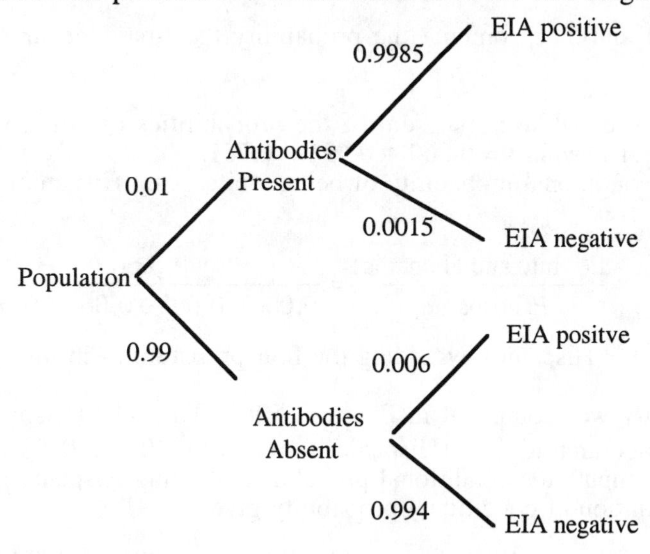

b) Let A be the event that antibodies are present and is B be the event that EIA is positive. Then

$$\begin{aligned} P(B) &= P(B|A)P(A) + P(B|A^c)P(A^c) \\ &= (0.9985)(.01) + (.006)(.99) \\ &= 0.01592 \end{aligned}$$

c) You are asked to find $P(A|B)$. This is an application of Baye's rule

$$\begin{aligned} P(A|B) &= \frac{P(B|A)P(A)}{P(B|A)P(A) + P(B|A^c)P(A^c)} \\ &= \frac{(0.9985)(0.01)}{(0.9985)(.01) + (.006)(.99)} \\ &= 0.627 \end{aligned}$$

Thus, $P(A^C|B)$, the probability of the antibody being absent but the test being positive is 1 – 0.627 = 0.373. or the probability of a false positive is almost 40%.

5.91 a) Let O indicate a 1 on a toss O^C and indicates a 1 does not occur on a toss, so that $P(O)$ = 1/6 and $P(O^C)$ = 5/6. The probability that the first 1 occurs on the second toss is

$$P(O^C O) = P(O)P(O^C) = \left(\frac{5}{6}\right)\left(\frac{1}{6}\right)$$

where we have used the independence of the tosses.

b) The probability that the first 1 occurs on the third toss is

$$P(O^C O^C O) = P(O^C)P(O^C)P(O) = \left(\frac{5}{6}\right)^2 \left(\frac{1}{6}\right)$$

c) If the first 1 occurs on the kth toss, then the first k – 1 tosses are not a 1 and the kth toss is a 1. The general pattern is

$$P(\text{first 1 on } k\text{th toss}) = \left(\frac{5}{6}\right)^{k-1}\left(\frac{1}{6}\right)$$

You can replace k by 4 and 5 to compute the probability the first 1 occurs on the fourth and fifth toss, respectively.

5.93 a) The probability of white is the sum of the probabilities of being white and Hispanic and white and not Hispanic, or P(white) = 0.060 + 0.691 = 0.751.

b) You want the conditional probability of being white given Hispanic. Using the definition of conditional probability gives

$$P(\text{white} \mid \text{Hispanic}) = \frac{P(\text{white and Hispanic})}{P(\text{Hispanic})} = \frac{0.060}{0.000+0.003+0.060+0.062} = \frac{0.060}{0.125} = 0.48$$

where we have computed P(Hispanic) by adding the four probabilities in this column.

5.95 a) This probability was computed in Exercise 5.93. To find P(Hispanic), you must add the four probabilities in this column to get P(Hispanic) = $0.000+0.003+0.060+0.062$ = 0.125.

b) You need to compute the conditional probability of being Hispanic given the person chosen is black. Using the definition of conditional probability gives

$$P(\text{Hispanic} \mid \text{black}) = \frac{P(\text{Hispanic and black})}{P(\text{black})} = \frac{0.003}{0.003+0.121} = \frac{0.003}{0.124} = 0.024$$

5.97 You need to first define the events of interest and determine which probabilities are given. Let I be the event that a person buys an iMac and F be the event that they are a first-time computer buyer. We are told $P(I)$ = 0.05, $P(F|I)$ = 0..32 and $P(F|I^C)$ = 0.40. You are asked to find $P(I|F)$. This is an example of Baye's rule and

$$\begin{aligned} P(I|F) &= \frac{P(F|I)P(I)}{P(F|I)P(I)+P(F|I^c)P(I^c)} \\ &= \frac{(0.32)(0.05)}{(0.32)(0.05)+(0.40)(0.95)} \\ &= 0.0404 \end{aligned}$$

Approximately 4% of first-time computer buyers bought an iMac.

CHAPTER 6

INTRODUCTION TO INFERENCE

SECTION 6.1

OVERVIEW

A **confidence interval** provides an estimate of an unknown parameter of a population or process along with an indication of how accurate this estimate is and how **confident** we are that the interval is correct. Confidence intervals have two parts. One is an interval computed from our data. This interval typically has the form

estimate ± margin of error.

The other part is the **confidence level,** which states the probability that the method used to construct the interval will give a correct answer. For example, if you use a 95% confidence interval repeatedly, in the long run 95% of the intervals you construct will contain the correct parameter value. Of course, when you apply the method only once you do not know whether your interval gives a correct value. Confidence refers to the probability that the method gives a correct answer in repeated use, not the correctness of any particular interval we compute from data.

Suppose we wish to estimate the unknown mean μ of a normal population with known standard deviation σ based on an SRS of size n. A level C confidence interval for μ is

$$\bar{x} \pm z^* \frac{\sigma}{\sqrt{n}}$$

where z^* is such that the probability is C that a standard normal random variable lies between $-z^*$ and z^* and is obtained from the bottom row in Table D.

The margin of error $z^* \frac{\sigma}{\sqrt{n}}$ of a confidence interval decreases when any of the following occur:

- the confidence level C decreases
- the sample size n increases
- the population standard deviation σ decreases.

The sample size needed to obtain a confidence interval for a normal mean of the form

estimate ± margin of error.

with a specified margin of error m is

$$n = \left(\frac{z^* \sigma}{m}\right)^2$$

where z^* is the critical point for the desired level of confidence. Many times, the n you will find will not be an integer. If it is not, round up to the next larger integer.

The formula for any specific confidence interval is a recipe that is correct under specific conditions. The most important conditions concern the methods used to produce the data. Many methods (including those discussed in this section) assume that our data were collected by random sampling. Other conditions, such as the actual distribution of the population, are also important.

APPLY YOUR KNOWLEDGE

6.1 If we take a random sample of size 100, the standard deviation for $\bar{x}$ is $\$200/\sqrt{100} = \$200/10 = \$20$.

6.3 The answer in the blank should be 2 × (standard deviation for $\bar{x}$) = 2 × ($20) = $40.

6.5 We are told that the mean assets, $\bar{x}$, are $220 dollars and that we can assume the population standard deviation is $\sigma = \$161$ million. The sample size is $n = 110$. For a 99% confidence interval, we see from Table D that $z^* = 2.576$, so our margin of error is

$$z^* \frac{\sigma}{\sqrt{n}} = \$2.054 \frac{161}{\sqrt{110}} \text{ million} = \$31.53 \text{ million.}$$

A 99% confidence interval for the mean assets of all community banks is therefore

$$\bar{x} \pm z^* \frac{\sigma}{\sqrt{n}} = \$220 \pm \$39.54 \text{ million} = (\$180.46 \text{ million}, \$259.54 \text{ million}).$$

6.7 We are told that an estimate of σ is $8000 and that we need to have a margin of error $m = \$500$. For a 95% confidence interval, Table D gives $z^* = 1.960$. Thus, we need a sample size of $n = \left(\frac{z^* \sigma}{m}\right)^2 = \left(\frac{1.96 \times \$8000}{\$500}\right)^2 = 983.45$. Because 983 observations will give a slightly wider interval than desired and 984 observations a slightly narrower interval, we should use $n = 984$.

6.9 a) 13,000 surveys were sent but only 1468 usable responses were received. Thus, the response rate is 1468/13000 = 0.1129, or converting this fraction to a percent, the response rate is 11.29%.

b) The response rate is low and the reasons organizations chose not to respond are not given. If there are systematic patterns in the organizations that did not respond, the survey results may be biased. Thus, it is not clear that the 1468 usable responses can plausibly be regarded as a simple random sample of the population. The small margin of error is probably not a good measure of the accuracy of the survey's results.

SECTION 6.1 EXERCISES

6.11 a) We are told that the mean time spent studying, $\bar{x}$, is 110 minutes and that we can assume the population standard deviation is $\sigma = 40$ minutes. The sample size is $n = 25$. For a 95% confidence interval, we see from Table D that $z^* = 1.96$, so our margin of error is

$$z^* \frac{\sigma}{\sqrt{n}} = 1.96 \frac{40}{\sqrt{25}} = 15.68 \text{ minutes.}$$

A 95% confidence interval for the mean time spent studying statistics by students in this class is therefore

$$\bar{x} \pm z^* \frac{\sigma}{\sqrt{n}} = 110 \pm 15.68 \text{ minutes.} = (94.32 \text{ minutes}, 125.68 \text{ minutes}).$$

b) It is not true that 95% of the students in the class have weekly study times that lie in the interval (94.32 minutes, 125.68 minutes). The confidence coefficient of 95% the probability that the method we used will give an interval containing the correct value of the population mean study time. It does not tell us about individual study times. To determine an interval that contains 95% of the student study times, we would need to know the distribution of student study times.

6.13 a) The mean weight of the runners in kg is $\bar{x}$ = 61.7917. To convert this to pounds we must multiply this by 2.2. Thus, the mean weight of runners in pounds is 2.2 × 61.7917 = 135.94.

b) The standard deviation of the mean weight in kg is $4.5/\sqrt{24}$ = 0.9186 kg. To convert this to pounds, we must multiply this by 2.2. Thus, the standard deviation of the mean weight of runners in pounds is 2.2 × 0.9186 = 2.0208.

c) A 95% confidence interval for the mean weight, in kg, of the population is

$$\bar{x} \pm z^* \frac{\sigma}{\sqrt{n}} = 61.7917 \pm 1.96 \frac{4.5}{\sqrt{24}} = 61.7917 \pm 1.8004 = (63.5921,\ 59.9913).$$

Converting this interval to pounds, we get (2.2 × 63.5921, 2.2 × 59.9913) = (139.9026, 131.9809) for our 95% confidence interval.

6.15 We are told that the mean time, $\bar{x}$, that the 114 general managers had spent with their current company was 11.8 years and that we can assume the population standard deviation is σ = 3.2 years. The sample size is n = 114. For a 99% confidence interval, we see from Table D that z^* = 2.576, so our margin of error is

$$z^* \frac{\sigma}{\sqrt{n}} = 2.576 \frac{3.2}{\sqrt{114}} = 0.77 \text{ years.}$$

A 99% confidence interval for the mean number of years general managers of major-chain hotels have spent with their current company is therefore

$$\bar{x} \pm z^* \frac{\sigma}{\sqrt{n}} = 11.8 \pm 0.77 \text{ years} = (11.03 \text{ years},\ 12.57 \text{ years}).$$

6.17 We compute the mean wages, $\bar{x}$, using statistical software and find that it is $17,528.90. We can assume the population standard deviation is σ = $1900. The sample size is n = 15. For a 95% confidence interval, we see from Table D that z^* = 1.96 so our margin of error is

$$z^* \frac{\sigma}{\sqrt{n}} = 1.96 \frac{\$1900}{\sqrt{15}} = \$961.53.$$

A 95% confidence interval for the mean earnings of all black female hourly workers at National Bank is therefore

$$\bar{x} \pm z^* \frac{\sigma}{\sqrt{n}} = \$17528.90 \pm \$961.53 = (\$16567.37,\ \$18490.43).$$

6.19 We are told that an estimate of σ is $22.00 and that we need to have a margin of error m = $4.00. For a 95% confidence interval, Table D gives z^* = 1.960. Thus, we need a sample size of n = $\left(\frac{z^*\sigma}{m}\right)^2 = \left(\frac{1.96 \times \$22.00}{\$4.00}\right)^2$ = 116.2. Because 116 observations will give a slightly wider interval than desired and 117 observations a slightly narrower interval, we should use n = 117.

6.21 a) We cannot be certain that the true population percent falls in this interval. The confidence level for this interval is 95% and this number is the probability that the method will produce an interval containing the population percentage. When we apply the method once, we do not know whether or not our interval correctly includes the population percent.

b) The announced result, 43%±3%, is a 95% confidence interval. That means that this particular interval was produced by a method that will give an interval that contains the true percent of the population that believe mergers are good for the economy 95% of the time. When we apply the method once, we do not know whether or not our interval correctly includes the population percent. Because the method yields a correct result 95% of the time, we say we are 95% confident that this is one of the correct intervals.

c) This is a 95% confidence interval. If we assume the sample percent has approximately a normal sampling distribution, z^* = 1.96 and 3% = $1.96\sigma_{\text{estimate}}$. Thus, σ_{estimate} = 3%/1.96 = 1.53%.

d) The announced margin of error includes only sampling error and other random effects. It does not include errors due to practical problems, such as undercoverage and nonresponse.

6.23 The estimated median income for four-person families in Michigan is based on the same sample survey that was used to estimate the median household income for the entire United States. Families from Michigan comprise only a fraction of all households that completed the survey. Thus, the number of households (sample size) used to compute the estimated median income for Michigan is smaller than the number used to compute median income for the entire United States. Assuming the standard deviations for the distribution of household income are the same for both Michigan and the United States, the margin of error for the estimate for Michigan will be larger than that for the entire United States because the estimate for Michigan is based on a smaller sample size.

6.25 a) 95% confidence means that this particular interval was produced by a method that will give an interval that contains the true percent of the population that will vote for Ringel 95% of the time. When we apply the method once, we do not know whether or not our interval correctly includes the true population percent. Thus, 95% refers to the method, not to any particular interval produced by the method.

b) The margin of error is 3%, so the 95% confidence interval is 52% ± 3% = (49%, 55%). The interval includes 50%, so we can not be "confident" that the true percent is 50% or even slightly less than 50%. Hence, the election is too close to call from the results of the poll.

6.27 The numbers given are the values obtained using the formula $\bar{x} \pm z^* \frac{\sigma}{\sqrt{n}}$. No arithmetic mistakes have been made. However, the results are not trustworthy. The formula for a confidence interval is relevant if our sample is an SRS (or can plausibly be considered an SRS). Phone-in polls are not SRS's. Only people who listen to the radio station are aware of the poll, and generally only listeners with strong opinions tend to call in. It is very unlikely that the listeners are representative of the entire population, and it is even less likely that those who feel strongly enough to call can be considered an SRS of all citizens.

SECTION 6.2

OVERVIEW

Tests of significance and confidence intervals are the two most widely used types of formal statistical inference. A test of significance is done to assess the evidence against the **null hypothesis** H_0 in favor of an **alternative hypothesis** H_a. Typically, the alternative hypothesis is the effect that the researcher is trying to demonstrate, and the null hypothesis is a statement that the effect is not present. The alternative hypothesis can be either **one** or **two-sided.**

Tests are usually carried out by first computing a **test statistic.** The test statistic is used to compute a ***P*-value,** which is the probability of getting a test statistic at least as extreme as the one observed, where the probability is computed when the null hypothesis is true. The *P*-value provides a measure of how incompatible our data is with the null hypothesis, or how unusual it would be to get data like ours if the null hypothesis were true. Because small *P*-values indicate data that is unusual or difficult to explain under the null hypothesis, we typically reject the null hypothesis in these cases. In this case, the alternative hypothesis provides a better explanation for our data.

Significance tests of the null hypothesis H_0: $\mu = \mu_0$ with either a one or two-sided alternative are based on the test statistic

$$z = \frac{\bar{x} - \mu_0}{\sigma / \sqrt{n}}$$

The use of this test statistic assumes that we have an SRS from a normal population with known standard deviation σ. When the sample size is large, the assumption of normality is less critical because the sampling distribution of $\bar{x}$ is approximately normal. P-values for the test based on z are computed using Table A.

When the P-value is below a specified value α, we say the results are statistically significant at level α, or we reject the null hypothesis at level α. Tests can be carried out at a fixed significance level by obtaining the appropriate critical value z^* from the bottom row in Table D.

Apply Your Knowledge

6.29 A mean of 0 or larger corresponds to opinions that the new form is no better (does not take less time) than the old form. Thus, a mean μ of 0 corresponds to the statement of "no effect" or "no difference," and such a statement is usually taken to be the null hypothesis. If the new form is an improvement, the mean response μ should be negative because negative values correspond to opinions that the new form takes less time than the old form. This is the statement that we hope or suspect is true, instead of the null hypothesis, and such a statement is usually taken to be the alternative hypothesis. Thus, the null and alternative hypotheses that provide a framework for examining whether or not the new form is an improvement are

$$H_0: \mu = 0$$
$$H_a: \mu < 0$$

6.31 a) We know that $\bar{x}$ = 28.6%, that the standard deviation σ = 9.6%, that the sample size is n = 40, and that under the null hypothesis μ = 31%. Thus the test statistic $z = \dfrac{\bar{x}-\mu}{\sigma/\sqrt{n}} = \dfrac{28.6\% - 31\%}{9.6\%/\sqrt{40}} =$ −1.58. The alternative hypothesis is two sided and the P-value is the probability (when H_0 is true) that $\bar{x}$ takes a value at least as far from 0 as the value actually observed A standard normal curve with the area that corresponds to the P-value follows.

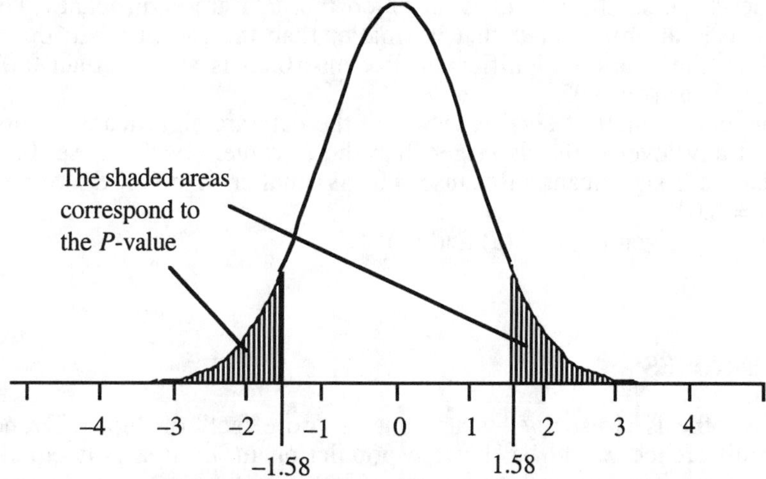

b) The P-value is the area under a standard normal curve to the left of −1.58 plus the area to the right of 1.58, as pictured in part (a). According to Table A, this area is 0.0571 + 0.0571 = 0.1142. This would probably not be considered strong evidence that Cleveland differs from the national average.

6.33 If the homebuilders have no idea whether Cleveland residents spend more or less than the national average, then they are not sure whether μ is larger or smaller than 31%. The appropriate hypotheses are

$$H_0: \mu = 31\%$$
$$H_a: \mu \neq 31\%$$

not the hypotheses the analyst tests.

6.35 Values in the extreme 0.005 area of the standard normal curve are significant at the 0.005 level. Because we are testing a two-sided alternative, this area is equally divided between the two tails of the curve. Thus, z is significant if it lies in the extreme 0.0025 area at either end. From Table D, we see that the extreme 0.0025 area in the right tail starts at $z^* = 2.807$. In other words, the z-values that are significant at the $\alpha = 0.005$ level are $z > 2.807$ and $z < -2.807$.

6.37 For a two-sided test, the significance level corresponding to $z^* = 2$ is the area under a standard normal curve to the right of 2 plus the area to the left of –2. From Table A, these areas are both 0.0228, so their sum is 0.0456. The significance level corresponding to $z^* = 3$ is the area under a standard normal curve to the right of 3 plus the area to the left of –3. From Table A, these areas are both 0.0013, so their sum is 0.0026. Thus, for a two-sided test, the significance level corresponding to $z^* = 2$ is 0.0456 and the significance level corresponding to $z^* = 3$ is 0.0026.

6.39 a) If the alternative is $H_a: \mu > \mu_0$, then the P-value is — using Table A — $P(Z \geq 1.8) = 1 - P(Z < 1.8) = 1 - 0.9641 = 0.0359$.

b) If the alternative is $H_a: \mu < \mu_0$, then the P-value is — using Table A — $P(Z \leq 1.8) = 0.9641$.

c) If the alternative is $H_a: \mu \neq \mu_0$, then the P-value is — using Table A — $P(Z \leq -1.8) + P(Z \geq 1.8) = 2P(Z \geq 1.8) = 2(0.0359) = 0.0718$.

6.41 a) A level 0.05 two-sided significance test rejects $H_0: \mu = 10$ exactly when 10 falls outside the 95% confidence interval for μ. With a P-value of 0.06, we would not reject $H_0: \mu = 10$ and thus 10 would not fall outside the 95% confidence interval.

b) A level 0.10 two-sided significance test rejects $H_0: \mu = 10$ exactly when 10 falls outside the 90% confidence interval for μ. With a P-value of 0.06, we would reject $H_0: \mu = 10$ and thus 10 would fall outside the 90% confidence interval.

6.43 a) The P-value is the smallest level α at which the data are significant. Thus, we would not reject the null hypothesis at any level α that is smaller than the P-value. In this case, 0.078 is the smallest level at which the data are significant. Because 0.05 is smaller than 0.078, we would not reject the null hypothesis at $\alpha = 0.05$.

b) The P-value is the smallest level α at which the data are significant. Thus, we would reject the null hypothesis at any level α that is larger than the P-value. In this case, 0.078 is the smallest level at which the data are significant. Because 0.01 is smaller than 0.078, we would not reject the null hypothesis at $\alpha = 0.01$.

c) Explanations are given in parts (a) and (b).

SECTION 6.2 EXERCISES

6.45 a) The null hypothesis is usually a statement of "no effect" or "no difference." In this case, the statement of no difference would be that the population mean area μ is equal to the advertised value of 1250 square feet. The alternative hypothesis is the statement that we hope or suspect is true instead of the null hypothesis. In this case, this is the tenant group's belief that the population mean area μ is less than the advertised value of 1250 square feet. Thus,

$$H_0: \mu = 1250$$
$$H_a: \mu < 1250$$

b) The null hypothesis is usually a statement of "no effect" or "no difference." In this case, the statement of no difference would be that the mean highway gas mileage μ is equal to the mileage Larry used to get with his old motor oil, namely 32 miles per gallon. The alternative hypothesis is the statement that we hope or suspect is true instead of the null hypothesis. In this case, this is whether Larry's mean highway gas mileage μ is more than the 32 miles per gallon Larry used to get with his old motor oil. Thus,

$$H_0: \mu = 32$$
$$H_a: \mu > 32$$

c) The null hypothesis is usually a statement of "no effect" or "no difference." In this case, the statement of no difference would be that the mean diameter of the spindle μ is equal to the target value of 5 mm. The alternative hypothesis is the statement that we hope or suspect is true instead of the null hypothesis. In this case, this is whether the mean spindle diameter μ has moved away from (differs from) the target value of 5 mm. Thus,

$$H_0: \mu = 5$$
$$H_a: \mu \neq 5$$

6.47 a) The parameter of interest is the correlation, ρ, between income and the percent of disposable income that is saved by employed young adults. The null hypothesis is a statement of "no effect" or "no difference." In this case, the statement of no effect would be that the correlation, ρ is (less than or) equal to 0. The alternative hypothesis is the statement that we hope or suspect is true instead of the null hypothesis. In this case, this is whether the correlation, ρ, is positive. Thus,

$$H_0: \rho = 0$$
$$H_a: \rho > 0$$

b) The parameters of interest are the percent of males, say p_M, and the percent of females, say p_F, in the population that will name economics as their favorite subject. The null hypothesis is a statement of "no effect" or "no difference." In this case, the statement of no effect would be that the two percents, p_M and p_F, are equal. The alternative hypothesis is the statement that we hope or suspect is true instead of the null hypothesis. In this case, this is whether the percent of males p_M, who will name economics as their favorite subject is larger than the percent of females, pF, who will name economics as their favorite subject Thus,

$$H_0: p_M = p_F$$
$$H_a: p_M > p_F$$

c) Let μ_A be the mean score on the test of basketball skills for the population of all sixth grade students if all were treated as those in Group A. Let μ_B be the mean score on the test of basketball skills for the population of all sixth grade students if all were treated as those in Group B. μ_A and μ_B are the parameters of interest. The null hypothesis is a statement of "no effect" or "no difference." In this case, the statement of no effect would be that teacher attitudes do not make a difference in learning, so that μ_A and μ_B are equal The alternative hypothesis is the statement that we hope or suspect is true intend of the null hypothesis. In this case, this is that positive teacher attitudes do improve learning so that μ_A will be larger than μ_B. Thus,

$$H_0: \mu_A = \mu_B$$
$$H_a: \mu_A > \mu_B$$

6.49 The conclusion might be interpreted as follows. Suppose that taking a calcium supplement actually has no effect on seated systolic blood pressure. In other words, the effect of the calcium supplement is the same as that of the placebo. If this is true, are the results we obtained (in support of calcium supplements) likely to occur simply by chance or are they unusual? If the results are unusual, this would be an indication that the assumption that there is no difference between the calcium supplement and the placebo is inconsistent with the data and hence strong evidence against such an assumption.

To be more specific, what is the probability that we would observe, simply by chance, results that are as strongly or more strongly in support of the calcium supplement if it is really no more effective than the placebo ? This probability is the P-value. In this case, the P-value is 0.008, which is very small. Because it is unlikely that we would obtain data this strongly in support of the calcium supplement by chance, this is strong evidence against the assumption that the effect of the calcium supplement is the same as that of the placebo. This is the basis for the conclusion given.

6.51 a) The null hypothesis is generally the hypothesis of "no effect." In words, we would formulate the null hypothesis here as

H_0: exercise has no effect on how students perform on their final exam in statistics

The alternative hypothesis is the statement that we hope or suspect is true instead of the null hypothesis. It is not clear whether the researchers wanted to show that exercise improved performance, reduced performance, or simply had an effect but without any idea as to the direction of the effect. Thus, any of the following alternative hypotheses is possible (depending on the purpose of the study).

H_a: exercise results in a higher performance by students on their final exam in statistics
or H_a: exercise results in a lower performance by students on their final exam in statistics
or H_a: exercise results in a change in performance by students on their final exam in statistics

b) The P-value is large, so we would not reject the null hypothesis. If we assume the appropriate alternative hypothesis is

H_a: exercise results in a higher performance by students on their final exam in statistics

in plain language, we conclude that there is little evidence that exercise results in a higher performance by students on their final exam in statistics. Evidence as or more strongly in favor of exercise improving performance than that actually observed could happen simply by chance (if exercise has no effect) with probability 0.87.

c) Some questions we might ask include: What was the actual study design? Was this a randomized comparative study? How many students were involved?

6.53 We assume that the number of new words in a sonnet follows approximately a normal distribution and can be considered a random sample of works from some population (either the population of unknown works by our poet or the population of unknown works by some other poet). We are told that this distribution has standard deviation $\sigma = 2.7$, that a sample of size $n = 5$ was used, and that the mean number of new words in these five sonnets is $\bar{x} = 11.2$. Under the null hypothesis $\mu = 6.9$, so the test statistic is

$$z = \frac{\bar{x} - \mu}{\sigma/\sqrt{n}} = \frac{11.2 - 6.9}{2.7/\sqrt{5}} = 3.56.$$

Because the alternative hypothesis is H_a: $\mu > 6.9$, the P-value is

$$P\text{-value} = P(Z \geq 3.56) = 1 - P(Z < 3.56)$$

which is less than 0.002 (3.56 is not included in Table A and all we can say is that $P(Z < 3.56) > 0.9998$). We would conclude that there is strong evidence that these five sonnets come from a population with a mean number of new words that is larger than 6.9, and we have evidence that the new sonnets are not by our poet.

6.55 We assume that the distribution of corn yields is approximately normal, and we are told that $\sigma = 10$, that a sample of size $n = 40$ was used, and that the ample mean yield of corn is $\bar{x} = 138.8$ bushels per acre. Under the null hypothesis $\mu = 135$, so the test statistic is

$$z = \frac{\bar{x} - \mu}{\sigma/\sqrt{n}} = \frac{138.8 - 135}{10/\sqrt{40}} = 2.40.$$

Because the alternative hypothesis is H_a:$\mu \neq 135$, which is two-sided, the P-value is

$$P\text{-value} = 2P(Z \geq 2.40) = 2(1 - P(Z < 2.40)) = 2(1 - 0.9918) = 2(0.0082) = 0.0164.$$

This is strong evidence against the null hypothesis that the population mean corn yield is 135.

Our results are based on the mean of a sample of 40 observations, and such a mean may vary approximately according to a normal distribution (by the central limit theorem) even if the population is not normal Because it is the sampling distribution of $\bar{x}$ rather than the distribution of the individual observations that determines the distribution of our test statistic, our conclusions are probably still valid.

6.57 a) Because the alternative hypothesis is two-sided, the *P*-value corresponding to a z statistic with value –1.37 is

$$P\text{-value} = 2P(Z \geq |-1.37|) = 2P(Z \geq 1.37) = 2(1 - P(Z < 1.37)) = 2(1 - 0.9147) = 0.1706.$$

This is greater than 0.05, so the result is not significant at the 5% level.

b) Because the *P*-value is 0.1706, which is greater than 0.01, the result is not significant at the 1% level.

6.59 The approximate *P*-value for this-two sided test with $z = 3.3$ is

$$P\text{-value} = 2P(Z \geq 3.3) = 2(1 - P[Z < 3.3]) = 2(1 - 0.9995) = 2(0.0005) = 0.001.$$

6.61 a) For the alternative H_a: $\mu > 0$, Table D tells us that the critical value at the 5% level is 1.645. Thus, we would reject H_0 for any value of z greater than 1.645.

b) If the alternative was H_a: $\mu \neq 0$, we would reject H_0 if either z is too large or z is too small. According to Table D, the upper tail probability for 1.96 is 0.025. This means that the lower tail probability for –1.96 is also 0.025. The sum of these two is 0.05, so we would reject H_0 at the 5% level if $z > 1.96$ or $z < -1.96$.

c) In part (a), we reject only for large values of z because such values are evidence against H_0 in favor of H_a: $\mu > 0$. In part (b), we reject H_0 in favor of H_a: $\mu \neq 0$ if either z is too large or z is too small because both extremes are evidence against H_0 in favor of H_a: $\mu \neq 0$. Guaranteeing that the overall probability is 5% of rejecting H_0 when the null hypothesis is true determines the precise range of values for which we reject in both parts.

6.63 a) The $n = 12$ readings have mean $\bar{x} = 104.133$ (we computed this using statistical software). We are told that $\sigma = 9$. For a 95% confidence interval, we see from Table D that $z^* = 1.96$ so our margin of error is

$$z^* \frac{\sigma}{\sqrt{n}} = 1.96 \frac{9}{\sqrt{12}} = 5.092.$$

A 95% confidence interval for the mean time spent studying statistics by students in this class is therefore

$$\bar{x} \pm z^* \frac{\sigma}{\sqrt{n}} = 104.133 \pm 5.092 = (99.041,\ 109.225).$$

b) The hypotheses we are testing are

$$H_0: \mu = 105$$
$$H_a: \mu \neq 105$$

The 95% confidence interval in part (a) contains 105, so we would not reject H_0 at the 5% level.

6.65 a) Let μ represent the mean amount (in milligrams) of the sugar in the hindguts of the population of all cockroaches of the same type used in the study under these conditions. The hypotheses we are testing are

$$H_0: \mu = 7$$
$$H_a: \mu \neq 7$$

The 95% confidence interval is $4.2 \pm 2.3 = (1.9, 6.5)$. 7 is not in this interval, so we would reject H_0 at the 5% level. Thus, the paper gives evidence that the mean amount of sugar in the hindguts under these conditions is not equal to 7 mg.

b) 5 is in the 95% confidence interval, so we would not reject the hypothesis that $\mu = 5$ at the 5% level in favor of a two-sided alternative.

SECTION 6.3

OVERVIEW

When describing the outcome of a hypothesis test, it is more informative to give the P-value than just the reject or not decision at a particular significance level α. The traditional levels of 0.01, 0.05, and 0.10 are arbitrary and serve as rough guidelines.

When testing hypotheses with a very large sample, the P-value can be very small for effects that may not be of practical interest. Do not confuse small P-values with large or important effects. Plot the data to display the effect you are trying to show, and also give a confidence interval that says something about the size of the effect. Statistical inference is not valid for data from badly designed surveys or experiments.

Just because a test is not statistically significant does not imply that the null hypothesis is true. This may occur when the test is based on a small sample size and has low power. Finally, if you run enough tests, you will invariably find statistical significance for one of them. Be careful in interpreting the results when testing many hypotheses on the same data.

APPLY YOUR KNOWLEDGE

6.67 a) We are told that $n = 100$, $\sigma = 100$, and $\bar{x} = 541.4$. Under the null hypothesis $\mu = 525$, so the z statistic for testing the hypotheses given is

$$z = \frac{\bar{x} - \mu}{\sigma/\sqrt{n}} = \frac{541.4 - 525}{100/\sqrt{100}} = 1.64.$$

According to Table D, the critical value for the one-sided alternative given is 1.645 and therefore we reject H_0 at the 5% if $z > 1.645$. Thus, in this case the result is not significant at the 5% level.

b) In this case, the z statistic for testing the hypotheses given is

$$z = \frac{\bar{x} - \mu}{\sigma/\sqrt{n}} = \frac{541.5 - 525}{100/\sqrt{100}} = 1.65.$$

According to Table D, the critical value for the one-sided alternative given is 1.645 and therefore we reject H_0 at the 5% if $z > 1.645$. Thus, in this case the result is significant at the 5% level.

6.69 Any convenience sample (phone-in or write-in polls, surveys of acquaintances only) would produce data for which statistical inference is not appropriate. To find a real example, look for man-on-the-street surveys or listen to radio stations in your area to see if any are conducting phone-in surveys. Poorly designed experiments also provide examples of data for which statistical inference is not valid. The EESEE supplement contains some examples so you might look through EESEE for an example.

SECTION 6.3 EXERCISES

6.71 A test of significance answers question (b) "Is the observed effect due to chance?" Whether or not an observed effect is important is something that the designers of a study must determine on practical rather than statistical grounds. Determining whether a sample or experiment is properly designed requires judgment using the principles described in Chapter 3.

6.73 a) We are told that $n = 100$, $\sigma = 100$, and $\bar{x} = 518$. Under the null hypothesis $\mu = 515$, so the z statistic for testing the hypotheses given is

$$z = \frac{\bar{x} - \mu}{\sigma/\sqrt{n}} = \frac{518 - 515}{100/\sqrt{100}} = 0.3.$$

Because the alternative hypothesis is H_a: $\mu > 515$, the P-value is

$$P\text{-value} = P(Z \geq 0.3) = 1 - P(Z < 0.3) = 1 - 0.6179 = 0.3821.$$

b) Now $n = 1000$, $\sigma = 100$, and $\bar{x} = 518$. Under the null hypothesis $\mu = 515$, so the z statistic for testing the hypotheses given is

$$z = \frac{\bar{x} - \mu}{\sigma/\sqrt{n}} = \frac{518 - 515}{100/\sqrt{1000}} = 0.95.$$

Because the alternative hypothesis is H_a: $\mu > 515$, the P-value is

$$P\text{-value} = P(Z \geq 0.95) = 1 - P(Z < 0.95) = 1 - 0.8289 = 0.1711.$$

c) Now $n = 10{,}000$, $\sigma = 100$, and $\bar{x} = 518$. Under the null hypothesis $\mu = 515$, so the z statistic for testing the hypotheses given is

$$z = \frac{\bar{x} - \mu}{\sigma/\sqrt{n}} = \frac{518 - 515}{100/\sqrt{10000}} = 3.$$

Because the alternative hypothesis is H_a: $\mu > 515$, the P-value is

$$P\text{-value} = P(Z \geq 3) = 1 - P(Z < 3) = 1 - 0.9987 = 0.0013.$$

6.75 The station's arithmetic is correct (1921/2372 = 0.81 and the reported margin of error is correct) but its conclusion is not justified. Statistical inference is not valid for badly designed surveys or experiments. This is a call-in poll and such polls are not random samples. Results are often biased (those who listen to Channel 13 and those who feel strongly enough to call in may differ systematically from the population at large) so inference based on these results are likely to be misleading.

6.77 Using the Bonferroni procedure for $k = 6$ tests with $\alpha = 0.05$, we should require a P-value of $\alpha/k = 0.05/6 = 0.0083$ (or less) for statistical significance for each test. Of the six P-values given, only two, 0.008 and 0.001 are below 0.05/6 = 0.0083 and so only these two are statistically significant under the Bonferroni procedure.

6.79 a) There are 77 independent tests, each with probability 0.05 of being significant at the 0.05 level (assuming all 77 null hypotheses are true). These tests form a sequence of 77 independent Bernoulli trials, each with probability of "success" 0.05 (equating "success" with statistical significance). Thus, if X is the total number that are significant, then X has a binomial distribution with $n = 77$ and $p = 0.05$.

b) The probability that two or more are significant is (using the formulas for binomial probabilities)

$$\begin{aligned} P(X \geq 2) &= 1 - P(X < 2) = 1 - P(X = 0) - P(X = 1) \\ &= 1 - \binom{77}{0}(0.05)^0(0.95)^{77} - \binom{77}{1}(0.05)^1(0.95)^{76} \\ &= 1 - (1)(1)(0.01926) - (77)(0.05)(0.02028) \\ &= 1 - 0.01926 - 0.07808 \\ &= 0.90266. \end{aligned}$$

SECTION 6.4

OVERVIEW

The **power** of a significance test is always calculated at a specific alternative hypothesis and is the probability that the test will reject H_0 when that alternative is true. This calculation requires knowledge of the sampling distribution under the specific alternative hypothesis of the test statistic used. Power is usually interpreted as the ability of a test to detect an alternative hypothesis or as the sensitivity of a test to an alternative hypothesis. The power of a test can be increased by increasing the sample size when the significance level remains fixed.

To compute the power of a significance test about a mean of a normal population, we need to

- state H_0, H_a (the particular alternative we want to detect), and the significance level α,
- find the values of $\bar{x}$ that will lead us to reject H_0, and
- calculate the probability of observing these values of $\bar{x}$ when the alternative is true.

Statistical inference can be regarded as giving rules for making decisions in the presence of uncertainty. From this **decision theory** point of view, H_0 and H_a are just two statements of equal status that we must decide between. Decision analysis chooses a rule for deciding between H_0 and H_a on the basis of the probabilities of the two types of errors that we can make. A **Type I error** occurs if H_0 is rejected when it is true. A **Type II error** occurs if H_0 is accepted when in fact H_a is true.

There is a clear relation between α-level significance tests and testing from the decision-making point of view. The significance level α is the probability of a Type I error, and the power of the test against a specific alternative is 1 minus the probability of a Type II error for that alternative.

SECTION 6.3 EXERCISES

6.81 a) We follow the steps outlined in Section 6.4.

Step 1. State H_0, H_a, the particular alternative we want to detect, and the significance level α.

The hypotheses we are testing are

$$H_0: \mu = 300$$
$$H_a: \mu < 300$$

at the 5% level. The problem tells us that the particular alternative we want to detect is $\mu = 298$.

Step 2. Find the values of $\bar{x}$ that will lead us to reject H_0.

We reject H_0 if $z \le -1.645$. We know that $z = \dfrac{\bar{x}-300}{3/\sqrt{6}}$, hence we require $\dfrac{\bar{x}-300}{3/\sqrt{6}} \le -1.645$. Solving for $\bar{x}$, this implies $\bar{x} \le -1.645(3/\sqrt{6}) + 300 = 297.985$. Thus, the values of $\bar{x}$ that will lead us to reject H_0 are $\bar{x} \le 297.985$.

Step 3. Calculate the probability of observing these values of $\bar{x}$ when the alternative is true. This is the power.

If $\mu = 298$,

$$P(\bar{x} \le 297.985) = P\left(\frac{\bar{x}-298}{3/\sqrt{6}} \le \frac{297.985-298}{3/\sqrt{6}}\right) = P(Z \le -0.01) = 0.4960.$$

Thus, the power of the test against the alternative $\mu = 298$ is 0.4960.

b) The power is higher. The alternative $\mu = 295$ is further away from $\mu_0 = 300$ (the value of μ under the null hypothesis) than $\mu = 298$. Values of μ that are in H_a but are closer to the hypothesized value μ_0 under the null hypothesis are harder to detect (have lower power) than values of μ that are further from μ_0.

6.83 Step 1. State H_0, H_a, the particular alternative we want to detect, and the significance level α.

The hypotheses we are testing are

$$H_0: \mu = 300$$
$$H_a: \mu < 300$$

at the 5% level. The problem tells us that the particular alternative we want to detect is $\mu = 298$.

Step 2. Find the values of $\bar{x}$ that will lead us to reject H_0.

We reject H_0 if $z \le -1.645$. We know that $z = \dfrac{\bar{x} - 300}{3/\sqrt{20}}$, hence we require $\dfrac{\bar{x} - 300}{3/\sqrt{20}} \le -1.645$. Solving for $\bar{x}$, this implies $\bar{x} \le -1.645(3/\sqrt{20}) + 300 = 298.897$. Thus, the values of $\bar{x}$ that will lead us to reject H_0 are $\bar{x} \le 298.897$.

Step 3. Calculate the probability of observing these values of $\bar{x}$ when the alternative is true. This is the power.

If $\mu = 298$,

$$P(\bar{x} \le 298.897) = P(\frac{\bar{x} - 298}{3/\sqrt{20}} \le \frac{298.897 - 298}{3/\sqrt{20}}) = P(Z \le 1.34) = 0.9099.$$

Thus, the power of the test against the alternative $\mu = 298$ is 0.9099. This is quite a bit larger than the power of 0.4960 that we found in Exercise 6.81.

6.85 The probability of a Type I error in Exercise 6.81 is the probability that we reject H_0 when, in fact, H_0 is true. This is equal to the significance level of the test, which in Exercise 6.81 was $\alpha = 0.05$.

The probability of a Type II error at $\mu = 298$ in Exercise 6.81 is the probability that we accept H_0 when $\mu = 298$, and so H_a is true. This is equal to 1 minus the power of the test at $\mu = 298$. In Exercise 6.81, we found that this power was 0.4960, so the probability of a Type II error is 1 – 0.4960 = 0.5040.

6.87 a) The decision to be made is whether or not the patient needs to see a doctor. Thus, we might formulate our two hypotheses to be (taking the null hypothesis to be the hypothesis of no effect)

H_0: patient does not need to see a doctor
H_a: patient does need to see a doctor

The two types of error that the program can make are

1. The patient is told to see a doctor when, in fact, the patient does not need to see a doctor. This is a "false-positive" in that the patient tests "positive" for needing to see a doctor but really does not need to see one.

2. The patient is diagnosed as not needing to see a doctor when, in fact, the patient does need to see one. This is a "false-negative" in that the patient test "negative" for needing to see a doctor, but really does need to see one.

b) The error probability one chooses to control usually depends on which error is considered most serious. In most cases, a false-negative is considered the more serious. If you have an illness and it is not detected, the consequences can be serious (serious health problems, possibly death). Thus, in most cases one would want to control the probability of a false-negative. However, if the disease being screened for is not serious, and the cost of a false-positive is high (expensive medical bills, painful treatment) one might argue that controlling the probability of a false positive is more important. Which one chooses to control (the probability of a false-negative or the probability of a false-positive) is to some extent a matter of judgment.

CHAPTER 6 REVIEW EXERCISES

6.89 This study tells us the following. Industries with SHRUSED values above the median were found to have cash flow elasticities less than those for industries with lower SHRUSED value. The probability is less than 0.05 that we would observe a difference as or more extreme than this by chance if, in fact, on average the cash flow elasticities for the two types of industries are the same. This probability is quite low, so it is unlikely that the observed difference is merely accidental. This is strong evidence that the difference is real rather than accidental. Thus, if the study was well-designed, we have good reason to think that an active used-equipment market really does change the relationship between cash flow and investment.

6.91 a) A stemplot of the data follows.

```
2|0 3 4
3|0 1 1 2 4 6
4|3
```

As this stemplot shows the data are roughly symmetric.

b) We compute the mean DMS odor threshold for the 10 students to be $\bar{x}$ = 30.4, and we are told that we can assume the population standard deviation is $\sigma = 7$. The sample size is $n = 10$. For a 95% confidence interval, we see from Table D that $z^* = 1.96$, so our margin of error is

$$z^* \frac{\sigma}{\sqrt{n}} = 1.96 \frac{7}{\sqrt{10}} = 4.34.$$

A 95% confidence interval for the mean DMS odor threshold among all beginning oenology students is therefore

$$\bar{x} \pm z^* \frac{\sigma}{\sqrt{n}} = 30.40 \pm 4.34 \ \mu\text{g/l} = (26.06 \ \mu\text{g/l}, 34.74 \ \mu\text{g/l}).$$

c) Let μ denote the mean DMS odor threshold among all beginning oenology students. To test whether the mean odor threshold μ for beginning oenology students is higher than the published threshold, we test the hypotheses

$$H_0: \mu = 25$$
$$H_a: \mu > 25$$

Notice that the null hypothesis is the hypothesis of no difference (from the published threshold) and the alternative hypothesis is what we hope to show. We know that $n = 10$, $\sigma = 7$, and $\bar{x}$ = 30.4. Under the null hypothesis $\mu = 25$, so the z statistic for testing the hypotheses given is

$$z = \frac{\bar{x} - \mu}{\sigma / \sqrt{n}} = \frac{30.4 - 25}{7 / \sqrt{10}} = 2.44.$$

The P-value for our one-sided alternative H_a: $\mu > 25$ is

$$P\text{-value} = P(Z \geq 2.44) = 1 - P(Z < 2.44) = 1 - 0.9927 = 0.0073.$$

This is below 0.01 and therefore we reject H_0 at the 1%. This is strong evidence against the null hypothesis and in favor of the claim that the mean odor threshold for beginning oenology students is higher than the published threshold of 25 μg/l.

6.93 In Exercise 6.92, the 90% confidence interval for the 1999 median annual household income is \$40,816 ± \$314 = (\$40,502, \$41,130). To convert this interval to its equivalent form expressed as weekly incomes, we must divide annual incomes by the number of weeks per year, which is 52.14. By doing so, our 90% confidence interval becomes

(\$40,816/52.14) ± (\$314/52.14) = (\$40,502/52.14, \$41,130/52.14)

or

\$782.82 ± \$6.02 = (\$776.80, \$788.84).

Thus, a 90% confidence interval for the 1999 median weekly household income is \$782.82 ± \$6.02 = (\$776.80, \$788.84).

6.95 a) The authors probably want to draw conclusions about the population of all adult Americans because this would be the group of people that will purchase nonprescription medications. The authors selected their sample from the Indianapolis telephone directory, so the population to which their conclusions most clearly apply are all people listed in the Indianapolis telephone directory.

b) Taking the sample standard deviations to be the population standard deviations, noticing that the sample size is $n = 201$, and recalling that $z^* = 1.96$ for 95% confidence, the 95% confidence intervals are summarized in the table below.

Store type	95% confidence interval
Food stores	$\bar{x} \pm z^* \frac{\sigma}{\sqrt{n}} = 18.67 \pm 1.96(\frac{24.95}{\sqrt{201}}) = 18.67 \pm 3.45$
Mass merchandisers	$\bar{x} \pm z^* \frac{\sigma}{\sqrt{n}} = 32.38 \pm 1.96(\frac{33.37}{\sqrt{201}}) = 32.38 \pm 4.61$
Pharmacies	$\bar{x} \pm z^* \frac{\sigma}{\sqrt{n}} = 48.60 \pm 1.96(\frac{35.62}{\sqrt{201}}) = 48.60 \pm 4.92$

c) None of these intervals overlap (in fact, the intervals are well separated), which suggests the observed differences are likely to be real. Because pharmacies have the highest rating, this would seem to be reasonably strong evidence that consumers (at least those listed in the Indianapolis phone directory) think pharmacies offer higher performance than other types of stores.

6.97 a) Increasing the size n of a sample will decrease the width of a level C confidence interval. If we examine the formula for such an interval in the case where we have normal data, namely $\bar{x} \pm z^* \frac{\sigma}{\sqrt{n}}$, the fact that n is in the denominator means that as n increases the margin of error $z^* \frac{\sigma}{\sqrt{n}}$ decreases and so the width of the interval decreases.

b) Increasing the size n of a sample will decrease the P-value of a test when H_0 is false and all facts about the population remain unchanged. In general, if H_0 is false, a larger sample size will give us more information and enable us to detect smaller departures from the null hypothesis. In other words, the same data will appear to be a more extreme departure from the null hypothesis and the probability (P-value) of observing such an extreme value by chance if H_0 were true decreases. To make this more explicit, consider the case of normal data. The test statistic is $z = \frac{\bar{x} - \mu}{\sigma / \sqrt{n}}$ and we see that as n increases, z increases, and hence the probability of observing a value as or more extreme than z under the null hypothesis decreases.

c) Increasing the sample size n will increase the power of a fixed level α test when the alternative hypothesis, the particular alternative at which we compute the power, and all facts about the population remain unchanged. In general, a larger sample size will give us more information about the population and enable us to detect smaller departures from the null hypothesis.

6.99 This is not correct. The null hypothesis is either true or false, i.e., it is true with either probability 1 or 0. Statistical significance at the 0.05 level means that if the null hypothesis is true, the probability we will obtain data that leads us to incorrectly reject H_0 is 0.05. The level of significance refers to the performance of our method for testing H_0 in repeated use, not to the truth of H_0.

6.101a) "A significant difference ($P < 0.01$)" means the following. Assume that in the population of all mothers with young children those who would choose to attend the training program and those who would not actually remain on welfare at the same rate. If this assumption is true, the probability is less than 0.01 that we would observe a difference as or more extreme than that actually observed. This probability is quite low and so it is unlikely that the observed difference is merely accidental. This is strong evidence that the difference is real rather than accidental.

b) 95% confidence means that the method used to construct the interval 21% ±4 % will produce an interval that contains the true difference 95% of the time. Because the method is reliable 95% of the time, we say that we are 95% confident that this particular interval is accurate. Technically, the interval is either correct (contains the true difference) or incorrect. Our confidence refers to the reliability of the method in repeated use, not whether this particular interval is correct.

c) Although the P-value is small and the confidence interval suggests that the difference is large, the study is not good evidence that requiring job training of all welfare mothers will greatly reduce the percent who remain on welfare for several years. The reason that the study does not provide good evidence is the design of the study. Mothers chose to participate in the training program; they were not assigned using randomization. Thus, the effect of the program is confounded with the reasons some women chose to participate and some did not. Perhaps those who chose to participate were more highly motivated to get off welfare and would have done so even if they had not attended the training. Statistical inference is not valid for badly designed studies.

6.103 We used statistical software to conduct the simulation. The results are summarized below.

1 Sample Mean = 17.577825 z statistic = –1.083; Fail to reject at Alpha = 0.05; P = 0.2787
2. Sample Mean = 19.877607 z statistic = –0.0547; Fail to reject at Alpha = 0.05; P = 0.9563
3. Sample Mean = 21.717800 z statistic = 0.7682; Fail to reject at Alpha = 0.05; P = 0.4424
4. Sample Mean = 16.135595 z statistic = –1.728; Fail to reject at Alpha = 0.05; P = 0.0840
5. Sample Mean = 17.118525 z statistic = –1.289; Fail to reject at Alpha = 0.05; P = 0.1975
6. Sample Mean = 21.605743 z statistic = 0.7181; Fail to reject at Alpha = 0.05; P = 0.4727
7. Sample Mean = 16.172664 z statistic = –1.712; Fail to reject at Alpha = 0.05; P = 0.0870
8. Sample Mean = 23.297126 z statistic = 1.475; Fail to reject at Alpha = 0.05; P = 0.1403
9. Sample Mean = 23.908858 z statistic = 1.748; Fail to reject at Alpha = 0.05; P = 0.0804
10. Sample Mean = 17.745180 z statistic = –1.008; Fail to reject at Alpha = 0.05; P = 0.3133
11. Sample Mean = 17.910399 z statistic = –0.9345; Fail to reject at Alpha = 0.05; P = 0.3500
12. Sample Mean = 20.065768 z statistic = 0.0294; Fail to reject at Alpha = 0.05; P = 0.9765
13. Sample Mean = 19.174106 z statistic = –0.3694; Fail to reject at Alpha = 0.05; P = 0.7119
14. Sample Mean = 18.415171 z statistic = –0.7088; Fail to reject at Alpha = 0.05; P = 0.4785
15. Sample Mean = 20.951042 z statistic = 0.4253; Fail to reject at Alpha = 0.05; P = 0.6706
16. Sample Mean = 22.713500 z statistic = 1.214; Fail to reject at Alpha = 0.05; P = 0.2249
17. Sample Mean = 18.138763 z statistic = –0.8324; Fail to reject at Alpha = 0.05; P = 0.4052
18. Sample Mean = 18.964147 z statistic = –0.4632; Fail to reject at Alpha = 0.05; P = 0.6432
19. Sample Mean = 17.558578 z statistic = –1.092; Fail to reject at Alpha = 0.05; P = 0.2749
20. Sample Mean = 23.031043 z statistic = 1.356; Fail to reject at Alpha = 0.05; P = 0.1753
21. Sample Mean = 19.354264 z statistic = –0.2888;l Fail to reject at Alpha = 0.05; P = 0.7727
22. Sample Mean = 22.653787 z statistic = 1.187; Fail to reject at Alpha = 0.05; P = 0.2353
23. Sample Mean = 19.927163 z statistic = –0.0326; Fail to reject at Alpha = 0.05; P = 0.9740
24. Sample Mean = 19.904690 z statistic = –0.0426; Fail to reject at Alpha = 0.05; P = 0.9660
25. Sample Mean = 20.684493 z statistic = 0.3061; Fail to reject at Alpha = 0.05; P = 0.7595
26. Sample Mean = 22.118202 z statistic = 0.9473; Fail to reject at Alpha = 0.05; P = 0.3435
27. Sample Mean = 17.337963 z statistic = –1.190; Fail to reject at Alpha = 0.05; P = 0.2339
28. Sample Mean = 19.001105 z statistic = –0.4467; Fail to reject at Alpha = 0.05; P = 0.6551
29. Sample Mean = 23.060083 z statistic = 1.369; Fail to reject at Alpha = 0.05; P = 0.1712
30. Sample Mean = 20.443916 z statistic = 0.1985; Fail to reject at Alpha = 0.05; P = 0.8426
31. Sample Mean = 17.254774 z statistic = –1.228; Fail to reject at Alpha = 0.05; P = 0.2196
32. Sample Mean = 17.973636 z statistic = –0.9062; Fail to reject at Alpha = 0.05; P = 0.3648
33. Sample Mean = 17.536305 z statistic = –1.102; Fail to reject at Alpha = 0.05; P = 0.2705
34. Sample Mean = 20.417951 z statistic = 0.1869; Fail to reject at Alpha = 0.05; P = 0.8517
35. Sample Mean = 21.151701 z statistic = 0.5151; Fail to reject at Alpha = 0.05; P = 0.6065
36. Sample Mean = 20.070363 z statistic = 0.0315; Fail to reject at Alpha = 0.05; P = 0.9749
37. Sample Mean = 21.358099 z statistic = 0.6074; Fail to reject at Alpha = 0.05; P = 0.5436
38. Sample Mean = 21.262138 z statistic = 0.5644; Fail to reject at Alpha = 0.05; P = 0.5725
39. Sample Mean = 16.766305 z statistic = –1.446; Fail to reject at Alpha = 0.05; P = 0.1481

40. Sample Mean = 19.714630 z statistic = –0.1276; Fail to reject at Alpha = 0.05; P = 0.8984
41. Sample Mean = 21.028823 z statistic = 0.4601; Fail to reject at Alpha = 0.05; P = 0.6454
42. Sample Mean = 18.919166 z statistic = –0.4834; Fail to reject at Alpha = 0.05; P = 0.6288
43. Sample Mean = 17.426017 z statistic = –1.151; Fail to reject at Alpha = 0.05; P = 0.2497
44. Sample Mean = 17.067582 z statistic = –1.311; Fail to reject at Alpha = 0.05; P = 0.1897
45. Sample Mean = 19.483805 z statistic = –0.2308; Fail to reject at Alpha = 0.05; P = 0.8174
46. Sample Mean = 21.398537 z statistic = 0.6254; Fail to reject at Alpha = 0.05; P = 0.5317
47. Sample Mean = 20.403869 z statistic = 0.1806; Fail to reject at Alpha = 0.05; P = 0.8567
48. Sample Mean = 16.662104 z statistic = –1.493; Fail to reject at Alpha = 0.05; P = 0.1355
49. Sample Mean = 19.849644 z statistic = –0.0672; Fail to reject at Alpha = 0.05; P = 0.9464
50. Sample Mean = 15.569270 z statistic = –1.981; Reject at Alpha = 0.05; P = 0.0475
51. Sample Mean = 21.632234 z statistic = 0.7300; Fail to reject at Alpha = 0.05; P = 0.4654
52. Sample Mean = 16.979065 z statistic = –1.351; Fail to reject at Alpha = 0.05; P = 0.1767
53. Sample Mean = 19.318209 z statistic = –0.3049; Fail to reject at Alpha = 0.05; P = 0.7604
54. Sample Mean = 20.117020 z statistic = 0.0523; Fail to reject at Alpha = 0.05; P = 0.9583
55. Sample Mean = 21.474592 z statistic = 0.6595; Fail to reject at Alpha = 0.05; P = 0.5096
56. Sample Mean = 19.263090 z statistic = –0.3296; Fail to reject at Alpha = 0.05; P = 0.7417
57. Sample Mean = 20.452935 z statistic = 0.2026; Fail to reject at Alpha = 0.05; P = 0.8395
58. Sample Mean = 16.874319 z statistic = –1.398; Fail to reject at Alpha = 0.05; P = 0.1622
59. Sample Mean = 20.763393 z statistic = 0.3414; Fail to reject at Alpha = 0.05; P = 0.7328
60. Sample Mean = 20.835662 z statistic = 0.3737; Fail to reject at Alpha = 0.05; P = 0.7086
61. Sample Mean = 22.405945 z statistic = 1.076; Fail to reject at Alpha = 0.05; P = 0.2819
62. Sample Mean = 16.307375 z statistic = –1.651; Fail to reject at Alpha = 0.05; P = 0.0987
63. Sample Mean = 16.509392 z statistic = –1.561; Fail to reject at Alpha = 0.05; P = 0.1185
64. Sample Mean = 21.140779 z statistic = 0.5102; Fail to reject at Alpha = 0.05; P = 0.6099
65. Sample Mean = 22.708451 z statistic = 1.211; Fail to reject at Alpha = 0.05; P = 0.2258
66. Sample Mean = 21.289483 z statistic = 0.5767; Fail to reject at Alpha = 0.05; P = 0.5642
67. Sample Mean = 24.526428 z statistic = 2.024; Reject at Alpha = 0.05; P = 0.0429
68. Sample Mean = 19.165256 z statistic = –0.3733; Fail to reject at Alpha = 0.05; P = 0.7089
69. Sample Mean = 20.233763 z statistic = 0.1045; Fail to reject at Alpha = 0.05; P = 0.9167
70. Sample Mean = 23.789297 z statistic = 1.695; Fail to reject at Alpha = 0.05; P = 0.0901
71. Sample Mean = 19.558681 z statistic = –0.1974; Fail to reject at Alpha = 0.05; P = 0.8435
72. Sample Mean = 17.626449 z statistic = –1.061; Fail to reject at Alpha = 0.05; P = 0.2885
73. Sample Mean = 20.284339 z statistic = 0.1272; Fail to reject at Alpha = 0.05; P = 0.8988
74. Sample Mean = 22.993913 z statistic = 1.3391; Fail to reject at Alpha = 0.051 P = 0.1806
75. Sample Mean = 19.418630 z statistic = –0.2600; Fail to reject at Alpha = 0.05; P = 0.7949
76. Sample Mean = 18.417701 z statistic = –0.7076; Fail to reject at Alpha = 0.05; P = 0.4792
77. Sample Mean = 19.804492 z statistic = –0.0874; Fail to reject at Alpha = 0.05; P = 0.9303
78. Sample Mean = 20.218534 z statistic = 0.0977; Fail to reject at Alpha = 0.05; P = 0.9221
79. Sample Mean = 18.993190 z statistic = –0.4503; Fail to reject at Alpha = 0.05; P = 0.6525
80. Sample Mean = 15.459194 z statistic = –2.031; Reject at Alpha = 0.05; P = 0.0423
81. Sample Mean = 20.975678 z statistic = 0.4363; Fail to reject at Alpha = 0.05; P = 0.6626
82. Sample Mean = 20.804681 z statistic = 0.3599; Fail to reject at Alpha = 0.05; P = 0.7189
83. Sample Mean = 19.942797 z statistic = –0.0256; Fail to reject at Alpha = 0.05; P = 0.9796
84. Sample Mean = 21.360327 z statistic = 0.6084; Fail to reject at Alpha = 0.05; P = 0.5430
85. Sample Mean = 20.352050 z statistic = 0.1574; Fail to reject at Alpha = 0.05; P = 0.8749
86. Sample Mean = 16.533325 z statistic = –1.550; Fail to reject at Alpha = 0.05; P = 0.1211
87. Sample Mean = 21.057047 z statistic = 0.4727; Fail to reject at Alpha = 0.05; P = 0.6364
88. Sample Mean = 17.785820 z statistic = –0.9902; Fail to reject at Alpha = 0.05; P = 0.3221
89. Sample Mean = 24.729923 z statistic = 2.115; Reject at Alpha = 0.05; P = 0.0344
90. Sample Mean = 21.030679 z statistic = 0.4609; Fail to reject at Alpha = 0.05; P = 0.6448
91. Sample Mean = 23.149961 z statistic = 1.409; Fail to reject at Alpha = 0.05; P = 0.1589
92. Sample Mean = 21.525307 z statistic = 0.6821; Fail to reject at Alpha = 0.05; P = 0.4952
93. Sample Mean = 21.055966 z statistic = 0.4722; Fail to reject at Alpha = 0.05; P = 0.6368
94. Sample Mean = 19.196987 z statistic = –0.3591; Fail to reject at Alpha = 0.05; P = 0.7195
95. Sample Mean = 21.387189 z statistic = 0.6204; Fail to reject at Alpha = 0.05; P = 0.5350
96. Sample Mean = 18.477575 z statistic = –0.6808; Fail to reject at Alpha = 0.05; P = 0.4960
97. Sample Mean = 19.962349 z statistic = –0.0168; Fail to reject at Alpha = 0.05; P = 0.9866
98. Sample Mean = 14.298872 z statistic = –2.550; Reject at Alpha = 0.05; P = 0.0108
99. Sample Mean = 20.847586 z statistic = 0.3791; Fail to reject at Alpha = 0.05; P = 0.7046
100. Sample Mean = 20.700917 z statistic = 0.3135; Fail to reject at Alpha = 0.05; P = 0.7539

Notice that only in cases 50, 67, 80, 89, and 98 did we reject H_0. Thus, the number of times we rejected H_0 was five times and the proportion of times was 0.05. Of course, this is consistent with the meaning of a 0.05 significance level. It is the probability that we would reject H_0 when, in fact, H_0 is true and so in 100 trials where H_0 was true, we would expect the proportion of times we reject to be about 0.05.

6.105 a) We used statistical software to conduct the simulations. The sample means for each of the 25 simulations are given below.

232.48764	247.55669	241.07254	238.81784	230.95930
257.42161	244.84676	242.09581	258.96179	238.76006
269.73055	249.22459	241.70413	204.54853	224.08684
271.19073	263.08773	250.25793	238.37828	251.21616
238.44008	245.16715	243.18843	226.99779	250.02756

b) This is a reasonable assumption because each sample mean is the average of an SRS of size 100. The sample size is large and the central limit theorem suggests that it is probably reasonable to assume the sampling distribution for the sample means is approximately normal.

c) We are told that the sample size is n = 100, σ = 150, and that we want a 50% confidence interval so from Table D z^* = 0.674. Therefore, the margin of error is

$$m = z^* \frac{\sigma}{\sqrt{n}} = 0.674\left(\frac{150}{\sqrt{100}}\right) = 10.11.$$

d) The 25 50% confidence intervals from our simulation follow.

(222.37029, 242.60499)	(237.43934, 257.67403)	(230.95519, 251.18989)
(228.70049, 248.93519)	(220.84196, 241.07665)	(247.30427, 267.53896)
(234.72941, 254.96410)	(231.97847, 252.21316)	(248.84444, 269.07913)
(228.64272, 248.87741)	(259.61321, 279.84790)	(239.10725, 259.34194)
(231.58678, 251.82148)	(194.43119, 214.66588)	(213.96950, 234.20419)
(261.07338, 281.30808)	(252.97039, 273.20508)	(240.14059, 260.37528)
(228.26093, 248.49562)	(241.09881, 261.33350)	(228.32274, 248.55743)
(235.04980, 255.28449)	(233.07108, 253.30577)	(216.88044, 237.11514)
(239.91022, 260.14491)		

Notice that the sample mean for the first simulation was 232.48764. The margin of error was found to be 10.11 in part (c), so the 50% confidence interval for the first simulation is 232.48764 ± 10.11 = (222.37764, 242.59764), which agrees with our simulation result up to round-off error.

e) We see that simulations 1, 2, 3, 4, 5, 7, 8, 10, 12, 13, 19, 21, 22, 23, and 25 contain μ = 240. Thus, 15 of the 25, or 60%, of the simulations contain μ = 240. If we repeated the simulation, we would not expect to get exactly the same number of intervals to contain μ = 240. This is because each simulation is random and so results will vary from one simulation to the next. This is the meaning of "random" (although there will be certain regular behavior in the long run). The probability that any given simulation contains μ = 240 is 0.50, so in a very large number of simulations we would expect about 50% to contain μ = 240.

CHAPTER 7

INFERENCE FOR DISTRIBUTIONS

SECTION 7.1

OVERVIEW

Confidence intervals and significance tests for the mean μ of a normal population are based on the sample mean $\bar{x}$ of an SRS. When the sample size n is large, the central limit theorem suggests that these procedures are approximately correct for other population distributions. In Chapter 6, we considered the (unrealistic) situation in which we knew the population standard deviation σ. In this section, we consider the more realistic case where σ is not known and we must estimate σ from our SRS by the sample standard deviation s. In Chapter 6, we used the **one-sample *z* statistic**

$$z = \frac{\bar{x} - \mu}{\sigma/\sqrt{n}}$$

which has the $N(0,1)$ distribution. Replacing σ by s, we now use the **one-sample *t* statistic**

$$t = \frac{\bar{x} - \mu}{s/\sqrt{n}}$$

which has the ***t* distribution** with $n - 1$ **degrees of freedom.** The quantity $\mathrm{SE}_{\bar{x}} = \frac{s}{\sqrt{n}}$ is the standard error of the sample mean.

For every positive value of k, there is a t distribution, with k degrees of freedom, denoted $t(k)$. All are symmetric, bell-shaped distributions, similar in shape to normal distributions but with greater spread. As k increases, $t(k)$ approaches the $N(0,1)$ distribution.

A level C **confidence interval for the mean** μ of a normal population when σ is unknown is

$$\bar{x} \pm t^* \frac{s}{\sqrt{n}}$$

where t^* is the upper $(1-C)/2$ critical value of the $t(n-1)$ distribution whose value can be found in Table D of the text or from statistical software. The quantity $t^* \frac{s}{\sqrt{n}}$ is the **margin of error.**

Significance tests of H_0: $\mu = \mu_0$ are based on the one-sample t statistic. Such tests are often referred to as **one-sample *t* tests.** P-values or fixed significance levels are computed from the $t(n-1)$ distribution using Table D or, more commonly in practice, using statistical software.

The power of the t test is calculated like that of the z test as described in Chapter 6, using an approximate value (perhaps based on past experience or a pilot study) for both σ and s.

One application of these one-sample t procedures is to the analysis of data from **matched pairs** studies. We compute the differences between the two values of a matched pair (often before and after measurements on the same unit) to produce a single sample value. The sample mean and standard deviation of these differences are computed. Depending on whether we are interested in a confidence interval or a test of significance concerning the difference in the population means of matched pairs, we use either the one-sample confidence interval or the one-sample significance test based on the t statistic.

For larger sample sizes, the t procedures are fairly **robust** against nonnormal populations. As a rule of thumb, t procedures are useful for nonnormal data when $n \geq 15$ unless the data show outliers or strong skewness, and for samples of size $n \geq 40$ t procedures can be used for even clearly skewed distributions. Data consisting of small samples from skewed populations can sometimes be analyzed by first applying a **transformation** (such as logarithms) to the data to obtain an approximately normally distributed variable. The t procedures can then be applied to the transformed data. When transformations are used, it is a good idea to examine stemplots, histograms, or normal quantile plots of the transformed data to verify that the transformed data appear to be approximately normally distributed.

Another procedure that can be used with smaller samples from a nonnormal population is the **sign test.** The sign test is most useful for testing for "no treatment effect" in matched pairs studies. As with the matched pairs t test, one computes the differences of the two values in each matched pair. Pairs with difference 0 are ignored and the number of trials n is the count of the remaining pairs. The test statistic is the count X of pairs with a positive difference. P-values for X are based on the binomial $B(n, 1/2)$ distribution. The sign test is less powerful than the t test in cases where the use of the t test is justified.

APPLY YOUR KNOWLEDGE

7.1 a) The standard error of the mean is $SE_{\bar{x}} = \frac{s}{\sqrt{n}} = \frac{80}{\sqrt{10}} = 25.30$.

b) The degrees of freedom for a one-sample t statistic are $n - 1 = 10 - 1 = 9$.

7.3 A 95% confidence interval for the mean monthly rent for unfurnished one-bedroom apartments in this community can be calculated from the sample of 10 apartments. We use the formula for a t interval, namely $\bar{x} \pm t^* \frac{s}{\sqrt{n}}$. In this problem, $\bar{x}$ = \$531, s = \$82.792, n = 10, and t^* is the upper $(1 - 0.95)/2 = 0.025$ critical value for the $t(9)$ distribution. From Table D, we see that $t^* = 2.262$. Thus, the 95% confidence interval is

$$531 \pm 2.262\frac{82.792}{\sqrt{10}} = 531 \pm 59.22 = (\$471.78, \$590.22)$$

7.5 We wish to know if there is evidence that the mean rent is greater than \$500 per month. This corresponds to the hypotheses

$$H_0: \mu = 500$$
$$H_a: \mu > 500$$

Recall from Exercise 7.3 that $\bar{x}$ = \$531, s = \$82.792, and n = 10. We compute

$$t = \frac{\bar{x} - \mu}{s/\sqrt{n}} = \frac{531 - 500}{26.18} = 1.18$$

According to the row corresponding to $n - 1 = 9$ in Table D,

df = 9

p	.15	.10
t^*	1.100	1.383

the P-value lies between 0.10 and 0.15. Using statistical software, we find the P-value is 0.13. We conclude that there is not much evidence that the mean rent of all advertised apartments exceeds \$500.

7.7 a) Because we are interested in whether the average sales are different from last month (no direction of the difference is specified), we wish to test the hypotheses

$$H_0: \mu = 0$$
$$H_a: \mu \neq 0$$

b) We compute

$$t = \frac{\bar{x} - \mu}{s/\sqrt{n}} = \frac{4.8 - 0}{15/\sqrt{50}} = 2.26$$

According to the row corresponding to $n - 1 = 49$ in Table D (we use the row for 50 as an approximation), .

df = 50

p	.01	.02
t^*	2.403	2.109

the P-value lies between $2 \times 0.01 = 0.02$ and $2 \times 0.02 = 0.04$. Using statistical software the P-value is 0.028. There is strong evidence that the average sales have increased.

c) The mean change is 4.8% and the standard deviation is 15%, so there are certainly stores with a percent change that is negative. Although there is strong evidence that the mean has increased, there are still stores for which the sales have decreased.

7.9 a) Using statistical software and df = 5000 – 1 = 4999, the absolute value of the t statistic must exceed 2.5768 to be significant at the 1% level.

b) This result can be found in Table D in the row corresponding to z^*. This illustrates that for large degrees of freedom, there is little difference between critical values for the t and the normal distribution.

7.11 The five differences (Before – After) are 53, 52, 57, 52, and 61 milligrams per 100 grams of blend, dry basis. These represent the loss in vitamin C. A 95% confidence interval for the mean loss in vitamin C can be calculated from this sample of five differences. We use the formula for a t interval, namely $\bar{x} \pm t^* \frac{s}{\sqrt{n}}$, where averages and standard deviations are computed using the differences because this is a matched pairs analysis. In this problem, $\bar{x} = 55$, $s = 3.94$, $n = 5$, and t^* is the upper $(1 - 0.95)/2 = 0.025$ critical value for the $t(4)$ distribution. From Table D, we see that $t^* = 2.776$. Thus, the 95% confidence interval is

$$55 \pm 2.776 \frac{3.94}{\sqrt{5}} = 55 \pm 4.89 = (50.11, 59.89)$$

Instead of using the differences, it would also be possible to look at the percentage of vitamin C lost in cooking.

7.13 a) The stemplot of the data is given below. The stemplot uses split stems to better reveal the characteristics of the data. The data are clearly skewed to the right and have several high outliers.

```
1 | 01233344
1 | 5566667778999999
2 | 00124444
2 | 5555566667
3 | 244
3 | 5
4 | 1
4 | 8
5 |
5 |
6 | 3
6 |
7 |
7 | 9
```

b) We use the formula for a t interval, namely $\bar{x} \pm t^* \frac{s}{\sqrt{n}}$, where $\bar{x} = 23.56$, $s = 12.52$, $n = 50$, and t^* is the upper $(1 - 0.95)/2 = 0.025$ critical value for the $t(49)$ distribution. From Table D we see that t^* is approximately 2.009, where the row corresponding to 50 df was used. Using statistical software, we get $t^* = 2.010$. Thus, the 95% confidence interval is

$$23.56 \pm 2.010\frac{12.52}{\sqrt{50}} = 23.56 \pm 3.56 = (20.00, 27.12)$$

which agrees quite well with the bootstrap intervals. The lesson is that for large sample sizes the t procedures are very robust.

7.15 We wish to determine the power of the t test against the alternative $\mu = 2.3$ when the significance level is 0.05 and $n = 20$. We use 3 as an estimate of both the population standard deviation σ and s in future samples. The t test with 20 observations rejects H_0: $\mu = 0$ (this is the null hypothesis that there is no improvement in the executives' comprehension of spoken French) if the t statistic

$$t = \frac{\bar{x} - 0}{s/\sqrt{20}}$$

exceeds the upper 5% point of $t(19)$, which is 1.729 from Table D. Using $s = 3$, we reject H_0 when

$$t = \frac{\bar{x}}{3/\sqrt{20}} \geq 1.729$$

or equivalently when

$$\bar{x} \geq 1.729\frac{3}{\sqrt{20}} = 1.16.$$

The power is the probability that $\bar{x} \geq 1.16$ when $\mu = 2.3$. Taking $\sigma = 3$, this probability is found by standardizing $\bar{x}$,

$$P(\bar{x} \geq 1.16 \text{ when } \mu = 2.3) = P\left(\frac{\bar{x} - 2.3}{3/\sqrt{20}} \geq \frac{1.16 - 2.3}{3/\sqrt{20}}\right)$$

$$= P(Z \geq -1.70) = 1 - 0.0446 = 0.9554.$$

Thus the power is 0.9554.

SECTION 7.1 EXERCISES

7.17 a) The degrees of freedom are $n - 1 = 12 - 1 = 11$. t^* is the upper $(1 - 0.90)/2 = 0.05$ critical value for the $t(11)$ distribution. From Table D, we see that $t^* = 1.796$.

b) The degrees of freedom are $n - 1 = 30 - 1 = 29$. t^* is the upper $(1 - 0.95)/2 = 0.025$ critical value for the $t(29)$ distribution. From Table D, we see that $t^* = 2.045$.

c) The degrees of freedom are $n - 1 = 18 - 1 = 17$. t^* is the upper $(1 - 0.80)/2 = 0.10$ critical value for the $t(17)$ distribution. From Table D, we see that $t^* = 1.333$.

7.19 a) The degrees of freedom for the statistic are $n - 1 = 15 - 1 = 14$.

b) The two values of t^* that bracket $t = 1.97$ are 1.761 and 2.145, using the row in Table D corresponding to 14 degrees of freedom.

c) The right tail probability corresponding to $t^* = 1.761$ is 0.05 and to $t^* = 2.145$ is 0.025.

d) The P-value lies between 0.025 and 0.05.

e) It is significant at the 5% level because $t = 1.97$ exceeds $t^* = 1.761$, but not significant at the 1% level because $t = 1.97$ does not exceed $t^* = 2.624$. Alternatively, because the P-value is less than

0.05 it is significant at the 5% level, and because the P-value is larger than 0.01 it is not significant at the 1% level.

f) Using statistical software, the P-value is 0.0345.

7.21 a) The degrees of freedom for the statistic are $n - 1 = 12 - 1 = 11$.

b) According to the row corresponding to $n - 1 = 11$ in Table D,

df = 11

p	.01	.02
t^*	2.718	2.328

the P-value lies between 0.01 and 0.02 because these values of t^* bracket $t = 2.45$.

c) Using statistical software, the P-value is 0.0161.

7.23 a) The stemplot and normal quantile plot are given below. In the stemplot, the stems are \$10,000 and the leaves are \$1000. The stemplot was produced by Minitab, and we note that the salaries were automatically truncated rather than rounded — the 14 in the stemplot corresponds to the observation \$14,698. Software packages adopt various conventions when analyzing data, and as you become accustomed to a particular package you will learn more about these. Neither the stemplot nor the normal quantile plot show serious departures from normality. There is no serious skewness and there are no outliers present.

```
1 | 45
1 | 777
1 | 88999
2 | 01
2 | 2
2 | 445
2 | 66
2 | 8
3 | 1
```

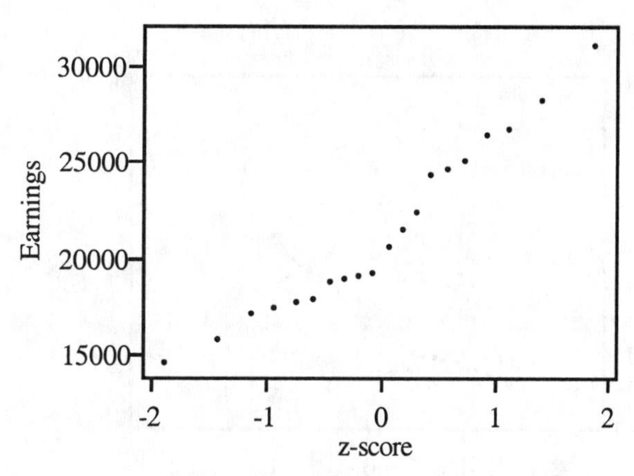

b) A 95% confidence interval for the mean annual earnings of hourly paid white female workers at this bank can be calculated from the sample of 20 white female workers. We use the formula for a t interval, namely $\bar{x} \pm t^* \frac{s}{\sqrt{n}}$. In this problem, $\bar{x}$ = \$21485, s = \$4543.1, n = 20, hence t^* is the upper $(1 - 0.95)/2 = 0.025$ critical value for the $t(19)$ distribution. From Table D, we see that $t^* = 2.093$. Thus, the 95% confidence interval is

$$21{,}485 \pm 2.093\frac{4543.1}{\sqrt{20}} = 21{,}485 \pm 2127 = (\$19{,}358, \$23{,}612)$$

7.25 We wish to know if there is evidence that the mean of annual earnings is greater than \$20,000. This corresponds to the hypotheses

$$H_0: \mu = 20000$$
$$H_a: \mu > 20000$$

Recall from Exercise 7.23 that $\bar{x}$ = \$21,485, s = \$4543.1, and n = 20. We compute

$$t = \frac{\bar{x} - \mu}{s/\sqrt{n}} = \frac{21,485 - 20,000}{1015.9} = 1.46$$

According to the row corresponding to n – 1 = 19 in Table D,

df = 19

p	.10	.05
t^*	1.328	1.729

the P-value lies between 0.05 and 0.10. Using statistical software, we find the P-value is 0.08. We conclude that there is weak evidence that the mean annual earnings exceed \$20,000.

7.27 a) The amounts spent have been rounded to the nearest dollar and then a stemplot has been drawn. The normal quantile plot is provided as well. Although there are no strong outliers, the distribution of the amount spent is skewed to the right and clearly nonNormal.

```
0 | 399
1 | 1345677889
2 | 000123455668888
3 | 25699
4 | 1345579
5 | 0359
6 | 1
7 | 0
8 | 366
9 | 3
```

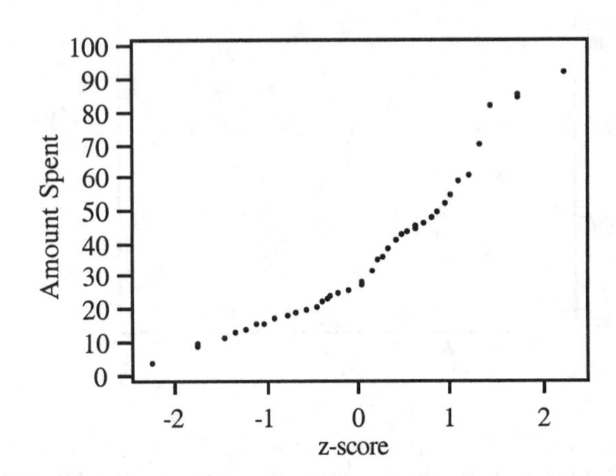

b) In this problem, $\bar{x}$ = \$34.70, s = \$21.70 and the standard error of the mean is $SE_{\bar{x}} = \frac{s}{\sqrt{n}} = \frac{21.70}{\sqrt{50}} = 3.07$.

c) A 95% confidence interval for the mean amount spent can be calculated from the sample of 50 shoppers. We use the formula for a t interval, namely $\bar{x} \pm t^* \frac{s}{\sqrt{n}}$. Because n = 50, t^* is the upper (1 – 0.95)/2 = 0.025 critical value for the $t(49)$ distribution. From Table D, we see that t^* is approximately 2.009 using the row corresponding to 50 degrees of freedom. Using statistical software, t^* = 2.010. Thus, the 95% confidence interval is

$$34.70 \pm 2.010\frac{21.70}{\sqrt{50}} = 34.70 \pm 6.17 = (\$28.53, \$40.87)$$

7.29 In Exercise 7.24, a 95% confidence interval for μ, the mean monthly cost of Internet access in August 2000, was obtained. The interval, based on 50 observations, was ($18.73, $23.07). If there are N people in a population with mean μ, then the population total is $N\mu$. In this case, there are 44 million users, so the total spent on the Internet is 44,000,000μ. Once you have a confidence interval for μ, it can be converted into a confidence interval for the total by multiplying both endpoints by 44 million. That is, if we are 95% confident that μ is in the interval ($18.73, $23.07), we are also 95% confident that 44,000,000μ is in the interval ($824,120,000, $1,015,080,000), where $824,120,000 = 44,000,000 × $18.73 and $1,015,080,000 = 44,000,000 × $23.07.

7.31 A 95% confidence interval for the mean price received by farmers for corn sold in October is based on the formula for a t interval, namely $\bar{x} \pm t^*\frac{s}{\sqrt{n}}$. In this problem, $\bar{x}$ = $2.08, the standard error $\frac{s}{\sqrt{n}}$ = $0.176, and t^* is the upper (1 – 0.95)/2 = 0.025 critical value for the $t(21)$ distribution. From Table D, we see that t^* = 2.080. Thus the 95% confidence interval is

$$2.08 \pm 2.080(0.176) = 2.08 \pm 0.37 = (\$1.71, \$2.45)$$

7.33 A 95% confidence interval for the mean score on the question "Feeling welcomed at Purdue" is based on the formula for a t interval, namely $\bar{x} \pm t^*\frac{s}{\sqrt{n}}$. In this problem, $\bar{x}$ = 3.9, s = 0.98, n = 1406, and t^* is the upper (1 – 0.99)/2 = 0.005 critical value for the $t(1405)$ distribution. From Table D, we see that t^* is approximately 2.581 using the row corresponding to 1000 degrees of freedom. Using statistical software, t^* = 2.579. Thus, the 99% confidence interval is

$$3.9 \pm 2.579\frac{0.98}{\sqrt{1406}} = 3.9 \pm 0.07 = (3.83, 3.97)$$

7.35 We wish to know if there is evidence that piano lessons improve the spatial-temporal reasoning of preschool children. This corresponds to the hypotheses

$$H_0: \mu = 0$$
$$H_a: \mu > 0$$

where μ represents the average improvement in scores over six months for preschool children. Recall from Exercise 7.34 that based on a sample of 34 children the changes in reasoning scores had $\bar{x}$ = 3.618, s = 3.055, and n = 34. We compute

$$t = \frac{\bar{x} - \mu}{s/\sqrt{n}} = \frac{3.618 - 0}{.524} = 6.90$$

According to the row corresponding to either 30 or 40 degrees of freedom in Table D, a t value of 6.90 is well beyond the largest critical value corresponding to a tail area of 0.0005. The P-value is very close to zero, so there is extremely strong evidence that the scores improved over 6 months. This is in agreement with the confidence interval (2.55, 6.88) obtained in Exercise 7.34. The confidence interval also indicates the mean improvement exceeded zero.

7.37 A 95% confidence interval for the mean amount of D-glucose in cockroach hindguts under these conditions is based on the formula for a t interval, namely $\bar{x} \pm t^*\frac{s}{\sqrt{n}}$. In this problem, $\bar{x}$ = 44.44, s = 20.74, n = 5, and t^* is the upper (1 – 0.95)/2 = 0.025 critical value for the $t(4)$ distribution. From Table D, we see that t^* = 2.776. Thus, the 95% confidence interval is

$$44.44 \pm 2.776\frac{20.74}{\sqrt{5}} = 44.44 \pm 25.75 = (18.69 \text{ micrograms}, 70.19 \text{ micrograms})$$

The confidence interval is extremely wide as it is based on a very small sample size and there was a great deal of variability in the five observations obtained.

7.39 a) We wish to know if there is evidence that vitamin C content is destroyed as a result of storage and shipment of the product. This is paired data. We take the difference in vitamin C content (Haiti – factory) as the change, and this corresponds to the hypotheses

$$H_0: \mu = 0$$
$$H_a: \mu < 0$$

where μ represents the average difference in vitamin C between the measurement five months later in Haiti and the factory measurement. Since Vitamin C is not going to be created, a confidence interval to determine the amount of Vitamin C lost is probably of more interest than the hypothesis test.

b) Since this is paired data we first compute the 34 differences (Haiti – Factory), giving

```
 -4    -13     -9     -9     -7     -6    -12    -13     -5     -7      8
-12     -2     -4     -6     -6     -6      3      4    -14     -7      1
  3     -4     -8     -1     -8
```

Computing the mean and standard deviation of these differences gives $\bar{x} = -5.33$, $s = 5.59$ and $n = 27$. We compute

$$t = \frac{\bar{x} - \mu}{s/\sqrt{n}} = \frac{-5.33 - 0}{1.076} = -4.95$$

According to the row corresponding to $n - 1 = 26$ in Table D, $t^* = 3.707$ corresponds to an upper tail probability $p = 0.0005$. The P-value is less than 0.0005, so there is very strong evidence that vitamin C is destroyed as a result of storage and shipment.

c) You are asked to find three confidence intervals, one for the mean at the factory, one for the mean 5 months later, and one for the change.

Mean at the factory — $\bar{x} = 42.852$, $s = 4.793$, $n = 27$, and t^* is the upper $(1 - 0.95)/2 = 0.025$ critical value for the $t(26)$ distribution. From Table D, we see that $t^* = 2.056$. Thus, the 95% confidence interval is

$$42.852 \pm 2.056\frac{4.793}{\sqrt{27}} = 42.852 \pm 1.896 = (40.96, 44.75)$$

Mean after 5 months — $\bar{x} = 37.519$, $s = 2.440$, $n = 27$, and $t^* = 2.056$. Thus, the 95% confidence interval is

$$37.519 \pm 2.056\frac{2.440}{\sqrt{27}} = 37.519 \pm 0.965 = (36.55, 38.48)$$

Mean change — $\bar{x} = -5.33$, $s = 5.59$, $n = 27$, and $t^* = 2.056$. Thus, the 95% confidence interval is

$$-5.33 \pm 2.056\frac{5.59}{\sqrt{27}} = -5.33 \pm 2.21 = (-7.54, -3.12)$$

This last confidence interval furnishes information on how much vitamin C is lost in storage and shipment, as opposed to the test which just answers the question whether any is lost.

7.41 Taking the 25 differences (right – left) we get the mean and standard deviation of the differences as $\bar{x} = -13.32$, $s = 22.94$, $n = 25$, and t^* is the upper $(1 - 0.90)/2 = 0.05$ critical value for the $t(24)$ distribution. From Table D, we see that $t^* = 1.711$. Thus, the 90% confidence interval is

$$-13.32 \pm 1.711\frac{22.94}{\sqrt{25}} = -13.32 \pm 7.85 = (-21.17, -5.47)$$

Computing the means, $\bar{x}_R = 104.12$, $\bar{x}_L = 117.44$, and $\bar{x}_R / \bar{x}_L = 88.7\%$, so those using the right-handed threads complete the task in about 90% of the time it takes those using the left-handed threads. (As an alternative, if for each subject we first take the ratio [right-thread] / [left-thread] and then average these ratios, we get 91.7%, which is almost the same answer.)

7.43 This is a matched pairs experiment for which the mean and standard deviation of the 10 differences (Variety A – Variety B) are $\bar{x} = 0.34$ and $s = 0.83$. To determine if there is evidence that variety A has the higher yield corresponds to the hypotheses

$$H_0: \mu = 0$$
$$H_a: \mu > 0$$

for the mean of the differences. With $n = 10$, we compute

$$t = \frac{\bar{x} - \mu}{s/\sqrt{n}} = \frac{0.34 - 0}{0.262} = 1.30$$

According to the row corresponding to $n - 1 = 9$ in Table D,

df = 9

p	.15	.10
t^*	1.100	1.383

the P-value lies between 0.10 and 0.15. Using statistical software, we find the P-value is 0.11. We conclude that there is very weak evidence that variety A is better.

7.45 a) This is a single sample of 20 measurements taken with the new method.

b) This would be two independent samples with 10 observations using each method. There is no indication that the observations on methods are paired in any way.

7.47 a) Using software, the t critical value for $\alpha = 0.01$ and $n = 50$ is $t^* = 2.405$. If you do not have access to software, use the row in Table D corresponding to 50 degrees of freedom instead of 49.

b) We use \$108 as an estimate of both the population standard deviation σ and s in future samples. The t test with 50 observations rejects $H_0: \mu = 0$ if the t statistic

$$t = \frac{\bar{x}}{108/\sqrt{50}} \geq 2.405$$

or equivalently when

$$\bar{x} \geq 2.405\frac{108}{\sqrt{50}} = 36.73.$$

c) The power is the probability that $\bar{x} \geq 36.73$ when $\mu = 100$. Taking $\sigma = 108$, this probability is found by standardizing $\bar{x}$,

$$P(\bar{x} \geq 36.73 \text{ when } \mu = 100) = P(\frac{\bar{x} - 100}{108/\sqrt{50}} \geq \frac{36.73 - 100}{108/\sqrt{50}})$$
$$= P(Z \geq -4.14) = 1 - 0.000 = 1.000.$$

Thus, the power is approximately 1. There is no need for increasing the sample size beyond 50.

7.49 a) In terms of the median for the population of differences

(time to complete task with left-hand thread) – (time to complete task with right-hand thread)

the hypotheses are

$$H_0\text{: population median} = 0 \text{ and } H_a\text{: population median} > 0$$

because these hypotheses test whether there is a tendency for the times using the left-hand thread to be longer than the times using the right-hand thread.

In terms of the probability p of completing the task faster with the *right-hand* thread, the hypotheses would be

$$H_0: p = 1/2 \text{ and } H_a: p > 1/2$$

since these hypotheses test whether we are more likely to complete the task faster with the right-hand thread.

b) There is only 1 pair with difference 0 in the data, thus $n = 24$. The number of pairs with a positive difference in our data is 19. Thus, the P-value is

$$P(X \geq 19) = 1 - P(X \leq 18)$$

where X has the binomial $B(24, 1/2)$ distribution. We use the normal approximation to the binomial to compute this probability. X has mean

$$\mu = np = (24)(1/2) = 12$$

and standard deviation

$$\sigma = \sqrt{np(1-p)} = \sqrt{(24)(1/2)(1/2)} = \sqrt{6} = 2.45.$$

The z-score of 19 is thus

$$z = \frac{19-12}{2.45} = 2.86$$

and so our P-value is

$$P(X \geq 19) = P(z \geq 2.86) = 0.0021.$$

The exact P-value using the binomial distribution is 0.0033. There is strong evidence that the right-hand threads are quicker than the left-hand threads.

SECTION 7.2

OVERVIEW

One of the most commonly used significance tests is the **comparison of two population means,** μ_1 and μ_2. In this setting, we have two distinct, independent SRSs from two populations or two treatments on two samples. The procedures are based on the difference $\bar{x}_1 - \bar{x}_2$. When the populations are not normal, the results obtained using the methods of this section are approximately correct for sufficiently large sample sizes due to the central limit theorem.

Tests and confidence intervals for the difference in the population means, $\mu_1 - \mu_2$, are based on the **two-sample *t* statistic,**

$$t = \frac{(\bar{x}_1 - \bar{x}_2) - (\mu_1 - \mu_2)}{\sqrt{\frac{s_1^2}{n_1} + \frac{s_2^2}{n_2}}}$$

Despite the name, this test statistic does *not* have an exact t distribution. However, there are good approximations to its distribution, which allow us to carry out valid significance tests. Conservative procedures use the $t(k)$ distribution as an approximation where the degrees of freedom k is taken to be the smaller of $n_1 - 1$ and $n_2 - 1$. More accurate procedures use the data to estimate the degrees of freedom k. This is the procedure that is followed by most statistical software.

To carry out a significance test for $H_0: \mu_1 = \mu_2$, use the two-sample t statistic

$$t = \frac{(\bar{x}_1 - \bar{x}_2)}{\sqrt{\frac{s_1^2}{n_1} + \frac{s_2^2}{n_2}}}$$

The P-value is found by using the approximate distribution $t(k)$, where k is estimated from the data by the **Satterthwaite approximation** when using statistical software, or can be taken to be the smaller of $n_1 - 1$ and $n_2 - 1$ for a conservative procedure.

An approximate confidence C level **confidence interval** for $\mu_1 - \mu_2$ is given by

$$(\bar{x}_1 - \bar{x}_2) \pm t^* \sqrt{\frac{s_1^2}{n_1} + \frac{s_2^2}{n_2}}$$

where t^* is the upper $(1 - C)/2$ critical value for $t(k)$, where k is estimated from the data when using statistical software, or can be taken to be the smaller of $n_1 - 1$ and $n_2 - 1$ for a conservative procedure. The procedures are most robust to failures in the assumptions when the sample sizes are equal.

The **pooled two-sample *t* procedures** are used when we can safely assume that the two populations have equal variances. The modifications in the procedure are the use of the pooled estimator of the common unknown variance

$$s_p^2 = \frac{(n_1 - 1)s_1^2 + (n_2 - 1)s_2^2}{n_1 + n_2 - 2}$$

and critical values obtained from the $t\,(n_1 + n_2 - 2)$ distribution.

APPLY YOUR KNOWLEDGE

7.51 a) You would use a two-sided significance test for this problem because you just want to know if there is evidence of a difference in the two designs, without a specified direction.

b) The sample sizes are $n_1 = 30$ and $n_2 = 30$, and using Table D the degrees of freedom can be taken to be the smaller of $n_1 - 1 = 29$ and $n_2 - 1 = 29$. Using 29 degrees of freedom, $t^* = 2.045$ for a two-sided test at the 5% level.

c) According to the row corresponding to $n - 1 = 29$ in Table D,

df = 29

p	.0025	.001
t^*	3.038	3.396

the P-value lies between $2 \times 0.0025 = 0.005$ and $2 \times 0.001 = 0.002$, where the upper tail probability must be doubled because it is a two-sided test. Using statistical software, we find the P-value is $2 \times 0.0019 = 0.0038$. We conclude that there is strong evidence of a difference in daily sales for the two designs.

7.53 This is a matched pairs experiment with the pairs being the two measurements on each day of the week. The matching has been done because we believe that sales differ for the different days of the week. To use a two-sample t, you would have to assign design A to half the days and design B to half the days as in Exercise 7.51.

7.55 Randomization makes the two groups similar except for the treatment. Although it may not seem that bias would exist by giving the next 10 employees who need new screens flat screens and the next 10 the standard monitor, it could be introduced if employees talked about their preferences with each other. Randomization is the best way to insure that no bias is introduced.

7.57 a) All four packages report the two means, with different numbers of significant digits. Typically, if you want more significant digits, this can be controlled as an option in the software package. SAS reports the two group means in the column labeled mean that is part of the confidence interval portion of the output.

b) Excel reports the two variances, SPSS and Minitab report both the standard deviation and the standard error of the mean for each group, while SAS reports only the standard error of the mean for each group.

c) The Excel output contains the pooled two-sample t procedure that is described later in this section. The t using Satterthwaite degrees of freedom is often referred to as the t with "equal variances not assumed" or "unequal variances". Both SPSS and SAS provide the t with Satterthwaite degrees of freedom as well as the pooled t. Minitab provides the t with Satterthwaite degrees of freedom. Minitab rounds the Satterthwaite degrees of freedom to an integer in the output, although it does the calculations for the P-values without rounding. SAS and SPSS both report degrees of freedom to several decimal places. All report P-values, to various accuracies.

d) All packages with the exception of Excel report the confidence interval for the mean difference as part of the output.

e) Excel has the least information, while the other three are equally good. SAS seems to provide the most information as it includes more information about the two groups individually, and SAS also presents the results in a simple format.

7.59 The first step in the calculations is to determine the pooled estimator of σ^2

$$s_p^2 = \frac{(n_1-1)s_1^2+(n_2-1)s_2^2}{n_1+n_2-2} = \frac{(45-1)(.19)^2+(90-1)(.22)^2}{45+90-2} = 0.0443.$$

and $s_p = 0.211$. The numerical value of the pooled two-sample t statistic is

$$t = \frac{\bar{x}_1-\bar{x}_2}{s_p\sqrt{\dfrac{1}{n_1}+\dfrac{1}{n_2}}} = \frac{3.61-2.95}{0.211\sqrt{\dfrac{1}{45}+\dfrac{1}{90}}} = 17.13$$

which agrees quite closely with the answer in Example 7.13. When the two standard deviations are close, the numerical values of the two-sample t statistic and the pooled two-sample t statistic are quite similar, as are the conclusions.

The P-value is found by comparing the computed value of t to critical values for the $t(n_1+n_2-2) = t(133)$ distribution, and then doubling the upper tail probability because the alternative is two-sided. The P-value is very close to zero, so there is strong evidence of a difference in mean wheat prices for these two months. The conclusions are essentially the same as with the unpooled analysis.

7.61 When the sample sizes are equal, the same value will be obtained for the pooled and unpooled t statistics, although the degrees of freedom will differ. When the sample sizes are unequal, both the values of the statistics and the degrees of freedom will differ. Algebraically, when $n_1 = n_2 = n$, the pooled variance is just the average of the two sample variances because

$$s_p^2 = \frac{(n_1-1)s_1^2+(n_2-1)s_2^2}{n_1+n_2-2} = \frac{(n-1)s_1^2+(n-1)s_2^2}{n+n-2} = \frac{s_1^2+s_2^2}{2}$$

Plugging this value of s_p into the formula for the pooled t gives

$$t = \frac{\bar{x}_1 - \bar{x}_2}{s_p\sqrt{\frac{1}{n_1} + \frac{1}{n_2}}} = \frac{\bar{x}_1 - \bar{x}_2}{\sqrt{\frac{s_1^2 + s_2^2}{2}}\sqrt{\frac{1}{n} + \frac{1}{n}}} = \frac{\bar{x}_1 - \bar{x}_2}{\sqrt{\frac{s_1^2}{n} + \frac{s_2^2}{n}}}$$

which is just the formula for the unpooled t when $n_1 = n_2 = n$.

SECTION 7.2 EXERCISES

7.63 The 95% confidence interval is $\bar{x}_1 - \bar{x}_2 \pm t^*\sqrt{\frac{s_1^2}{n_1} + \frac{s_2^2}{n_2}}$, where t^* is the upper $(1 - C)/2 = 0.025$ critical value for the t distribution with degrees of freedom equal to the smaller of $n_1 - 1 = 9$ and $n_2 - 1 = 9$. Using 9 degrees of freedom, we find the value of $t^* = 2.262$, and the confidence interval is

$$609 - (531) \pm 2.262\sqrt{\frac{89.312^2}{10} + \frac{82.792^2}{10}} = 78 \pm 87.11$$

The Minitab statistical software using Satterthwaite's approximation to the degrees of freedom gives the interval (–3, 159). As expected, this interval is somewhat narrower than that obtained by the conservative approach.

7.65 The following output was obtained using the Minitab software. The time of 0 corresponds to immediately after baking. The hypotheses are

$$H_0: \mu_0 = \mu_3$$
$$H_a: \mu_0 > \mu_3$$

which are written out by Minitab as "`T-Test mu 0 = mu 3 (vs >).`"

```
Two-sample T for Vit C
Time  N      Mean       StDev    SE Mean
0     2     48.71       1.53       1.1
3     2    21.795      0.771      0.55

90% C.I. for mu 0 - mu 3: (19.2,   34.58)
T-Test mu 0 = mu 3 (vs >): T = 22.16   P = 0.014   DF = 1
```

a) The numerical value of the two-sample t statistic is 22.16. The degrees of freedom are reported as 1, but Minitab truncates the degrees of freedom in the printout. The degrees of freedom are actually 1.47 and this is what is being used by the software when calculating the P-value. The P-value is 0.014, which indicates there is strong evidence that the vitamin C content has decreased after three days. Minitab provides the option of a one-sided test as you can see from the greater than sign in the output "`T-Test mu 0 = mu 3 (vs >)`". If you are using software that performs only two-sided tests, then you need to divide the reported P-value by two in this case. Finally, you might be surprised that the P-value is not closer to zero given the size of the t statistic, but with very small sample sizes large differences are required for statistical significance.

b) Minitab gives the option of choosing the confidence coefficient, and in this case we have selected 90%. The 90% confidence interval is (19.2, 34.58) and it is fairly wide due to the small sample sizes involved.

7.67 The following output was obtained using the Minitab software. The time of 0 corresponds to immediately after baking. The hypotheses are

$$H_0: \mu_0 = \mu_3$$
$$H_a: \mu_0 > \mu_3$$

which are written out by Minitab as "`T-Test mu 0 = mu 3 (vs >).`"

```
Two-sample T for Vit E
Time   N        Mean        StDev     SE Mean
0      2      95.300        0.990        0.70
3      2       95.85         2.19         1.5

90% C.I. for mu 0  - mu 3:  (-11.29,    10.2)
T-Test mu 0 = mu 3 (vs >): T = -0.32   P = 0.60   DF = 1
```

a) The value of the two-sample t statistic is –0.32, and the P-value is 0.60. For the one-sided alternative, a P-value larger than 0.5 occurs when the sample means are in the opposite direction from that hypothesized. In this case, the sample mean of the two loaves after three days is slightly higher than the mean obtained immediately after baking. There is no evidence of a loss of vitamin E.

b) The 90% confidence interval is (–11.29, 10.2) and it is fairly wide due to the small sample sizes involved.

7.69 a) When the software reports a two-sided P-value of 0.06, it is adding the area to the right of 2.07 and the area to the left of –2.07. Because the t distribution is symmetric, the area to the right of 2.07 must be 0.03 and the area to the left of –2.07 must also be 0.03. Assuming the software has taken the first mean minus the second mean in the calculation of the t statistic, the P-value is $P(t > 2.07) = 0.03$ and we reject with $\alpha = 0.05$ because the P-value is below 0.05. It is typically the case that the reported P-value must be divided by two when testing a one-sided alternative, but this is not always the case as is seen in part (b) of this exercise.

b) The P-value is $P(t > -2.07)$ because we are testing the same hypothesis as in part (a). Now,

$$P(t > -2.07) = 1 - P(t < -2.07) = 1 - 0.03 = 0.97$$

What has happened in this example is that the means are in the reverse order to what was hypothesized, which is why the t statistic is negative rather than positive. Because the P-value is 0.97, we do not reject with $\alpha = 0.05$ as the P-value is larger than 0.05. There is no evidence that the first mean is greater than the second mean. It is a common mistake to automatically divide the P-value by two without paying attention to the sign of the t statistic. The moral of this exercise is that it is up to the user to understand what the software is doing.

7.71 The sample mean improvement for the 34 children who took piano lessons was 3.618 with standard deviation 3.055, and for the 44 children who did not the sample mean improvement was 0.386 with standard deviation 2.423. We wish to compare the two groups by constructing a confidence interval for the difference in the mean improvement. The 95% confidence interval is $\bar{x}_1 - \bar{x}_2 \pm t^* \sqrt{\frac{s_1^2}{n_1} + \frac{s_2^2}{n_2}}$, where t^* is the upper $(1 - C)/2 = 0.025$ critical value for the t distribution with degrees of freedom equal to the smaller of $n_1 - 1 = 33$ and $n_2 - 1 = 43$. Using 33 degrees of freedom, we find the value of $t^* = 2.035$ and the confidence interval is

$$3.618 - 0.386 \pm 2.035\sqrt{\frac{3.055^2}{34} + \frac{2.423^2}{44}} = 3.232 \pm 1.300 = (1.932, 4.532)$$

The Minitab statistical software using Satterthwaite's approximation to the degrees of freedom gives the interval (1.95, 4.51). This interval is narrower than that obtained by the conservative approach, although the difference is quite small as there is little difference in the values of t^* used in the two intervals when the degrees of freedom is large.

7.73 a) When means and standard deviations for the two samples are given, it is relatively easy to compute the value of the two-sample t statistic. With raw data as in this exercise, the details of the computations are best left to statistical software. However, although you may let the computer do the tedious computations, it is still important to know what all the numbers in the output mean and how to interpret the results. The output below is reproduced from Minitab. The output is fairly standard between the different software packages.

```
Two-sample T for Low vs High
        N      Mean     StDev    SE Mean
Low    14     4.640     0.690     0.18
High   14     6.429     0.430     0.12

T-Test mu Low = mu High (vs not =): T = -8.23 P = 0.0000 DF = 21
```

The output begins with summary information for the two samples. For the Low group, $n_1 = 14$, $\bar{x}_1 = 4.640$, $s_1 = 0.690$, and the SE Mean is $s_1 / \sqrt{n_1} = 0.18$. For the High group, $n_2 = 14$, $\bar{x}_2 = 6.429$, $s_2 = 0.430$, and the SE Mean is $s_2 / \sqrt{n_2} = 0.12$. The next line gives the hypotheses, test statistic, and P-value:

`mu Low = mu High` corresponds to H_0: $\mu_{Low} = \mu_{High}$, and `(vs not =)` corresponds to H_a: $\mu_{Low} \neq \mu_{High}$.

Next is $t = \dfrac{\bar{x}_1 - \bar{x}_2}{\sqrt{\dfrac{s_1^2}{n_1} + \dfrac{s_2^2}{n_2}}} = \dfrac{4.640 - 6.429}{\sqrt{\dfrac{(0.690)^2}{14} + \dfrac{(0.430)^2}{14}}} = -8.23$ and the P-value = 0.0000.

The degrees of freedom are found using the formula

$$\frac{\left(\dfrac{s_1^2}{n_1} + \dfrac{s_2^2}{n_2}\right)^2}{\dfrac{1}{n_1 - 1}\left(\dfrac{s_1^2}{n_1}\right)^2 + \dfrac{1}{n_2 - 1}\left(\dfrac{s_2^2}{n_2}\right)^2} = \frac{\left(\dfrac{0.690^2}{14} + \dfrac{0.430^2}{14}\right)^2}{\dfrac{1}{14-1}\left(\dfrac{0.690^2}{14}\right)^2 + \dfrac{1}{14-1}\left(\dfrac{0.430^2}{14}\right)^2} = 21.77$$

Although truncated in the printed output to 21, the value of 21.77 is used in computing the P-value. You should make sure you understand how all the numbers in the computer output are obtained. Because the P-value in the output is given as 0.0000, it is smaller than both 5% and 1%. So we reject H_0 at both these significance levels.

b) The individuals in the study are not random samples from low-fitness and high-fitness groups of middle-aged men. There are two ways in which they might be systematically different. The first is that all subjects in the study are college faculty. The "ego strength" personality factor for low-fitness and high-fitness middle-aged college faculty members might differ from the general population, and so might the mean difference between the groups. In addition, we do not have random samples of college faculty members. We are using volunteers in the study, and they might differ from both college faculty and general population. It is hard to say in which direction this might bias the results, but the possibility of bias is definitely present.

c) This is observational data and we cannot conclude a cause-and-effect relationship.

7.75 This would now be a matched pairs design, because we have before and after measurements on the same men. The analysis would proceed by taking the change in score and computing the mean and standard deviation of the changes. These values would be used to compute the t statistic.

7.77 a) The 95% confidence interval is $\bar{x}_1 - \bar{x}_2 \pm t^* \sqrt{\dfrac{s_1^2}{n_1} + \dfrac{s_2^2}{n_2}}$, where t^* is the upper $(1 - C)/2 = 0.025$ critical value for the t distribution with degrees of freedom equal to the smaller of $n_1 - 1 = 75 - 1 = 74$ and $n_2 - 1 = 53 - 1 = 52$. Using 52 degrees of freedom and statistical software, we find the value of $t^* = 2.007$. The 95% confidence interval is

$$52 - 49 \pm 2.007\sqrt{\frac{13^2}{75} + \frac{11^2}{53}} = 3 \pm 4.3 = (-1.3, 7.3)$$

b) The confidence interval is wide and includes both positive and negative values. The width indicates we are not very certain by how much the sales have changed. The positive values correspond to an increase in mean sales, while the negative values correspond to a decrease in mean sales. If the true difference in means were –1, which is included in the confidence interval, then the percent sales would have decreased, not increased by 6%.

7.79 a) The two histograms below are drawn using the same scale for ease of comparison. Since there are different numbers of men and women, the vertical axis uses percents rather than frequency. With the scores ranging in the fixed interval 0 to 200, there are no high or low outliers. The men's histogram looks skewed to the right and the women's is less so. The normal quantile plots are fairly linear, and in spite of the skewness, with the sum of the sample sizes close to 40, the t procedures can be used. With the histograms on the same scale it is easy to see that the variability is slightly larger for men, and that the men's scores tend to be lower.

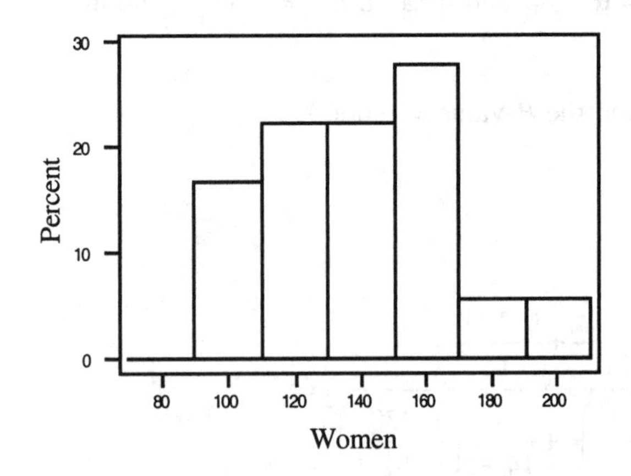

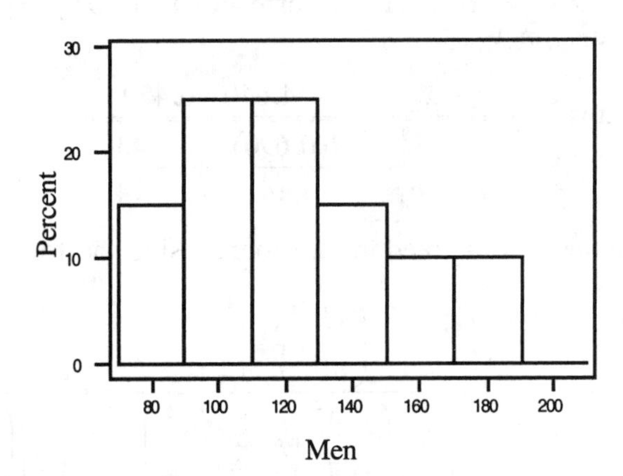

b) The output below was obtained using Minitab. A one-sided test was requested as well as the 90% confidence interval. The value of the t statistic is 2.06 with a P-value of 0.024. This gives fairly strong evidence that the mean SSHA score is lower for men than for women at this college.

```
Two-sample T for women vs men
        N        Mean      StDev     SE Mean
women  18       141.1       26.4         6.2
men    20       121.2       32.9         7.3

90% C.I. for mu women - mu men: ( 3.5,  36.1)
T-Test mu women = mu men (vs >): T = 2.06   P = 0.024   DF = 35
```

c) The confidence interval for the difference is (3.5, 36.1). Although we have good evidence that the women's scores are higher, the interval is quite wide suggesting that we haven't pinned the difference down very well. A larger sample size would be required to reduce the sampling variability and get a narrower interval.

7.81 The 21 observations in the treatment group have a mean of 51.48 and a standard deviation of 11.01, while the 23 observations in the control group have a mean of 41.52 and a standard deviation of 17.15 The t statistic is computed as

$$t = \frac{\bar{x}_1 - \bar{x}_2}{\sqrt{\frac{s_1^2}{n_1} + \frac{s_2^2}{n_2}}} = \frac{51.48 - 41.52}{\sqrt{\frac{(11.01)^2}{21} + \frac{(17.15)^2}{23}}} = 2.31$$

and the approximate degrees of freedom are found using the formula

$$\frac{\left(\frac{s_1^2}{n_1}+\frac{s_2^2}{n_2}\right)^2}{\frac{1}{n_1-1}\left(\frac{s_1^2}{n_1}\right)^2+\frac{1}{n_2-1}\left(\frac{s_2^2}{n_2}\right)^2}=\frac{\left(\frac{11.01^2}{21}+\frac{17.15^2}{23}\right)^2}{\frac{1}{21-1}\left(\frac{11.01^2}{21}\right)^2+\frac{1}{23-1}\left(\frac{17.15^2}{23}\right)^2}=37.859$$

7.83 a) The sample sizes are $n_1 = 2$ and $n_2 = 2$, and using Table D the degrees of freedom can be taken to be the smaller of $n_1 - 1 = 1$ and $n_2 - 1 = 1$. Using 1 degree of freedom, $t^* = 12.71$ for a two-sided test at the 5% level.

b) For the pooled t statistic, the degrees of freedom are $n_1 + n_2 - 2 = 2$, and using 2 degrees of freedom $t^* = 4.303$ for a two-sided test at the 5% level.

c) The increase in degrees of freedom means that the value of the t statistic needs to be less extreme for the pooled t statistic in order to find a statistically significant difference.

SECTION 7.3

OVERVIEW

There are formal inference procedures to compare the standard deviations of two normal populations as well as the two means. The validity of the procedures is seriously affected by nonnormality, and they are not recommended for regular use. The procedures are based on the **F statistic,** which is the ratio of the two sample variances

$$F = \frac{s_1^2}{s_2^2}.$$

If the data consist of independent simple random samples of sizes n_1 and n_2 from two normal populations, then the F statistic has the F distribution, $F(n_1 - 1, n_2 - 1)$, if the two population standard deviations σ_1 and σ_2 are equal.

The power of the pooled two-sample t test can be found using the noncentral t-distribution or a normal approximation to it. The critical value for the significance test, the degrees of freedom and the noncentrality parameter for alternatives of interest are all required for the computation. Power calculations can be useful when comparing alternative designs and assumptions.

SECTION 7.3 EXERCISES

7.85 a) The F distribution $F(n_1 - 1, n_2 - 1) = F(9, 20)$ is appropriate. The upper 5% critical value is $F^* = 2.39$.

b) We know that because $F = 2.45$, we have taken the ratio of the larger variance divided by the smaller variance. To test at the 10% level of significance, you need to find the upper 5% critical value, because the upper tail probability must be doubled. From part (a), this value is $F^* = 2.39$ and $F = 2.45$ is significant at the 10% level. At the 5% level, $F^* = 2.84$ (the upper 0.025 critical value), so we do not reject at the 5% level. The P-value is between 0.05 and 0.10.

7.87 The hypotheses of interest are

$$H_0: \sigma_1 = \sigma_2$$
$$H_a: \sigma_1 \neq \sigma_2$$

For the 34 children who took piano lessons, the sample standard deviation of the improvement scores is 3.055, and the sample variance is $(3.055)^2 = 9.333$. For the 44 children who did not take piano lessons, the sample standard deviation is 2.423, and the sample variance is $(2.423)^2 = 5.871$. Computing the test statistic

$$F = \frac{\text{larger } s^2}{\text{smaller } s^2} = \frac{9.333}{5.871} = 1.59,$$

we can go to the $F(33, 43)$ distribution in Table E and find the critical value corresponding to 0.05/2 = 0.025 because the test is two-sided. The value is 1.94 for the $F(30, 40)$ distribution If we use statistical software to find the exact critical value, we find $F^* = 1.89$, so we do not reject at the 5% level. Using statistical software, the exact upper tail probability (probability of exceeding 1.59 for an $F(33, 43)$ distribution) is 0.0764 and the P-value = 2 x 0.0764 = 0.1528, so there is little evidence of a difference in variances between the two groups.

7.89 a) You need to go to the $F(1, 1)$ distribution in Table E and find the critical value corresponding to 0.05/2 = 0.025 because the test is two-sided. The value in the table is 647.79. The ratio of the larger to smaller variance would need to exceed 647.79 to reject the null hypothesis of equal standard deviations. This suggests the power of the test will be very low, as this is unlikely to occur unless the population standard deviations are very far apart.

b) For the data collected immediately after baking, the sample standard deviation is 1.53 (mg/100 g) and the sample variance is $(1.53)^2 = 2.3409$. For the data collected three days after baking, the sample standard deviation is 0.771 (mg/100 g) and the sample variance is $(0.771)^2 = 0.5944$. Computing the test statistic

$$F = \frac{\text{larger } s^2}{\text{smaller } s^2} = \frac{2.3409}{0.5944} = 3.94,$$

we see the computed value does not exceed 647.79, so we fail to reject at the 5% level of significance. Using statistical software, the exact upper tail probability (probability of exceeding 3.94 for an $F(1, 1)$ distribution is 0.2981 and the P-value = 2 × 0.2981 = 0.5962.

7.91 a) The hypotheses of interest are

$$H_0\text{: } \sigma_M = \sigma_W$$
$$H_a\text{: } \sigma_M > \sigma_W$$

b) For the 20 men, the sample standard deviation is 32.9, and the sample variance is $(32.9)^2 = 1082.41$. For the 18 women, the sample standard deviation is 26.4 and the sample variance is $(26.4)^2 = 696.96$. Because the men are hypothesized to have the larger variance, the sample variance of the men must be put in the numerator of F. Computing the test statistic, we have

$$F = \frac{s_M^2}{s_W^2} = \frac{1082.41}{696.96} = 1.56$$

c) We need to use the $F(19, 17)$ distribution. Using statistical software, $P(F > 1.56) = 0.1807$, so there is little evidence that the males have larger variability in their scores.

7.93 The researcher thinks the true difference in mean birth weights might be about 300 grams and a difference this large is clinically important. The value $\mu_1 - \mu_2 = 300$ is an alternative considered important to detect.

The sample sizes are $n_1 = n_2 = 25$, 50, 75, 100, and 125. The type I error is $\alpha = 0.05$ and our guess for σ is going to be 650 grams.

For the different sample sizes involved, we need to find the value for the degrees of freedom, df $= n_1 + n_2 - 2$, the critical value t^*, and the noncentrality parameter $\delta = \frac{|\mu_1 - \mu_2|}{\sigma\sqrt{(1/n_1)+(1/n_2)}}$. These can be used to find the power using SAS and the noncentral t distribution, or the normal approximation. The necessary SAS function is called PROBT, and to use it you need to specify t^*, DF, and δ. The SAS "command" is POWER = 1 – PROBT(t^*, DF, δ). The following table summarizes the calculations using SAS and provides the answers using the normal approximation.

Sample size	df	t^*	δ	POWER(SAS)	Normal approximation
25	48	1.6772	1.6318	0.4855	0.4819
50	98	1.6606	2.3077	0.7411	0.7412
75	148	1.6546	2.8263	0.8787	0.8794
100	198	1.6526	3.2636	0.9460	0.9464
125	248	1.6510	3.6488	0.9769	0.9771

7.95 a) The researchers think a difference of about 0.5 points would be of interest, so the value $\mu_1 - \mu_2 = 0.5$ is an alternative considered important to detect.

The sample sizes are $n_1 = n_2 = 20$. The type I error is $\alpha = 0.01$ and our guess for σ is going to be 0.7. The degrees of freedom are $n_1 + n_2 - 2 = 38$ and the critical value t^* for a two-sided test is $t^*= 2.7116$. The noncentrality parameter is $\delta = \dfrac{|\mu_1 - \mu_2|}{\sigma\sqrt{(1/n_1)+(1/n_2)}} = 2.2588$. The SAS function is called PROBT, and to use it you need to specify t^*, DF, and δ. The required SAS "command" is POWER = 1 – PROBT(t^*, DF, δ) = 0.3390. There is slightly more than a 30% chance of detecting this alternative with sample sizes of 20 in each group.

b) Using 30 students and $\alpha = 0.05$, the sample sizes are $n_1 = n_2 = 30$, the degrees of freedom are $n_1 + n_2 - 2 = 58$, and the critical value t^* for a two-sided test is $t^* = 2.0017$. The noncentrality parameter is $\delta = \dfrac{|\mu_1 - \mu_2|}{\sigma\sqrt{(1/n_1)+(1/n_2)}} = 2.7664$. The SAS "command" gives POWER = 1-PROBT(t^*, DF, δ) = 0.7765, which is greater than the power found in (a).

CHAPTER 7 REVIEW EXERCISES

7.97 The two histograms for men and women are given below. They are drawn using the same scale for comparison purposes using the vertical axis as percent. Because there are different numbers of men and women, using frequency as the vertical axis would make comparisons difficult. As can be seen from the two histograms, there appears to be little difference between the salaries of the samples of men and women.

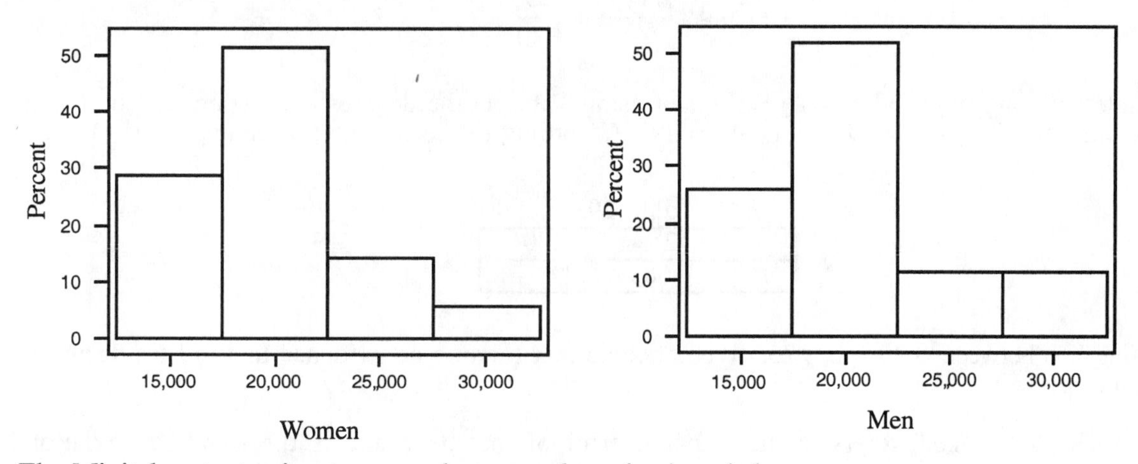

The Minitab output using two-sample t procedures is given below.

```
Two-sample T for men vs women
          N       Mean      StDev    SE Mean
men      27      20626       4196        808
women    35      19789       4121        697

95% C.I. for mu men - mu women: (-1301,  2975)
T-Test mu men = mu women (vs not =): T = 0.78   P = 0.44   DF = 55
```

The test has a P-value of 0.44, which indicates no evidence of a difference in mean salaries between men and women. As expected, the confidence interval includes zero, which corresponds to no difference. Even if a difference had been found, it would not necessarily be an indication of discrimination as this is observational data and there can be other factors besides gender (seniority, for example) that might account for the observed differences.

7.99 We would expect that those who evacuated all or some pets would tend to score higher on the scale that measured "commitment to adult animals," and we would like to see if the data supports this hypothesis. Letting group 1 correspond to those who evacuated all or some pets and group 2 correspond to those who did not evacuate any pets, the hypotheses of interest are

$$H_0: \mu_1 = \mu_2$$
$$H_a: \mu_1 > \mu_2$$

To carry out a significance test, use the two-sample t statistic

$$t = \frac{(\bar{x}_1 - \bar{x}_2)}{\sqrt{\frac{s_1^2}{n_1} + \frac{s_2^2}{n_2}}} = \frac{(7.95 - 6.26)}{\sqrt{\frac{3.62^2}{116} + \frac{3.56^2}{125}}} = 3.65$$

The sample sizes are $n_1 = 116$ and $n_2 = 125$, and the conservative degrees of freedom can be taken to be the smaller of $n_1 - 1 = 115$ and $n_2 - 1 = 124$, which is 115. The area to the right of 3.65 for a t distribution with 115 degrees of freedom is 0.0002 using statistical software. Not surprisingly, there is strong evidence that the group that evacuated all or some pets scored higher on the scale measuring commitment to adult animals.

7.101 To carry out a significance test for the hypotheses $H_0: \mu_N = \mu_C$ and $H_a: \mu_N < \mu_C$, use the two-sample t statistic

$$t = \frac{(\bar{x}_1 - \bar{x}_2)}{\sqrt{\frac{s_1^2}{n_1} + \frac{s_2^2}{n_2}}} = \frac{(7880 - 8112)}{\sqrt{\frac{1115^2}{30} + \frac{1250^2}{30}}} = -0.76$$

The sample sizes are $n_1 = 30$ and $n_2 = 30$, and using Table D the degrees of freedom can be taken to be the smaller of $n_1 - 1 = 29$ and $n_2 - 1 = 29$. According to the row corresponding to $n - 1 = 29$ in Table D,

df = 29

p	.25	.20
t^*	0.683	0.854

the P-value lies between 0.20 and 0.25. We conclude that there is no evidence that nitrites decrease amino acid uptake.

7.103 a) The term "s.e." refers to the standard error of the mean and is the standard deviation divided by $\sqrt{n}$. In the table below, the s.e. has been multiplied by $\sqrt{n}$, where n is 94 for the drivers and 83 for the conductors, respectively. This gives the standard deviations, which are in parentheses next to the means.

	Drivers	Conductors
Total Calories	2821 (435.578)	2844 (437.301)
Alcohol grams	0.24 (0.594)	0.39 (1.002)

b) To see if there is significant evidence that the conductors consume more calories than the drivers, you must consider the hypotheses $H_0: \mu_D = \mu_C$ and $H_a: \mu_D < \mu_C$. The computed value of the two-sample t statistic is

$$t = \frac{\bar{x}_1 - \bar{x}_2}{\sqrt{\frac{s_1^2}{n_1} + \frac{s_2^2}{n_2}}} = \frac{2844 - 2821}{\sqrt{\frac{(437.301)^2}{83} + \frac{(435.578)^2}{94}}} = 0.35$$

The sample sizes are $n_1 = 83$ and $n_2 = 94$, and using Table D the degrees of freedom can be taken to be the smaller of $n_1 - 1 = 82$ and $n_2 - 1 = 93$. The P-value is larger than 0.25, so there is little evidence that the conductors consume more calories than the drivers.

c) To see if there is significant evidence that the conductors and drivers have different mean alcohol consumption corresponds to the hypotheses $H_0: \mu_D = \mu_C$ and $H_a: \mu_D \neq \mu_C$. The computed value of the two-sample t statistic is

$$t = \frac{\bar{x}_1 - \bar{x}_2}{\sqrt{\frac{s_1^2}{n_1} + \frac{s_2^2}{n_2}}} = \frac{0.39 - 0.24}{\sqrt{\frac{(1.002)^2}{83} + \frac{(0.594)^2}{94}}} = 1.19$$

The sample sizes are $n_1 = 83$ and $n_2 = 94$, and the degrees of freedom can be taken to be the smaller of $n_1 - 1 = 82$ and $n_2 - 1 = 93$. Using degrees of freedom of 82 and statistical software, the area to the right of 1.19 is 0.1187 and the P-value is $2 \times 0.1187 = 0.2374$, so there is little evidence of a difference in mean alcohol consumption between conductors and drivers.

7.105 The standard deviations are quite similar in part (b), so the pooled two-sample t would be justified. The first step in the calculations is to determine the pooled estimator of σ^2

$$s_p^2 = \frac{(n_1 - 1)s_1^2 + (n_2 - 1)s_2^2}{n_1 + n_2 - 2} = \frac{(83 - 1)(437.301)^2 + (94 - 1)(435.578)^2}{83 + 94 - 2} = 190{,}432.9117.$$

and $s_p = 436.39$. The numerical value of the pooled two-sample t statistic is

$$t = \frac{\bar{x}_1 - \bar{x}_2}{s_p\sqrt{\frac{1}{n_1} + \frac{1}{n_2}}} = \frac{2844 - 2821}{436.39\sqrt{\frac{1}{83} + \frac{1}{94}}} = 0.35$$

which agrees with the answer in Exercise 7.103 to two decimal places. When the two standard deviations are close, the numerical values of the two-sample t statistic and the pooled two-sample t statistic are quite similar, as are the conclusions.

7.107 We have the populations of all 58 counties and when we average them we have the true mean μ for the average population of a California county. There is no reason to do statistical inference in this problem.

7.109 For this power calculation, the sample size is fixed and we are interested in the probability of detecting various alternatives (this is the power) with the test. The sample sizes are $n_1 = n_2 = 20$. The type I error is $\alpha = 0.05$ and our guess for σ is going to be 0.7. The degrees of freedom are $n_1 + n_2 - 2 = 38$ and the critical value t^* for a two-sided test is $t^* = 2.0017$. The noncentrality parameter is $\delta = \frac{|\mu_1 - \mu_2|}{\sigma\sqrt{(1/n_1) + (1/n_2)}}$, where we will specify different values of $|\mu_1 - \mu_2|$ and study the effect on the power. The following table gives the values of $|\mu_1 - \mu_2|$, δ and the corresponding power.

$\lvert\mu_1-\mu_2\rvert$	δ	POWER = 1-PROBT(t*, DF, δ)
.50	2.259	0.604
.75	3.388	0.914
1.00	4.518	0.993

There is about a 60% chance of detecting differences as small as 0.5, about a 90% chance of detecting a difference of 0.75, and the chance is close to 1 for differences of 1 or larger.

7.111 The margin of error for the one-sample t confidence interval is $t^*\dfrac{s}{\sqrt{n}}$. When fixing $s = 1$, the margin of error is $t^* / \sqrt{n}$. In the graph below, the margin of error has been computed for $n = 5$, 10, 15, ..., 100 and the points connected. Note that the value of t^* depends on n through the degrees of freedom $n - 1$, although the dependence becomes small as the degrees of freedom get larger (because $t^* \approx z$). The margin of error decreases quite rapidly for smaller values of n, and then the decrease becomes less pronounced. As n continues to become larger, the margin of error will continue to decrease toward zero.

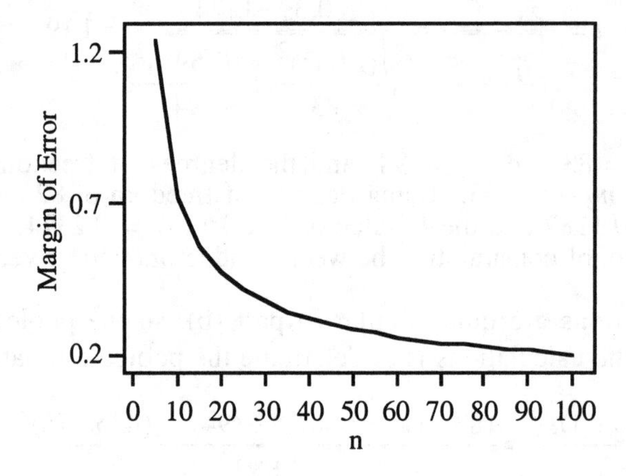

CHAPTER 8

INFERENCE FOR PROPORTIONS

SECTION 8.1

OVERVIEW

In this section, we consider inference about a population proportion p from an SRS of size n based on the **sample proportion** $\hat{p} = X/n$ and the **Wilson estimate** $\tilde{p} = (X + 2)/(n + 4)$, which is the sample proportion with two successes and two failures added to the data. In these formulas, X is the number of "successes" (occurrences of the event of interest) in the sample. If the population is at least 10 times as large as the sample, the individual observations will be approximately independent and X will have a distribution which is approximately binomial $B(n, p)$. While it is possible to develop procedures for inference about p based on the binomial $B(n, p)$ distribution, these can be awkward to work with because of the discrete nature of the binomial distribution. When n is large, we can treat $\hat{p}$ as having a distribution, which is approximately normal with mean $\mu = p$ and standard deviation $\sigma = \sqrt{p(1-p)/n}$, and $\tilde{p}$ as having a distribution, which is approximately normal with mean $\mu = p$ and standard deviation $\sigma = \sqrt{p(1-p)/(n+4)}$.

An **approximate level C confidence interval** for p is

$$\tilde{p} \pm z^* \text{SE}_{\tilde{p}}$$

where z^* is the upper $(1 - C)/2$ critical value of the standard normal distribution, and

$$\text{SE}_{\tilde{p}} = \sqrt{\frac{\tilde{p}(1-\tilde{p})}{n+4}}$$

is the **standard error** of $\tilde{p}$ and $z^* \text{SE}_{\tilde{p}}$ is the **margin of error.**

Tests of the hypothesis $H_0\text{: } p = p_0$ are based on the **z statistic**

$$z = \frac{\hat{p} - p_0}{\sqrt{\dfrac{p_0(1-p_0)}{n}}}$$

with P-values calculated from the $N(0, 1)$ distribution.

The **sample size** n required to obtain a confidence interval of approximate margin of error m for a proportion is

$$n + 4 = \left(\frac{z^*}{m}\right)^2 p^*(1 - p^*)$$

where p^* is a guessed value for the population proportion and z^* is the upper $(1 - C)/2$ critical value of the standard normal distribution. To guarantee that the margin of error of the confidence interval is less than or equal to m no matter what the value of the population proportion may be, use a guessed value of $p^* = 1/2$, which yields

$$n + 4 = \left(\frac{z^*}{2m}\right)^2.$$

APPLY YOUR KNOWLEDGE

8.1 The sample size is $n = 40$ and the count is $X = 11$. The Wilson estimate of the proportion of all our customers who would be willing to buy the $100 upgrade is

$$\tilde{p} = \frac{X+2}{n+4} = \frac{11+2}{40+4} = \frac{13}{44} = 0.295$$

The standard error is

$$SE_{\tilde{p}} = \sqrt{\frac{\tilde{p}(1-\tilde{p})}{n+4}} = \sqrt{\frac{0.295(1-0.295)}{44}} = 0.069$$

The z critical value for 95% confidence is $z^* = 1.96$, so the confidence interval is

$$\tilde{p} \pm z^* SE_{\tilde{p}} = 0.295 \pm (1.96)(0.069) = 0.295 \pm 0.135 = (0.160, 0.430)$$

8.3 a) The sample size is $n = 5$ and the count is $X = 5$. The Wilson estimate of the proportion of all her customers who would buy the new product is

$$\tilde{p} = \frac{X+2}{n+4} = \frac{5+2}{5+4} = \frac{7}{9} = 0.778$$

b) The standard error is

$$SE_{\tilde{p}} = \sqrt{\frac{\tilde{p}(1-\tilde{p})}{n+4}} = \sqrt{\frac{0.778(1-0.778)}{9}} = 0.139$$

The z critical value for 95% confidence is $z^* = 1.96$, so the margin of error is $m = z^* SE_{\tilde{p}} = 1.96(0.139) = 0.272$.

c) The results apply to all of your top salesperson's customers (because these are a random sample of her customers), but not to all of your sales force. Your top salesperson is not a random sample from your sales force.

8.5 a) The sample size is $n = 100$ and the count of those who answered no is $X = 32$. The Wilson estimate of the proportion who would answer no is

$$\tilde{p} = \frac{X+2}{n+4} = \frac{32+2}{100+4} = \frac{34}{104} = 0.327$$

The standard error is

$$SE_{\tilde{p}} = \sqrt{\frac{\tilde{p}(1-\tilde{p})}{n+4}} = \sqrt{\frac{0.327(1-0.327)}{104}} = 0.046$$

The z critical value for 95% confidence is $z^* = 1.96$, so the confidence interval is

$$\tilde{p} \pm z^* SE_{\tilde{p}} = 0.327 \pm (1.96)(0.046) = 0.327 \pm 0.090 = (0.237, 0.417)$$

In Example 8.1, we computed an estimate of the proportion who would answer yes and here we estimate the proportion who would answer no. The two responses are "opposites," and if we know how many answered yes, we automatically know how many answered no. In particular, we would expect the estimate of the proportion who answer no to be 1 minus the estimate of the proportion who would answer yes, and, as a result, we would expect the estimates to have the same margin of error. We see that this is the case.

b) We test the hypotheses

$$H_0: p = 0.25$$
$$H_a: p \neq 0.25$$

The expected number of "Yes" and "No" responses are $100 \times 0.75 = 75$ and $100 \times 0.25 = 25$ and both are greater than 10, so we can use the z test. We find $\hat{p} = 32/100 = 0.32$, so the test statistic is

$$z = \frac{\hat{p} - p_0}{\sqrt{\frac{p_0(1-p_0)}{n}}} = \frac{0.32 - 0.25}{\sqrt{\frac{0.25(1-0.25)}{100}}} = 1.62.$$

From Table A, we find $P(Z \geq 1.62) = 1 - P(Z < 1.62) = 1 - 0.9474 = 0.0526$. The P-value for our two-sided test is the area in both tails, which is equal to $2 \times P(Z \geq 1.62) = 2 \times (0.0526) = 0.1052$.

This agrees with the result in Example 8.2. Because the proportion who answer yes must always be 1 minus the proportion who answer no, testing whether the proportion who answered yes is equal to 0.75 is equivalent to testing whether the proportion who answered no is 0.25.

8.7 We do not have a guessed value for the true proportion p^* of customers who would be interested in the new product. Thus, for 95% confidence with a margin of error $m = 0.05$ or less, the desired sample size n must satisfy

$$n + 4 = \left(\frac{z^*}{2m}\right)^2 = \left(\frac{1.96}{2 \times 0.05}\right)^2 = 384.16$$

and we round up to get $n + 4 = 385$ or $n = 381$.

SECTION 8.1 EXERCISES

8.9 No. A person could have both lied about having a degree and about their major. For example, a person with a bachelor's degree could lie about having an advanced degree, such as a masters or Ph.D., and about their undergraduate major. Because lying about having a degree and lying about their major are not necessarily mutually exclusive events, we cannot automatically conclude that a total of 25 = 15 + 9 applicants lied about one or the other.

8.11 The sample size is $n = 100$ and the count of orders that were shipped on time is $X = 86$. The Wilson estimate of the proportion shipped on time is

$$\tilde{p} = \frac{X+2}{n+4} = \frac{86+2}{100+4} = \frac{88}{104} = 0.846$$

The standard error is

$$SE_{\tilde{p}} = \sqrt{\frac{\tilde{p}(1-\tilde{p})}{n+4}} = \sqrt{\frac{0.846(1-0.846)}{104}} = 0.035$$

The z critical value for 95% confidence is $z^* = 1.96$, so the confidence interval is

$$\tilde{p} \pm z^* SE_{\tilde{p}} = 0.846 \pm (1.96)(0.035) = 0.846 \pm 0.069 = (0.777, 0.915).$$

8.13 The sample size is $n = 200$ and the count of those who say that "being well-off financially" is an important goal is $X = 132$. The Wilson estimate of the proportion who say this goal is important is

$$\tilde{p} = \frac{X+2}{n+4} = \frac{132+2}{200+4} = \frac{134}{204} = 0.657$$

The standard error is

$$SE_{\tilde{p}} = \sqrt{\frac{\tilde{p}(1-\tilde{p})}{n+4}} = \sqrt{\frac{0.657(1-0.657)}{204}} = 0.033$$

The z critical value for 95% confidence is $z^* = 1.96$, so the confidence interval is

$$\tilde{p} \pm z^* SE_{\tilde{p}} = 0.657 \pm (1.96)(0.033) = 0.657 \pm 0.065 = (0.592, 0.722).$$

8.15 a) Of the $n = 1711$ bicyclists examined, $X = 542$ tested positive for alcohol. The proportion of the bicyclists ages 15 or older who were fatally injured in bicycle accidents between 1987 and 1991 and tested positive for alcohol is therefore

$$\hat{p} = X/n = 542/1711 = 0.3168$$

Expressed as a percentage, we can say that 31.68% of the bicyclists ages 15 or older who were fatally injured in bicycle accidents between 1987 and 1991 and were tested for alcohol tested positive. The proportion $\hat{p} = 0.3168$, the corresponding percentage of 31.68%, the sample size $n = 1711$, and the number $X = 542$ who tested positive are the basic summary statistics.

b) We assume that the 1711 records can be considered a random sample from the population of all bicyclists ages 15 or older who were fatally injured in bicycle accident and were tested for alcohol. The sample size is $n = 1711$ and the count of who tested positive for alcohol is $X = 542$. The Wilson estimate of the probability (proportion) that a tested bicycle rider is positive for alcohol is

$$\tilde{p} = \frac{X+2}{n+4} = \frac{542+2}{1711+4} = \frac{544}{1715} = 0.317$$

The standard error is

$$SE_{\tilde{p}} = \sqrt{\frac{\tilde{p}(1-\tilde{p})}{n+4}} = \sqrt{\frac{0.317(1-0.317)}{1715}} = 0.011$$

The z critical value for 95% confidence is $z^* = 1.96$, so the confidence interval is

$$\tilde{p} \pm z^* SE_{\tilde{p}} = 0.317 \pm (1.96)(0.011) = 0.317 \pm 0.022 = (0.295, 0.339).$$

c) Although the probability that a fatally injured bicyclist who is tested for alcohol tests positive seems high (nearly 1/3), we cannot conclude from this study that alcohol causes fatal accidents. First, this is an observational study, not a designed experiment. Observational studies generally do not provide a good basis for concluding causality. Second, this is not a random sample from the population of all bicyclists who were fatally injured in bicycle accidents. This is a nonrandom sample of bicyclists who were fatally injured and were tested for alcohol. Unless all fatally injured bicyclists are tested for alcohol, it may be that only those for whom there is reason to believe that they might have been drinking are tested. In this case one, can only say that bicyclists who are killed in bicycle accidents and for whom there is reason to believe they might have been drinking have a probability of about 0.317 of testing positive. This is not a statement of causality but rather one of association. Association does not imply causation.

8.17 a) We can safely use the z test when np_0 and $n(1-p_0)$ are both greater than 10. In this case, $n = 10$ and $p_0 = 0.4$ so that $np_0 = 10 \times 0.4 = 4$ and $n(1-p_0) = 10 \times (1-0.4) = 6$. Neither is greater than 10, so the sample is not large enough to permit safe use of the z test.

b) In this case, $n = 100$ and $p_0 = 0.6$, so that $np_0 = 100 \times 0.6 = 60$ and $n(1 - p_0) = 100 \times (1 - 0.6) = 40$. Both are greater than 10, so the sample is large enough to permit safe use of the z test.

c) In this case, $n = 1000$ and $p_0 = 0.996$, so that $np_0 = 1000 \times 0.996 = 996$ and $n(1 - p_0) = 1000 \times (1 - 0.996) = 4$. $n(1 - p_0)$ is less than 10, so the sample is not large enough to permit safe use of the z test.

d) In this case, $n = 500$ and $p_0 = 0.3$, so that $np_0 = 500 \times 0.3 = 150$ and $n(1 - p_0) = 500 \times (1 - 0.3) = 350$. Both are greater than 10, so the sample is large enough to permit safe use of the z test.

8.19 a) According to the census, 64% of Indiana households are in urban areas. To examine how well the sample represents the state in regard to rural versus urban residence , we test the hypothesis

$$H_0: p = 0.64$$
$$H_a: p \neq 0.64$$

where p is the proportion of (potential) urban respondents.

b) We can safely use the z test when np_0 and $n(1 - p_0)$ are both greater than 10. In this case, $n = 500$ and $p_0 = 0.64$, so that $np_0 = 500 \times 0.64 = 320$, $n(1 - p_0) = 500 \times (1 - 0.64) = 180$ and we can use the z test. We are told that $\hat{p} = 0.62$, so the test statistic is

$$z = \frac{\hat{p} - p_0}{\sqrt{\frac{p_0(1-p_0)}{n}}} = \frac{0.62 - 0.64}{\sqrt{\frac{0.64(1-0.64)}{500}}} = -0.93.$$

From Table A, we find $P(Z \leq -0.93) = 0.1762$. The P-value for our two-sided test is the area in both tails, which is equal to $2 \times P(Z \leq -0.93) = 2 \times (0.1762) = 0.3524$. This is not strong evidence against the null hypothesis that the sample represents the state in regard to rural versus urban residence in terms of the proportion of urban residents.

c) These results are consistent (same P-value) with the previous exercise. We would expect this to be the case because the observed proportions and value of p_0 in this exercise are 1 minus the values in the other. Because the values are functions of each other, the results should be the same. We conclude that there is not strong evidence against the hypothesis that the sample represents the state in either the proportion of rural residents or in the proportion of urban residents.

8.21 The sample size is $n = 75$ and the number of locations in which egg masses were found (locations infected) is $X = 13$. The Wilson estimate of the proportion of all possible locations that are infected is

$$\tilde{p} = \frac{X+2}{n+4} = \frac{13+2}{75+4} = \frac{15}{79} = 0.190$$

The standard error is

$$SE_{\tilde{p}} = \sqrt{\frac{\tilde{p}(1-\tilde{p})}{n+4}} = \sqrt{\frac{0.190(1-0.190)}{79}} = 0.044$$

The z critical value for 95% confidence is $z^* = 1.96$, so the confidence interval is

$$\tilde{p} \pm z^* SE_{\tilde{p}} = 0.190 \pm (1.96)(0.044) = 0.190 \pm 0.086 = (0.104, 0.276).$$

8.23 a) To see if this is significant evidence that Kerrich's coin comes up heads a different proportion of times than 0.5, we test the hypotheses

$$H_0: p = 0.5$$
$$H_a: p \neq 0.5$$

where p is the probability that Kerrich's coin comes up heads.

We can safely use the z test when np_0 and $n(1 - p_0)$ are both greater than 10. Because $n = 10{,}000$ and $p_0 = 0.5$, $np_0 = 10{,}000 \times 0.5 = 5{,}000$ and $n(1 - p_0) = 10{,}000 \times (1 - 0.5) = 5000$,so we can use the z test. We find that $\hat{p} = 5067/10{,}000 = 0.5067$, so the test statistic is

$$z = \frac{\hat{p} - p_0}{\sqrt{\frac{p_0(1-p_0)}{n}}} = \frac{0.5067 - 0.5}{\sqrt{\frac{0.5(1-0.5)}{10{,}000}}} = 1.34.$$

From Table A, we find $P(Z \geq 1.34) = 1 - P(Z < 1.34) = 1 - 0.9099 = 0.0901$. The P-value for our two-sided test is the area in both tails, which is equal to $2 \times P(Z \geq 1.34) = 2 \times (0.0901) = 0.1802$. This is larger than 0.05, so we would not reject H_0 at the 5% significance level. Thus, this is not strong evidence against the null hypothesis that the probability Kerrich's coin comes up heads is 0.5.

b) The sample size here is $n = 10{,}000$ and the number of heads is $X = 5067$. The Wilson estimate of the probability of heads is

$$\tilde{p} = \frac{X+2}{n+4} = \frac{5067+2}{10{,}000+4} = \frac{5069}{10{,}004} = 0.5067$$

The standard error is

$$\mathrm{SE}_{\tilde{p}} = \sqrt{\frac{\tilde{p}(1-\tilde{p})}{n+4}} = \sqrt{\frac{0.5067(1-0.5067)}{10004}} = 0.0050$$

The z critical value for 95% confidence is $z^* = 1.96$, so the confidence interval is

$$\tilde{p} \pm z^* \mathrm{SE}_{\tilde{p}} = 0.5067 \pm (1.96)(0.0050) = 0.5067 \pm 0.0098 = (0.4969, 0.5165).$$

This gives the range of probabilities of heads that are roughly consistent with Kerrich's result.

8.25 a) To see if Leroy's free-throw probability has remained the same as last year's value of 0.384 or if his work in the summer has resulted in a higher probability of success, we test the hypotheses

$$H_0: p = 0.384$$
$$H_a: p > 0.384$$

where p is Leroy's probability of making each free throw he shoots this season.

b) We can safely use the z test when np_0 and $n(1 - p_0)$ are both greater than 10. Because $n = 40$ (the number of attempts this season) and $p_0 = 0.384$, $np_0 = 40 \times 0.384 = 15.36$ and $n(1 - p_0) = 40 \times (1 - 0.387) = 24.64$, so we can use the z test. We find that $\hat{p} = 25/40 = 0.625$, so the z test statistic is

$$z = \frac{\hat{p} - p_0}{\sqrt{\frac{p_0(1-p_0)}{n}}} = \frac{0.625 - 0.384}{\sqrt{\frac{0.384(1-0.384)}{40}}} = 3.13$$

c) From Table A, we find $P(Z \geq 3.13) = 1 - P(Z < 3.13) = 1 - 0.9991 = 0.0009$. This is the P-value for our one-sided test. It is smaller than 0.05, so we would reject H_0 at the 0.05 significance level.

d) The sample size here is $n = 40$ and the number of free throws made is $X = 25$. The Wilson estimate of the probability of making a free throw is

$$\tilde{p} = \frac{X+2}{n+4} = \frac{25+2}{40+4} = \frac{27}{44} = 0.614$$

The standard error is

$$\mathrm{SE}_{\tilde{p}} = \sqrt{\frac{\tilde{p}(1-\tilde{p})}{n+4}} = \sqrt{\frac{0.614(1-0.614)}{44}} = 0.073$$

The z critical value for 90% confidence is $z^* = 1.645$, so the confidence interval is

$$\tilde{p} \pm z^* \mathrm{SE}_{\tilde{p}} = 0.614 \pm (1.645)(0.073) = 0.614 \pm 0.120 = (0.494, 0.734).$$

Last year's probability of 0.384 is well outside this range, so this would appear to be good evidence that Leroy is a better free-throw shooter than last season. However, see the answer to Part (e).

e) We need to assume that the 40 free throws this season are a random sample of all free throws he will take this season. We also need to assume that free throws are independent. If these are not true, then the inferences in the previous parts are not valid.

8.27 We have a guessed value of 0.2 for the true proportion p^* of the magazine's subscribers who have high incomes (incomes in excess of $100,000). Thus, for 95% confidence $z^* = 1.96$ and with a margin of error $m = 0.05$ or less, the desired sample size n must satisfy

$$n + 4 = \left(\frac{z^*}{m}\right)^2 p^*(1-p^*) = \left(\frac{1.96}{0.05}\right)^2 (0.2)(1 - 0.2) = 245.86$$

and we round up to get $n + 4 = 246$ or $n = 242$.

8.29 We have a guessed value of 0.7 for the true proportion p^* of the student body that is expected to respond favorably to the question of whether they would patronize the nightclub. Thus, for 90% confidence $z^* = 1.645$ and with a margin of error $m = 0.04$ or less, the desired sample size n must satisfy

$$n + 4 = \left(\frac{z^*}{m}\right)^2 p^*(1-p^*) = \left(\frac{1.645}{0.04}\right)^2 (0.7)(1 - 0.7) = 355.16$$

and we round up to get $n + 4 = 356$ or $n = 352$.

If we now take a such a sample, the sample size is $n = 352$. If 50% respond favorably, the number who respond favorably is $X = 0.5 \times 352 = 176$. The Wilson estimate of the proportion of the student body that would patronize the nightclub is

$$\tilde{p} = \frac{X+2}{n+4} = \frac{176+2}{352+4} = \frac{178}{356} = 0.5$$

The standard error is

$$\mathrm{SE}_{\tilde{p}} = \sqrt{\frac{\tilde{p}(1-\tilde{p})}{n+4}} = \sqrt{\frac{0.5(1-0.5)}{356}} = 0.026$$

The z critical value for 90% confidence is $z^* = 1.645$, so the margin of error for the 90% confidence interval is

$$z^* \mathrm{SE}_{\tilde{p}} = (1.645)(0.026) = 0.043.$$

8.31 If $\hat{p} = 0.1, 0.2, 0.3, 0.4, 0.5, 0.6, 0.7, 0.8$, and 0.9, then because $n = 100$ and $\hat{p} = X/n$, the observed counts are $X = 10, 20, 30, 40, 50, 60, 70, 80$, and 90. For each of these cases, we compute the Wilson estimates and find the following.

$$\hat{p} = 0.1: \ \tilde{p} = \frac{X+2}{n+4} = \frac{10+2}{100+4} = 0.115$$

$$\hat{p} = 0.2: \ \tilde{p} = \frac{X+2}{n+4} = \frac{20+2}{100+4} = 0.212$$

$$\hat{p} = 0.3: \ \tilde{p} = \frac{X+2}{n+4} = \frac{30+2}{100+4} = 0.308$$

$$\hat{p} = 0.4: \ \tilde{p} = \frac{X+2}{n+4} = \frac{40+2}{100+4} = 0.404$$

$$\hat{p} = 0.5\text{: } \tilde{p} = \frac{X+2}{n+4} = \frac{50+2}{100+4} = 0.500$$
$$\hat{p} = 0.6\text{: } \tilde{p} = \frac{X+2}{n+4} = \frac{60+2}{100+4} = 0.596$$
$$\hat{p} = 0.7\text{: } \tilde{p} = \frac{X+2}{n+4} = \frac{70+2}{100+4} = 0.692$$
$$\hat{p} = 0.8\text{: } \tilde{p} = \frac{X+2}{n+4} = \frac{80+2}{100+4} = 0.788$$
$$\hat{p} = 0.9\text{: } \tilde{p} = \frac{X+2}{n+4} = \frac{90+2}{100+4} = 0.885$$

Thus, in each case, the margins of error are for 95% confidence ($z^* = 1.96$)

$$\hat{p} = 0.1\text{: } m = z^*\text{SE}_{\tilde{p}} = (1.96)\sqrt{\frac{\tilde{p}(1-\tilde{p})}{n+4}} = (1.96)\sqrt{\frac{0.115(1-0.115)}{104}} = 0.061$$
$$\hat{p} = 0.2\text{: } m = z^*\text{SE}_{\tilde{p}} = (1.96)\sqrt{\frac{\tilde{p}(1-\tilde{p})}{n+4}} = (1.96)\sqrt{\frac{0.212(1-0.212)}{104}} = 0.079$$
$$\hat{p} = 0.3\text{: } m = z^*\text{SE}_{\tilde{p}} = (1.96)\sqrt{\frac{\tilde{p}(1-\tilde{p})}{n+4}} = (1.96)\sqrt{\frac{0.308(1-0.308)}{104}} = 0.089$$
$$\hat{p} = 0.4\text{: } m = z^*\text{SE}_{\tilde{p}} = (1.96)\sqrt{\frac{\tilde{p}(1-\tilde{p})}{n+4}} = (1.96)\sqrt{\frac{0.404(1-0.404)}{104}} = 0.094$$
$$\hat{p} = 0.5\text{: } m = z^*\text{SE}_{\tilde{p}} = (1.96)\sqrt{\frac{\tilde{p}(1-\tilde{p})}{n+4}} = (1.96)\sqrt{\frac{0.500(1-0.500)}{104}} = 0.096$$
$$\hat{p} = 0.6\text{: } m = z^*\text{SE}_{\tilde{p}} = (1.96)\sqrt{\frac{\tilde{p}(1-\tilde{p})}{n+4}} = (1.96)\sqrt{\frac{0.596(1-0.596)}{104}} = 0.094$$
$$\hat{p} = 0.7\text{: } m = z^*\text{SE}_{\tilde{p}} = (1.96)\sqrt{\frac{\tilde{p}(1-\tilde{p})}{n+4}} = (1.96)\sqrt{\frac{0.692(1-0.692)}{104}} = 0.089$$
$$\hat{p} = 0.8\text{: } m = z^*\text{SE}_{\tilde{p}} = (1.96)\sqrt{\frac{\tilde{p}(1-\tilde{p})}{n+4}} = (1.96)\sqrt{\frac{0.788(1-0.788)}{104}} = 0.079$$
$$\hat{p} = 0.9\text{: } m = z^*\text{SE}_{\tilde{p}} = (1.96)\sqrt{\frac{\tilde{p}(1-\tilde{p})}{n+4}} = (1.96)\sqrt{\frac{0.885(1-0.885)}{104}} = 0.061$$

Summarized in a table, we have

$\hat{p}$:	0.1	0.2	0.3	0.4	0.5	0.6	0.7	0.8	0.9
m:	0.061	0.079	0.089	0.094	0.096	0.094	0.089	0.079	0.061

SECTION 8.2

OVERVIEW

Confidence intervals and tests designed to compare two population proportions are based on the **difference in the sample proportions** $D = \hat{p}_1 - \hat{p}_2$ and the difference in **Wilson estimates** $\tilde{D} = \tilde{p}_1 - \tilde{p}_2$ where

$$\tilde{p}_1 = \frac{X_1 + 1}{n_1 + 2} \text{ and } \tilde{p}_2 = \frac{X_2 + 1}{n_2 + 2},$$

and X_1 and X_2 are the number of successes in each group.

The formula for the level C confidence interval is

$$(\tilde{p}_1 - \tilde{p}_2) \pm z^* \text{SE}_{\tilde{D}}$$

where z^* is the upper $(1 - C)/2$ standard normal critical value and $\text{SE}_{\tilde{D}}$ is the standard error for the difference in the two proportions computed as

$$\text{SE}_{\tilde{D}} = \sqrt{\frac{\tilde{p}_1(1-\tilde{p}_1)}{n_1 + 2} + \frac{\tilde{p}_2(1-\tilde{p}_2)}{n_2 + 2}}$$

Significance tests for the equality of the two proportions, H_0: $p_1 = p_2$, use a different standard error for the difference in the sample proportions, which is based on a **pooled estimate** of the common (under H_0) value of p_1 and p_2,

$$\hat{p} = \frac{X_1 + X_2}{n_1 + n_2}.$$

The test uses the z statistic

$$z = \frac{\hat{p}_1 - \hat{p}_2}{\text{SE}_{Dp}}$$

where

$$\text{SE}_{Dp} = \sqrt{\hat{p}(1-\hat{p})\left(\frac{1}{n_1} + \frac{1}{n_2}\right)}$$

and P-values are computed using Table A of the standard normal distribution.

Relative risk is the ratio of two sample proportions:

$$\text{RR} = \frac{\hat{p}_1}{\hat{p}_2}$$

Confidence intervals for relative risk are an alternative to confidence intervals for the difference when we want to compare two proportions.

APPLY YOUR KNOWLEDGE

8.33 a) The means and standard deviations of the two sample proportions are

For $\hat{p}_1$: mean $\mu_{\hat{p}_1} = p_1$, standard deviation $\sigma_{\hat{p}_1} = \sqrt{\dfrac{p_1(1-p_1)}{n_1}}$

For $\hat{p}_2$: mean $\mu_{\hat{p}_2} = p_2$, standard deviation $\sigma_{\hat{p}_2} = \sqrt{\dfrac{p_2(1-p_2)}{n_2}}$

b) The addition rule for means says that the mean of a difference is equal to the difference in the means. Thus,

$$\mu_D = \mu_{\hat{p}_1 - \hat{p}_2} = \mu_{\hat{p}_1} - \mu_{\hat{p}_2} = p_1 - p_2$$

c) The addition rule for the variance of the difference between two independent random variables says this variance is the sum of the variances of the random variables. Thus,

$$\sigma_D^2 = \sigma_{\hat{p}_1 - \hat{p}_2}^2 = \sigma_{\hat{p}_1}^2 + \sigma_{\hat{p}_2}^2 = \frac{p_1(1-p_1)}{n_1} + \frac{p_2(1-p_2)}{n_2}$$

8.35 There were $n_1 = 639$ patients who filed complaints and $X_1 = 54$ of these left the HMO voluntarily. There were $n_2 = 743$ patients who had not filed complaints and $X_2 = 22$ of these left the HMO voluntarily. The Wilson estimates for the two population proportions (the populations of complainers and of noncomplainers) are

$$\tilde{p}_1 = \frac{X_1+1}{n_1+2} = \frac{54+1}{639+2} = 0.086 \text{ and } \tilde{p}_2, = \frac{X_2+1}{n_2+2} = \frac{22+1}{743+2} = 0.031$$

Thus, the estimated difference is

$$\tilde{D} = \tilde{p}_1 - \tilde{p}_2 = 0.086 - 0.031 = 0.055$$

The standard deviation of $\tilde{D}$ is approximately

$$\text{SE}_{\tilde{D}} = \sqrt{\frac{\tilde{p}_1(1-\tilde{p}_1)}{n_1+2} + \frac{\tilde{p}_2(1-\tilde{p}_2)}{n_2+2}} = \sqrt{\frac{0.086(1-0.086)}{639+2} + \frac{0.031(1-0.031)}{743+2}} = 0.013$$

and so a 95% confidence interval ($z^* = 1.96$) for the difference in the two proportions is

$$(\tilde{p}_1 - \tilde{p}_2) \pm z^* \text{SE}_{\tilde{D}} = 0.055 \pm (1.96)(0.013) = 0.055 \pm 0.025 = (0.030, 0.080)$$

8.37 We expect a higher proportion of complainers to leave, so we should test the hypotheses

$$H_0: p_1 = p_2$$
$$H_a: p_1 > p_2$$

There were $n_1 = 639$ patients who filed complaints and $X_1 = 54$ of these left the HMO voluntarily. There were $n_2 = 743$ patients who had not filed complaints and $X_2 = 22$ of these left the HMO voluntarily. The pooled estimate of the common (under H_0) value of p_1 and p_2,

$$\hat{p} = \frac{X_1+X_2}{n_1+n_2} = \frac{54+22}{639+743} = 0.055$$

and the standard error of the pooled estimate is

$$SE_{Dp} = \sqrt{\hat{p}(1-\hat{p})\left(\frac{1}{n_1}+\frac{1}{n_2}\right)} = \sqrt{0.055(1-0.055)\left(\frac{1}{639}+\frac{1}{743}\right)} = 0.012$$

We find that

$$\hat{p}_1 = X_1/n_1 = 54/639 = 0.085 \text{ and } \hat{p}_2 = X_2/n_2 = 22/743 = 0.030$$

and thus the test statistic for testing the hypotheses is

$$z = \frac{\hat{p}_1 - \hat{p}_2}{SE_{Dp}} = \frac{0.085-0.030}{0.012} = 4.58$$

From Table A, we find $P(Z \geq 4.58)$ is approximately 0. This is the P-value for our one-sided test and this is strong evidence against the null hypothesis. We would conclude that there is strong evidence that a higher proportion of complainers than noncomplainers leave voluntarily.

SECTION 8.2 EXERCISES

8.39 a) There were $n_1 = 263 + 252 = 515$ farmers surveyed in Tippecanoe County and $X_1 = 263$ of these were in favor of the program. There were $n_2 = 260 + 377 = 637$ farmers surveyed in Benton County and $X_2 = 260$ of these were in favor of the program. Thus, the proportion of farmers in favor of the program in each of the two counties is

Tippecanoe County: $\hat{p}_1 = X_1/n_1 = 263/515 = 0.511$
Benton County: $\hat{p}_2 = X_2/n_2 = 260/637 = 0.408$

b) The Wilson estimates for the two population proportions are

$$\tilde{p}_1 = \frac{X_1+1}{n_1+2} = \frac{263+1}{515+2} = 0.511 \text{ and } \tilde{p}_2 = \frac{X_2+1}{n_2+2} = \frac{260+1}{637+2} = 0.408$$

Thus, the estimated difference is

$$\tilde{D} = \tilde{p}_1 - \tilde{p}_2 = 0.511 - 0.408 = 0.103$$

The standard deviation of $\tilde{D}$ is approximately

$$SE_{\tilde{D}} = \sqrt{\frac{\tilde{p}_1(1-\tilde{p}_1)}{n_1+2}+\frac{\tilde{p}_2(1-\tilde{p}_2)}{n_2+2}} = \sqrt{\frac{0.511(1-0.511)}{515+2}+\frac{0.408(1-0.408)}{637+2}} = 0.029$$

This is the standard error we need to compute a confidence interval for the difference in proportions.
c) A 99% confidence interval ($z^* = 2.576$) for the difference in the two proportions is

$$(\tilde{p}_1 - \tilde{p}_2) \pm z^* SE_{\tilde{D}} = 0.103 \pm (2.576)(0.029) = 0.103 \pm 0.075 = (0.028, 0.178)$$

The interval does not contain 0 and this is strong evidence that the opinions differed in the two counties.

8.41 a) Let p_1 denote the proportion of rural households that prefer natural trees and p_2 the proportion of urban households that prefer natural trees. We are simply interested in determining if these proportions are different, so we would test the hypotheses

$$H_0: p_1 = p_2$$
$$H_a: p_1 \neq p_2$$

b) There were $n_1 = 160$ rural respondents and $X_1 = 64$ of these preferred natural trees. There were $n_2 = 261$ urban respondents and $X_2 = 89$ of these preferred natural trees. We find

$$\hat{p}_1 = X_1/n_1 = 64/160 = 0.400 \text{ and } \hat{p}_2 = X_2/n_2 = 89/261 = 0.341$$

The pooled estimate of the common (under H_0) value of p_1 and p_2,

$$\hat{p} = \frac{X_1 + X_2}{n_1 + n_2} = \frac{64+89}{160+261} = 0.363$$

and the standard error of the pooled estimate is

$$\text{SE}_{Dp} = \sqrt{\hat{p}(1-\hat{p})\left(\frac{1}{n_1}+\frac{1}{n_2}\right)} = \sqrt{0.363(1-0.363)\left(\frac{1}{160}+\frac{1}{261}\right)} = 0.048$$

Thus, the test statistic for testing the hypotheses is

$$z = \frac{\hat{p}_1 - \hat{p}_2}{\text{SE}_{Dp}} = \frac{0.400-0.341}{0.048} = 1.23$$

From Table A,we find $P(Z \geq 1.23) = 1 - P(Z < 1.23) = 1 - 0.8907 = 0.1093$. Because we have a two-sided alternative, we need to double this (to account for the tail probabilities above and below 1.23 and −1.23, respectively), and this gives P-value = $2\times 0.1093 = 0.2186$. This is not strong evidence against the null hypothesis, and we would conclude that there is not strong evidence that there is a difference in preference for natural trees versus artificial trees between urban and rural households.

c) The Wilson estimates for the two population proportions are

$$\tilde{p}_1 = \frac{X_1+1}{n_1+2} = \frac{64+1}{160+2} = 0.401 \text{ and } \tilde{p}_2 = \frac{X_2+1}{n_2+2} = \frac{89+1}{261+2} = 0.342$$

Thus, the estimated difference is

$$\tilde{D} = \tilde{p}_1 - \tilde{p}_2 = 0.401 - 0.342 = 0.059$$

The standard deviation of $\tilde{D}$ is approximately

$$\text{SE}_{\tilde{D}} = \sqrt{\frac{\tilde{p}_1(1-\tilde{p}_1)}{n_1+2}+\frac{\tilde{p}_2(1-\tilde{p}_2)}{n_2+2}} = \sqrt{\frac{0.401(1-0.401)}{160+2}+\frac{0.342(1-0.342)}{261+2}} = 0.048$$

This is the standard error we need in order to compute a confidence interval for the difference in proportions. A 90% confidence interval ($z^* = 1.645$) for the difference in the two proportions is

$$(\tilde{p}_1 - \tilde{p}_2) \pm z^* \text{SE}_{\tilde{D}} = 0.059 \pm (1.645)(0.048) = 0.059 \pm 0.079 = (-0.020, 0.138)$$

8.43 a) Let p_1 denote the proportion of bolt-on shields removed in all older tractors and p_2 the proportion of flip-up shields removed. We are simply interested in determining if these proportions are different, so we would test the hypotheses

$$H_0\colon p_1 = p_2$$
$$H_a\colon p_1 \neq p_2$$

There were $n_1 = 83$ older tractors with bolt-on shields and $X_1 = 35$ of these had been removed. There were $n_2 = 136$ older tractors with flip-up shields and $X_2 = 15$ of these had been removed We find

$$\hat{p}_1 = X_1/n_1 = 35/83 = 0.422 \text{ and } \hat{p}_2 = X_2/n_2 = 15/136 = 0.110$$

The pooled estimate of the common (under H_0) value of p_1 and p_2 is

$$\hat{p} = \frac{X_1 + X_2}{n_1 + n_2} = \frac{35+15}{83+136} = 0.228$$

and the standard error of the pooled estimate is

$$SE_{Dp} = \sqrt{\hat{p}(1-\hat{p})\left(\frac{1}{n_1}+\frac{1}{n_2}\right)} = \sqrt{0.228(1-0.228)\left(\frac{1}{83}+\frac{1}{136}\right)} = 0.058$$

Thus, the test statistic for testing the hypotheses is

$$z = \frac{\hat{p}_1 - \hat{p}_2}{SE_{Dp}} = \frac{0.422 - 0.110}{0.058} = 5.40$$

From Table A, we find $P(Z \geq 5.40)$ is approximately 0. Because we have a two-sided alternative, we need to double this (to account for the tail probabilities above and below 5.40 and –5.40, respectively), and this gives a P-value = approximately 0. This is strong evidence against the null hypothesis, and we would conclude that there is strong evidence that there is a difference in the proportions of the two types of shields removed.

b) The Wilson estimates for the two population proportions are

$$\tilde{p}_1 = \frac{X_1+1}{n_1+2} = \frac{35+1}{83+2} = 0.424 \text{ and } \tilde{p}_2 = \frac{X_2+1}{n_2+2} = \frac{15+1}{136+2} = 0.116$$

Thus, the estimated difference is

$$\tilde{D} = \tilde{p}_1 - \tilde{p}_2 = 0.424 - 0.116 = 0.308$$

The standard deviation of $\tilde{D}$ is approximately

$$SE_{\tilde{D}} = \sqrt{\frac{\tilde{p}_1(1-\tilde{p}_1)}{n_1+2}+\frac{\tilde{p}_2(1-\tilde{p}_2)}{n_2+2}} = \sqrt{\frac{0.424(1-0.424)}{83+2}+\frac{0.116(1-0.116)}{136+2}} = 0.060$$

This is the standard error we need in order to compute a confidence interval for the difference in proportions. A 90% confidence interval ($z^* = 1.645$) for the difference in the two proportions is

$$(\tilde{p}_1 - \tilde{p}_2) \pm z^* SE_{\tilde{D}} = 0.308 \pm (1.645)(0.060) = 0.308 \pm 0.099 = (0.209, 0.407)$$

Zero is well outside this interval so that bolt-on shields appear to be removed more often than flip-up shields. We would recommend that flip-up shields be used on new tractors.

8.45 a) Let p_1 denote the proportion of females who were fatally injured in bicycle accidents and tested positive for alcohol and p_2 the proportion of males who were fatally injured in bicycle accidents and tested positive for alcohol. There were $n_1 = 191$ females who were fatally injured in bicycle accidents and were tested for alcohol, and $X_1 = 27$ of these tested positive. There were $n_2 = 1520$ males who were fatally injured in bicycle accidents and were tested for alcohol, and $X_2 = 515$ of these tested positive. The Wilson estimates for the two population proportions are

$$\tilde{p}_1 = \frac{X_1+1}{n_1+2} = \frac{27+1}{191+2} = 0.145 \text{ and } \tilde{p}_2, = \frac{X_2+1}{n_2+2} = \frac{515+1}{1520+2} = 0.339$$

Thus, the estimated difference is

$$\tilde{D} = \tilde{p}_1 - \tilde{p}_2 = 0.145 - 0.339 = -0.194$$

The standard deviation of $\tilde{D}$ is approximately

$$SE_{\tilde{D}} = \sqrt{\frac{\tilde{p}_1(1-\tilde{p}_1)}{n_1+2} + \frac{\tilde{p}_2(1-\tilde{p}_2)}{n_2+2}} = \sqrt{\frac{0.145(1-0.145)}{181+2} + \frac{0.339(1-0.339)}{1520+2}} = 0.029$$

This is the standard error we need in order to compute a confidence interval for the difference in proportions. A 90% confidence interval ($z^* = 1.645$) for the difference in the two proportions is

$$(\tilde{p}_1 - \tilde{p}_2) \pm z^* SE_{\tilde{D}} = -0.194 \pm (1.645)(0.029) = -0.194 \pm 0.048 = (-0.242, -0.146)$$

b) The term $\frac{\tilde{p}_1(1-\tilde{p}_1)}{n_1+2}$ contributes the largest amount to the standard error of the difference because $n_1 = 191$ is so much smaller than $n_2 = 1520$. Fewer females means a larger standard error for $\tilde{p}_1$ as compared to that for $\tilde{p}_2$ (the latter being based on a much larger sample).

8.47 Let p_1 denote the proportion of females who were fatally injured in bicycle accidents and tested positive for alcohol and p_2 the proportion of males who were fatally injured in bicycle accidents and tested positive for alcohol. There were $n_1 = 191$ females who were fatally injured in bicycle accidents and were tested for alcohol, and $X_1 = 27$ of these tested positive. There were $n_2 = 1520$ males who were fatally injured in bicycle accidents and were tested for alcohol, and $X_2 = 515$ of these tested positive. We are not given any information as to which proportion we expect to be larger, so we simply test to see if these proportions are different, i.e., we test the hypotheses

$$H_0: p_1 = p_2$$
$$H_a: p_1 \neq p_2$$

We find

$$\hat{p}_1 = X_1/n_1 = 27/191 = 0.141 \text{ and } \hat{p}_2 = X_2/n_2 = 515/1520 = 0.339$$

The pooled estimate of the common (under H_0) value of p_1 and p_2 is

$$\hat{p} = \frac{X_1 + X_2}{n_1 + n_2} = \frac{27+515}{191+1520} = 0.317$$

and the standard error of the pooled estimate is

$$SE_{Dp} = \sqrt{\hat{p}(1-\hat{p})\left(\frac{1}{n_1} + \frac{1}{n_2}\right)} = \sqrt{0.317(1-0.317)\left(\frac{1}{191} + \frac{1}{1520}\right)} = 0.036$$

Thus, the test statistic for testing the hypotheses is

$$z = \frac{\hat{p}_1 - \hat{p}_2}{SE_{Dp}} = \frac{0.141 - 0.339}{0.036} = -5.5$$

From Table A, we find $P(Z \leq -5.5)$ is approximately 0. Because we have a two-sided alternative, we need to double this (to account for the tail probabilities above and below 5.5 and –5.5, respectively), and this gives P-value = approximately 0. This is strong evidence against the null hypothesis, and we would conclude that there is strong evidence that there is a difference in the proportions of females and males who were in fatal bicycle accidents and were tested for alcohol and tested positive.

8.49 a) Let p_1 denote the probability that the Yankees win at home and p_2 the probability that the Yankees win away. There were $n_1 = 80$ games at home and $X_1 = 44$ of these were won. There were $n_2 = 81$ games away and $X_2 = 43$ of these were won. The Wilson estimates for the two probabilities are

$$\tilde{p}_1 = \frac{X_1 + 1}{n_1 + 2} = \frac{44 + 1}{80 + 2} = 0.549 \text{ and } \tilde{p}_2, = \frac{X_2 + 1}{n_2 + 2} = \frac{43 + 1}{81 + 2} = 0.530$$

b) The standard deviation of the difference in proportions $\tilde{D}$ is approximately

$$SE_{\tilde{D}} = \sqrt{\frac{\tilde{p}_1(1-\tilde{p}_1)}{n_1+2} + \frac{\tilde{p}_2(1-\tilde{p}_2)}{n_2+2}} = \sqrt{\frac{0.549(1-0.549)}{80+2} + \frac{0.530(1-0.530)}{81+2}} = 0.078$$

c) The estimated difference is

$$\tilde{D} = \tilde{p}_1 - \tilde{p}_2 = 0.549 - 0.530 = 0.019$$

Using the standard error computed in part (b), a 90% confidence interval ($z^* = 1.645$) for the difference in the two proportions is

$$(\tilde{p}_1 - \tilde{p}_2) \pm z^* SE_{\tilde{D}} = 0.019 \pm (1.645)(0.078) = 0.019 \pm 0.128 = (-0.109, 0.147)$$

This interval contains 0 and so provides no evidence that there is a significant difference in the probability of winning at home versus winning on the road. This evidence does not convince us that the 2000 Yankees were more likely to win at home.

8.51 a) Let p_1 denote the probability that the Yankees win at home and p_2 the probability that the Yankees win away. We are interested in whether the Yankees are more likely to win at home, so we test the hypotheses

$$H_0: p_1 = p_2$$
$$H_a: p_1 > p_2$$

b) There were $n_1 = 80$ games at home and $X_1 = 44$ of these were won. There were $n_2 = 81$ games away and $X_2 = 43$ of these were won. We find

$$\hat{p}_1 = X_1/n_1 = 44/80 = 0.550 \text{ and } \hat{p}_2 = X_2/n_2 = 43/81 = 0.531$$

The pooled estimate of the common (under H_0) value of p_1 and p_2 is

$$\hat{p} = \frac{X_1 + X_2}{n_1 + n_2} = \frac{44 + 43}{80 + 81} = 0.540$$

Thus, the proportion of all games played that the Yankees won is 0.540.

c) The standard error of the pooled estimate is

$$SE_{Dp} = \sqrt{\hat{p}(1-\hat{p})\left(\frac{1}{n_1} + \frac{1}{n_2}\right)} = \sqrt{0.540(1-0.540)\left(\frac{1}{80} + \frac{1}{81}\right)} = 0.079$$

and this is the standard error needed for testing our hypotheses.

d) The z test statistic for testing the hypotheses is

$$z = \frac{\hat{p}_1 - \hat{p}_2}{SE_{Dp}} = \frac{0.550 - 0.531}{0.079} = 0.24$$

From Table A, we find $P(Z \geq 0.24) = 1 - P(Z < 0.24) = 1 - 0.5948 = 0.4052$. This is the P-value. This is not strong evidence against the null hypothesis and we would conclude that there is not strong evidence that the Yankees are more likely to win at home than on the road.

8.53 Let p_1 denote the proportion of all Danish males born in Copenhagen with a normal male chromosome who have had criminal records and p_2 the proportion of all Danish males born in Copenhagen with an abnormal male chromosome who have had criminal records. We are interested in whether the men with abnormal male chromosomes are more likely to be criminals, so we test the hypotheses

$$H_0: p_1 = p_2$$
$$H_a: p_1 < p_2$$

There were $n_1 = 4096$ men with a normal male chromosome and $X_1 = 381$ of these had criminal records There were $n_2 = 28$ men with an abnormal male chromosome and $X_2 = 8$ of had criminal records. We find

$$\hat{p}_1 = X_1/n_1 = 381/4096 = 0.093 \text{ and } \hat{p}_2 = X_2/n_2 = 8/28 = 0.286$$

The pooled estimate of the common (under H_0) value of p_1 and p_2 is

$$\hat{p} = \frac{X_1 + X_2}{n_1 + n_2} = \frac{381+8}{4096+28} = 0.094$$

The standard error of the pooled estimate is

$$\text{SE}_{Dp} = \sqrt{\hat{p}(1-\hat{p})\left(\frac{1}{n_1}+\frac{1}{n_2}\right)} = \sqrt{0.094(1-0.094)\left(\frac{1}{4096}+\frac{1}{28}\right)} = 0.055$$

The z test statistic for testing the hypotheses is

$$z = \frac{\hat{p}_1 - \hat{p}_2}{\text{SE}_{Dp}} = \frac{0.093-0.286}{0.055} = -3.51$$

From Table A, we find the P-value $= P(Z \leq -3.51) < 0.0003$. This is strong evidence against the null hypothesis, and we would conclude that there is strong evidence that the proportion of males born in Copenhagen with abnormal male chromosomes who have criminal records is larger than that of males born in Copenhagen with normal male chromosomes. If males born in Copenhagen can be regarded as representative of all males, these conclusions would apply to all males.

8.55 a) Let p_1 denote the proportion of all German cockroaches that would die on glass and p_2 the proportion that would die on plasterboard. There were $n_1 = 18$ cockroaches placed on glass and $X_1 = 9$ of these died. There were $n_2 = 18$ cockroaches placed on plasterboard and $X_2 = 13$ of these died The Wilson estimates for the two probabilities are

$$\tilde{p}_1 = \frac{X_1+1}{n_1+2} = \frac{9+1}{18+2} = 0.5 \text{ and } \tilde{p}_2 = \frac{X_2+1}{n_2+2} = \frac{13+1}{18+2} = 0.7$$

The standard deviation of the difference in proportions $\tilde{D}$ is approximately

$$\text{SE}_{\tilde{D}} = \sqrt{\frac{\tilde{p}_1(1-\tilde{p}_1)}{n_1+2}+\frac{\tilde{p}_2(1-\tilde{p}_2)}{n_2+2}} = \sqrt{\frac{0.5(1-0.5)}{18+2}+\frac{0.7(1-0.7)}{18+2}} = 0.152$$

The estimated difference is

$$\tilde{D} = \tilde{p}_1 - \tilde{p}_2 = 0.5 - 0.7 = -0.2$$

A 90% confidence interval ($z^* = 1.645$) for the difference in the two proportions is

$$(\tilde{p}_1 - \tilde{p}_2) \pm z^* \text{SE}_{\tilde{D}} = -0.2 \pm (1.645)(0.152) = -0.2 \pm 0.25 = (-0.45, 0.05)$$

b) We are interested in whether the mortality rate on plasterboard will be greater than that for glass, so we test the hypotheses

$$H_0: p_1 = p_2$$
$$H_a: p_1 < p_2$$

We find

$$\hat{p}_1 = X_1/n_1 = 9/18 = 0.50 \text{ and } \hat{p}_2 = X_2/n_2 = 13/18 = 0.72$$

The pooled estimate of the common (under H_0) value of p_1 and p_2 is

$$\hat{p} = \frac{X_1 + X_2}{n_1 + n_2} = \frac{9 + 13}{18 + 18} = 0.61$$

The standard error of the pooled estimate is

$$SE_{Dp} = \sqrt{\hat{p}(1-\hat{p})\left(\frac{1}{n_1} + \frac{1}{n_2}\right)} = \sqrt{0.61(1-0.61)\left(\frac{1}{18} + \frac{1}{18}\right)} = 0.163$$

The z test statistic for testing the hypotheses is

$$z = \frac{\hat{p}_1 - \hat{p}_2}{SE_{Dp}} = \frac{0.50 - 0.72}{0.163} = -1.35$$

From Table A we find P-value $= P(Z \le -1.35) = 0.0885$. This is not strong evidence against the null hypothesis and we would conclude that there is not strong evidence that the proportion of German cockroaches that will die on glass is less than the proportion that will die on plasterboard.

8.57 a) Let p_1 denote the proportion of all German cockroaches that would die on glass and p_2 the proportion that would die on plasterboard. There were $n_1 = 36$ cockroaches placed on glass and $X_1 = 18$ of these died. There were $n_2 = 36$ cockroaches placed on plasterboard and $X_2 = 26$ of these died. All values are just double those in Exercise 8.55. In this case, we find

$$\hat{p}_1 = X_1/n_1 = 18/36 = 0.50 \text{ and } \hat{p}_2 = X_2/n_2 = 26/36 = 0.72$$

The pooled estimate of the common (under H_0) value of p_1 and p_2 is

$$\hat{p} = \frac{X_1 + X_2}{n_1 + n_2} = \frac{18 + 26}{36 + 36} = 0.61$$

The standard error of the pooled estimate is

$$SE_{Dp} = \sqrt{\hat{p}(1-\hat{p})\left(\frac{1}{n_1} + \frac{1}{n_2}\right)} = \sqrt{0.61(1-0.61)\left(\frac{1}{36} + \frac{1}{36}\right)} = 0.115$$

The z test statistic for testing the hypotheses is

$$z = \frac{\hat{p}_1 - \hat{p}_2}{SE_{Dp}} = \frac{0.50 - 0.72}{0.115} = -1.91$$

From Table A, we find P-value $= P(Z \le -1.91) = 0.0281$. This is reasonably strong evidence against the null hypothesis, and we would conclude that there is reasonably strong evidence that the proportion of German cockroaches that will die on glass is less than the proportion that will die on plasterboard.

b) In Exercise 8.55, the P-value is 0.0885 (not strong evidence against the null hypothesis), while here the P-value is 0.0281 (reasonably strong evidence against the null hypothesis). If the observed proportions remain the same, increasing the sample size gives a smaller P-value and results that were not statistically significant for smaller sample sizes can be statistically significant for larger sample sizes.

CHAPTER 8 REVIEW EXERCISES

8.59 a) Let p_1 denote the proportion of all Internet users who have completed college and p_2 the proportion of all nonusers who have completed college. We wish to test if users and nonusers differ significantly in the proportion of college graduates, so we test the hypotheses

$$H_0: p_1 = p_2$$
$$H_a: p_1 \neq p_2$$

There were $n_1 = 1132$ Internet users and $X_1 = 643$ of these completed college. There were $n_2 = 852$ nonusers and $X_2 = 349$ of these completed college. We find

$$\hat{p}_1 = X_1/n_1 = 643/1132 = 0.568 \text{ and } \hat{p}_2 = X_2/n_2 = 349/852 = 0.410$$

The pooled estimate of the common (under H_0) value of p_1 and p_2 is

$$\hat{p} = \frac{X_1 + X_2}{n_1 + n_2} = \frac{643 + 349}{1132 + 852} = 0.5$$

The standard error of the pooled estimate is

$$\mathrm{SE}_{Dp} = \sqrt{\hat{p}(1-\hat{p})\left(\frac{1}{n_1} + \frac{1}{n_2}\right)} = \sqrt{0.5(1-0.5)\left(\frac{1}{1132} + \frac{1}{852}\right)} = 0.023$$

The z test statistic for testing the hypotheses is

$$z = \frac{\hat{p}_1 - \hat{p}_2}{\mathrm{SE}_{Dp}} = \frac{0.568 - 0.410}{0.023} = 6.87$$

From Table A we find $P(Z \geq 6.87)$ is approximately 0. Because we have a two-sided alternative, we need to double this (to account for the tail probabilities above and below 6.87 and –6.87, respectively), and this gives a P-value = approximately 0. This is strong evidence against the null hypothesis and we would conclude that there is strong evidence that there is a difference in the proportions of users and nonusers who completed college.

b) The Wilson estimates for the two probabilities are

$$\tilde{p}_1 = \frac{X_1 + 1}{n_1 + 2} = \frac{643 + 1}{1132 + 2} = 0.568 \text{ and } \tilde{p}_2 = \frac{X_2 + 1}{n_2 + 2} = \frac{349 + 1}{852 + 2} = 0.410$$

The standard deviation of the difference in proportions $\tilde{D}$ is approximately

$$\mathrm{SE}_{\tilde{D}} = \sqrt{\frac{\tilde{p}_1(1-\tilde{p}_1)}{n_1 + 2} + \frac{\tilde{p}_2(1-\tilde{p}_2)}{n_2 + 2}} = \sqrt{\frac{0.568(1-0.568)}{1132 + 2} + \frac{0.410(1-0.410)}{852 + 2}} = 0.022$$

The estimated difference is

$$\tilde{D} = \tilde{p}_1 - \tilde{p}_2 = 0.568 - 0.410 = 0.158$$

A 95% confidence interval ($z^* = 1.96$) for the difference in the two proportions is

$$(\tilde{p}_1 - \tilde{p}_2) \pm z^* \mathrm{SE}_{\tilde{D}} = 0.158 \pm (1.96)(0.022) = 0.158 \pm 0.043 = (0.115, 0.201)$$

8.61 The total number of users and nonusers for the analysis of education is $n_1 = 1132$ and $n_2 = 852$, respectively. The total number of users and nonusers for the analysis of income is 493 + 378 = 871 and 477 + 200 = 677, respectively. The number of users who chose "Rather not say" is $X_1 = 1132 - 871 = 261$, and the number of nonusers who chose "Rather not say" is $X_2 = 852 - 677 = 175$.

Let p_1 be the proportion of all Internet users who would rather not give their incomes and p_2 the proportion of nonusers who would rather not give their income. We wish to test if users and nonusers differ significantly in the proportion who choose "Rather not say" for the income question, so we test the hypotheses

$$H_0: p_1 = p_2$$
$$H_a: p_1 \neq p_2$$

We find

$$\hat{p}_1 = X_1/n_1 = 261/1132 = 0.231 \text{ and } \hat{p}_2 = X_2/n_2 = 175/852 = 0.205$$

The pooled estimate of the common (under H_0) value of p_1 and p_2 is

$$\hat{p} = \frac{X_1 + X_2}{n_1 + n_2} = \frac{261+175}{1132+852} = 0.220$$

The standard error of the pooled estimate is

$$\text{SE}_{Dp} = \sqrt{\hat{p}(1-\hat{p})\left(\frac{1}{n_1}+\frac{1}{n_2}\right)} = \sqrt{0.220(1-0.220)\left(\frac{1}{1132}+\frac{1}{852}\right)} = 0.019$$

The z test statistic for testing the hypotheses is

$$z = \frac{\hat{p}_1 - \hat{p}_2}{\text{SE}_{Dp}} = \frac{0.231-0.205}{0.019} = 1.37$$

From Table A, we find $P(Z \geq 1.37) = 1 - P(Z < 1.37) = 1 - 0.9147 = 0.0853$. Because we have a two-sided alternative, we need to double this (to account for the tail probabilities above and below 1.37 and −1.37, respectively), and this gives a P-value = 2 × 0.0853 = 0.1706. This is not strong evidence against the null hypothesis, and we would conclude that there is not strong evidence that there is a difference in the proportions of users and nonusers who would choose "Rather not say" for the income question.

The Wilson estimates for the two probabilities are

$$\tilde{p}_1 = \frac{X_1+1}{n_1+2} = \frac{261+1}{1132+2} = 0.231 \text{ and } \tilde{p}_2 = \frac{X_2+1}{n_2+2} = \frac{175+1}{852+2} = 0.206$$

The standard deviation of the difference in proportions $\tilde{D}$ is approximately

$$\text{SE}_{\tilde{D}} = \sqrt{\frac{\tilde{p}_1(1-\tilde{p}_1)}{n_1+2}+\frac{\tilde{p}_2(1-\tilde{p}_2)}{n_2+2}} = \sqrt{\frac{0.231(1-0.231)}{1132+2}+\frac{0.206(1-0.206)}{852+2}} = 0.019$$

The estimated difference is

$$\tilde{D} = \tilde{p}_1 - \tilde{p}_2 = 0.231 - 0.206 = 0.025$$

A 95% confidence interval ($z^* = 1.96$) for the difference in the two proportions is

$$(\tilde{p}_1 - \tilde{p}_2) \pm z^* \text{SE}_{\tilde{D}} = 0.025 \pm (1.96)(0.019) = 0.025 \pm 0.037 = (-0.012, 0.062)$$

Although the difference in nonresponse between users and nonusers is small (and not statistically significant), the proportion of users and nonusers who did not respond is about 0.22. This is more than 1/5 of those surveyed and is large enough to be a serious limitation of the income study.

8.63 The number of die-hard fans is $n_1 = 134$ and the number of less loyal fans is $n_2 = 237$.

Let p_1 be the proportion of all die-hard fans who attend a Cubs game at least once a month and p_2 the proportion of less loyal fans who attend a Cubs game at least once a month. We wish to test if die-hard fans are more likely to attend a Cubs game at least once a month (i.e., are more loyal consumers) than less loyal fans, so we test

$$H_0: p_1 = p_2$$
$$H_a: p_1 > p_2$$

We are told that

$$\hat{p}_1 = X_1/n_1 = 0.67 \text{ and } \hat{p}_2 = X_2/n_2 = 0.20$$

This implies that $X_1 = 0.67(134) = 90$ and $X_2 = 0.20(237) = 47$ (after rounding off). The pooled estimate of the common (under H_0) value of p_1 and p_2 is

$$\hat{p} = \frac{X_1 + X_2}{n_1 + n_2} = \frac{90+47}{134+237} = 0.369$$

The standard error of the pooled estimate is

$$SE_{Dp} = \sqrt{\hat{p}(1-\hat{p})\left(\frac{1}{n_1}+\frac{1}{n_2}\right)} = \sqrt{0.369(1-0.369)\left(\frac{1}{134}+\frac{1}{237}\right)} = 0.052$$

The z test statistic for testing the hypotheses is

$$z = \frac{\hat{p}_1 - \hat{p}_2}{SE_{Dp}} = \frac{0.67-0.20}{0.052} = 9.04$$

From Table A, we find P-value = $P(Z \geq 9.04)$ is approximately 0. This is strong evidence against the null hypothesis, and we would conclude that there is strong evidence that the proportion of die-hard Cubs fans who attend a game at least once a month is larger than the proportion of less loyal Cubs fans who attend a game at least once a month.

The Wilson estimates for the two probabilities are

$$\tilde{p}_1 = \frac{X_1+1}{n_1+2} = \frac{90+1}{134+2} = 0.67 \text{ and } \tilde{p}_2, = \frac{X_2+1}{n_2+2} = \frac{47+1}{237+2} = 0.20$$

The standard deviation of the difference in proportions $\tilde{D}$ is approximately

$$SE_{\tilde{D}} = \sqrt{\frac{\tilde{p}_1(1-\tilde{p}_1)}{n_1+2}+\frac{\tilde{p}_2(1-\tilde{p}_2)}{n_2+2}} = \sqrt{\frac{0.67(1-0.67)}{134+2}+\frac{0.20(1-0.20)}{237+2}} = 0.048$$

The estimated difference is

$$\tilde{D} = \tilde{p}_1 - \tilde{p}_2 = 0.67 - 0.20 = 0.47$$

A 95% confidence interval ($z^* = 1.96$) for the difference in the two proportions is

$$(\tilde{p}_1 - \tilde{p}_2) \pm z^* SE_{\tilde{D}} = 0.47 \pm (1.96)(0.048) = 0.47 \pm 0.094 = (0.376, 0.564)$$

Not only is the difference in the proportions statistically significant, the magnitude of the difference is large.

8.65 According to Exercise 8.64, 76% of the 1025 people polled said they had at least one credit card. The number of people in the survey who said they had at least one credit card is $n = 0.76(1025) = 779$. Of these, 41% said they did not pay the full balance each month. Thus, the number who do not pay the full balance each month is $X = 0.41(779) = 319$. The Wilson estimate of the proportion of the $n = 779$ people who do not pay their balance in full is

$$\tilde{p} = \frac{X+2}{n+4} = \frac{319+2}{779+4} = 0.41$$

The standard error is

$$\mathrm{SE}_{\tilde{p}} = \sqrt{\frac{\tilde{p}(1-\tilde{p})}{n+4}} = \sqrt{\frac{0.41(1-0.41)}{783}} = 0.0176$$

The z critical value for 95% confidence is $z^* = 1.96$, so the 95% margin of error m for the proportion of credit card owners who do not pay their balance in full is

$$m = z^* \mathrm{SE}_{\tilde{p}} = 1.96(0.0176) = 0.0345, \text{ or } 3.45\%$$

8.67 The sample size here is $n = 248$ and the count is $X = 152$. The Wilson estimate of the proportion of all "heavy" players that are male is

$$\tilde{p} = \frac{X+2}{n+4} = \frac{152+2}{248+4} = 0.61$$

The standard error is

$$\mathrm{SE}_{\tilde{p}} = \sqrt{\frac{\tilde{p}(1-\tilde{p})}{n+4}} = \sqrt{\frac{0.61(1-0.61)}{252}} = 0.031$$

The z critical value for 95% confidence is $z^* = 1.96$, so a 95% confidence interval is

$$\tilde{p} \pm z^* \mathrm{SE}_{\tilde{p}} = 0.61 \pm (1.96)(0.031) = 0.61 \pm 0.061 = (0.549, 0.671)$$

This does not include 0.485, the proportion of U.S. adults who are male, so we would conclude that the proportion of heavy lottery players who are male is different from the proportion of U.S. adults who are male.

8.69 Let p be the proportion of products that will fail to conform to specifications in the modified process. We wish to see if the modified process reduces the proportion of noncomformities from the previous value of 0.11, so we test the hypotheses

$$H_0: p = 0.11$$
$$H_a: p < 0.11$$

In the trial run, the proportion of noncomforming items in the $n = 300$ produced was $\hat{p} = 16/300 = 0.053$. Thus, the z statistic for testing the hypotheses is

$$z = \frac{\hat{p} - p_0}{\sqrt{\frac{p_0(1-p_0)}{n}}} = \frac{0.053 - 0.110}{\sqrt{\frac{0.11(1-0.11)}{300}}} = -3.16$$

From Table A, we find the P-value $= P(Z \le -3.16) = 0.0008$. This is strong evidence against the null hypothesis, and thus strong evidence that the proportion of noncomforming items is less than 0.11, the former value. This conclusion is justified provided the trial run can be considered a random sample of all (future) runs from the modified process and that items produced by the process are independently conforming or noncomforming.

8.71 Let p_1 denote the proportion of male college students who engage in frequent binge drinking and p_2 denote the proportion of female college students who engage in frequent binge drinking. We are interested in whether there is a difference in the proportion of men and women who engage in frequent binge drinking, so we test the hypotheses

$$H_0: p_1 = p_2$$
$$H_a: p_1 \neq p_2$$

There were $n_1 = 7180$ men in the survey and $X_1 = 1630$ of these said they engaged in frequent binge drinking. There were $n_2 = 9916$ women in the survey and $X_2 = 1684$ of these indicated that they engaged in frequent binge drinking. We find

$$\hat{p}_1 = X_1/n_1 = 1630/7180 = 0.227 \text{ and } \hat{p}_2 = X_2/n_2 = 1684/9916 = 0.170$$

The pooled estimate of the common (under H_0) value of p_1 and p_2 is

$$\hat{p} = \frac{X_1 + X_2}{n_1 + n_2} = \frac{1630 + 1684}{7180 + 9916} = 0.194$$

The standard error of the pooled estimate is

$$\text{SE}_{Dp} = \sqrt{\hat{p}(1-\hat{p})\left(\frac{1}{n_1} + \frac{1}{n_2}\right)} = \sqrt{0.194(1-0.194)\left(\frac{1}{7180} + \frac{1}{9916}\right)} = 0.006$$

The z test statistic for testing the hypotheses is

$$z = \frac{\hat{p}_1 - \hat{p}_2}{\text{SE}_{Dp}} = \frac{0.227 - 0.170}{0.006} = 9.5$$

From Table A we find $P(Z \geq 9.5)$ is approximately 0. Because we have a two-sided alternative, we need to double this (to account for the tail probabilities above and below 9.5 and –9.5, respectively), and this gives P-value = approximately 0. This is strong evidence against the null hypothesis, and we would conclude that there is strong evidence that there is a difference in the proportions of male and female college students who engage in frequent binge drinking. Although statistically significant because of the large sample sizes, the actual difference in the proportions is 0.227 – 0.170 = 0.057. This is not very large and might not be considered a practically important difference.

8.73 a) The proportion of women in the class is $p_0 = 214/851 = 0.251$.

b) We are interested in whether the probability p that a woman will be among the top 30 students is larger than their proportion in the class would suggest. Thus, we want to test the hypotheses

$$H_0: p = 0.251$$
$$H_a: p > 0.251$$

We find that the proportion of the $n = 30$ top students who were female is $\hat{p} = 15/30 = 0.5$, so the test statistic is

$$z = \frac{\hat{p} - p_0}{\sqrt{\frac{p_0(1-p_0)}{n}}} = \frac{0.5 - 0.251}{\sqrt{\frac{0.251(1-0.251)}{30}}} = 3.15$$

From Table A, we find P-value = $P(Z \geq 3.15) = 1 - P(Z < 3.15) = 1 - 0.9992 = 0.0008$. This is strong evidence against the null hypothesis, and we would conclude that there is strong evidence that the probability that a women will be in the top 30 students is larger than 0.251, the proportion of females in the class. This conclusion is reasonable, provided the first-semester grades can be considered a random sample of grades for all semesters.

8.75 a) Let p_1 denote the proportion of men with low blood pressure who die from cardiovascular disease and p_2 the proportion of men with high blood pressure who die from cardiovascular disease. There were $n_1 = 2676$ men with low blood pressure and $X_1 = 21$ of these died from cardiovascular disease. There were $n_2 = 3338$ men with high blood pressure and $X_2 = 55$ of these died of cardiovascular disease. The Wilson estimates for the two population proportions are

$$\tilde{p}_1 = \frac{X_1+1}{n_1+2} = \frac{21+1}{2676+2} = 0.0082 \text{ and } \tilde{p}_2, = \frac{X_2+1}{n_2+2} = \frac{55+1}{3338+2} = 0.0168$$

Thus, the estimated difference is

$$\tilde{D} = \tilde{p}_1 - \tilde{p}_2 = 0.0082 - 0.0168 = -0.0086$$

The standard deviation of $\tilde{D}$ is approximately

$$SE_{\tilde{D}} = \sqrt{\frac{\tilde{p}_1(1-\tilde{p}_1)}{n_1+2} + \frac{\tilde{p}_2(1-\tilde{p}_2)}{n_2+2}} = \sqrt{\frac{0.0082(1-0.0082)}{2676+2} + \frac{0.0168(1-0.0168)}{3338+2}} = 0.0028$$

This is the standard error we need in order to compute a confidence interval for the difference in proportions. A 95% confidence interval ($z^* = 1.96$) for the difference in the two proportions is

$$(\tilde{p}_1 - \tilde{p}_2) \pm z^* SE_{\tilde{D}} = -0.0086 \pm (1.96)(0.0028) = -0.0086 \pm 0.0055 = (-0.0141, -0.0031)$$

This is consistent with the results in Example 8.8.

b) We are interested in whether death rates from cardiovascular disease are higher among men with high blood pressure, so we test the hypotheses

$$H_0: p_1 = p_2$$
$$H_a: p_1 < p_2$$

We find

$$\hat{p}_1 = X_1/n_1 = 21/2676 = 0.0078 \text{ and } \hat{p}_2 = X_2/n_2 = 55/3338 = 0.0165$$

The pooled estimate of the common (under H_0) value of p_1 and p_2 is

$$\hat{p} = \frac{X_1+X_2}{n_1+n_2} = \frac{21+55}{2676+3338} = 0.0126$$

The standard error of the pooled estimate is

$$SE_{Dp} = \sqrt{\hat{p}(1-\hat{p})\left(\frac{1}{n_1}+\frac{1}{n_2}\right)} = \sqrt{0.0126(1-0.0126)\left(\frac{1}{2676}+\frac{1}{3338}\right)} = 0.0029$$

The z test statistic for testing the hypotheses is

$$z = \frac{\hat{p}_1 - \hat{p}_2}{SE_{Dp}} = \frac{0.0078-0.0165}{0.0029} = -3$$

From Table A, we find P-value $= P(Z \le -3) = 0.0013$. This is strong evidence against the null hypothesis, and we would conclude that there is strong evidence that the proportion of men with low blood pressure who die from cardiovascular disease is less than the proportion of men with high blood pressure who die from cardiovascular disease.

8.77 The margin of error m for a 99% confidence interval for the difference in two proportions that are both estimated to be $\tilde{p}_1 = \tilde{p}_2 = 0.5$ and with sample sizes $n_1 = n_2 = n$ is (because $z^* = 2.576$ for 99% confidence)

$$m = z^* \text{SE}_{\tilde{D}} = (2.576)\sqrt{\frac{\tilde{p}_1(1-\tilde{p}_1)}{n_1+2}+\frac{\tilde{p}_2(1-\tilde{p}_2)}{n_2+2}} = (2.576)\sqrt{\frac{0.5(1-0.5)}{n+2}+\frac{0.5(1-0.5)}{n+2}}$$

$$= (2.576)\sqrt{\frac{0.5}{n+2}} = \frac{1.822}{\sqrt{n+2}}.$$

Applying these to the cases listed, we obtain the following table of results.

n:	10	30	50	100	200	500
m:	0.526	0.322	0.253	0.180	0.128	0.081

We see that the margin of error decreases as sample size increases, as we would expect.

8.79 The margin of error m for a 90% confidence interval for the difference in two proportions that are both estimated to be $\tilde{p}_1 = \tilde{p}_2 = 0.5$ and with sample sizes $n_1 = 20$ is (because $z^* = 1.645$ for 90% confidence)

$$m = z^* \text{SE}_{\tilde{D}} = (1.645)\sqrt{\frac{\tilde{p}_1(1-\tilde{p}_1)}{n_1+2}+\frac{\tilde{p}_2(1-\tilde{p}_2)}{n_2+2}} = (1.645)\sqrt{\frac{0.5(1-0.5)}{20+2}+\frac{0.5(1-0.5)}{n_2+2}}$$

$$= (1.645)\sqrt{0.0114+\frac{0.25}{n_2+2}} > (1.645)\sqrt{0.0114} = 0.176.$$

It is impossible to guarantee a margin of error less than or equal to 0.1 in this case. No matter how large n_2 is, the margin of error is larger than 0.176.

8.81 a) The proportion of eligible voters who were Mexican-Americans is p_0 = 143,611/181,535 = 0.791.

b) Because we are interested in whether Mexican-Americans are being discriminated against (less likely to be selected), we test the hypotheses

$$H_0: p = 0.791$$
$$H_a: p < 0.791$$

We find that the proportion of the $n = 870$ people selected for jury duty that were Mexican-American is $\hat{p}$ = 339/870 = 0.390, so the test statistic is

$$z = \frac{\hat{p}-p_0}{\sqrt{\frac{p_0(1-p_0)}{n}}} = \frac{0.390-0.791}{\sqrt{\frac{0.791(1-0.791)}{870}}} = -29.09.$$

From Table A, we find P-value = $P(Z \le -29.09)$ is approximately 0. This is strong evidence against the null hypothesis, and we would conclude that there is strong evidence that the probability that a randomly selected juror is Mexican-American is less than the proportion of eligible voters who were Mexican-American.

c) There were $n_1 = 870$ people selected for jury duty and $X_1 = 339$ of these were Mexican-American. There were n_2 = 181,535 – 870 = 180,665 people who were not selected for jury duty and X_2 = 143,611 – 339 = 143,272 of these were Mexican-American. We find

$$\hat{p}_1 = X_1/n_1 = 339/870 = 0.390 \text{ and } \hat{p}_2 = X_2/n_2 = 143{,}272/180{,}665 = 0.793$$

The pooled estimate of the common (under H_0) value of p_1 and p_2 is

$$\hat{p} = \frac{X_1 + X_2}{n_1 + n_2} = \frac{339 + 143272}{870 + 180665} = 0.791$$

The standard error of the pooled estimate is

$$SE_{Dp} = \sqrt{\hat{p}(1-\hat{p})\left(\frac{1}{n_1} + \frac{1}{n_2}\right)} = \sqrt{0.791(1-0.791)\left(\frac{1}{870} + \frac{1}{180665}\right)} = 0.0138$$

The z test statistic for testing the hypotheses is

$$z = \frac{\hat{p}_1 - \hat{p}_2}{SE_{Dp}} = \frac{0.390 - 0.793}{0.0138} = -29.20$$

From Table A, we find P-value = $P(Z \leq -29.20)$ is approximately 0. This is strong evidence against the null hypothesis, and we would conclude that there is strong evidence that the probability that a randomly selected juror is Mexican-American is less than the probability that a randomly selected nonjuror is Mexican-American. The z-statistic and P-value are almost the same as in part (b) and this conclusion is the same as that in part (b).

CHAPTER 9

INFERENCE FOR TWO-WAY TABLES

OVERVIEW

There are two common models that generate data that can be summarized in a two-way table. In the first model, independent SRSs are drawn from c populations, and each observation is classified according to a categorical variable that has r possible values. In the table, the c populations form the columns, and the categorical classification variable forms the rows. The null hypothesis is that the distributions of the row categorical variable are the same for each of c populations. In the second model, an SRS is drawn from a single population, and the observations are cross-classified according to two categorical variables having r and c possible values, respectively. In this model, the null hypothesis is that the row and column variables are independent. When one of the variables is an explanatory variable and the other is a response, the explanatory variable is used to form the columns of the table and the response forms the rows.

A test of the null hypothesis is carried out using the X^2 statistic in both models. Although the data are generated differently, the question in both cases is quite similar: Are the distributions of the row categorical variables the same? The cell counts are compared to the **expected cell counts** under the null hypothesis. The expected cell counts are computed using the formula

$$\text{expected count} = \frac{\text{row total} \times \text{column total}}{n}$$

where n is the total number of observations.

The **chi-square statistic** is used to test the null hypothesis by comparing the observed counts with the expected counts

$$X^2 = \sum \frac{(\text{observed - expected})^2}{\text{expected}}.$$

When the null hypothesis is true, the distribution of X^2 is approximately χ^2 with $(r - 1)(c - 1)$ degrees of freedom. The P-value is the probability of getting differences between observed and expected counts as large as we did and is computed as $P(\chi^2 > X^2)$. The use of the χ^2 distribution is an approximation that works well when the average expected count exceeds 5 and all of the individual expected counts are greater than 1. In the special case of the 2 × 2 table, all expected counts should exceed 5 before applying the approximation.

In addition to computing the X^2 statistic, tables or bar charts should be examined to describe the relationship between the two variables.

APPLY YOUR KNOWLEDGE

9.1 a) If all we record is whether or not the music is French and whether or not the wine purchased is French, then the data can be summarized in the 2 × 2 table below.

		Type of Music	
		French	Other
Type of Wine Purchased	French		
	Other		

If you are more specific about the type of wine or music (see Case 9.2), then there would be additional rows and columns.

b) The explanatory variable is the type of music, as we think this influences the type of wine purchased. The type of wine purchased is the response. When there is an explanatory and a response variable, the explanatory variable is usually associated with the columns of the table.

9.3 The percent of successful firms that offer exclusive territories is 108 / 123 = 87.8% and the percent of unsuccessful firms that offer exclusive territories is 34 / 47 = 72.3%. These row percentages can be read directly off all three outputs.

9.5 There are 28 firms that lack exclusive territories, and the percent of all firms that are unsuccessful is 47 / 170 = 27.6%. This last percentage can be read directly from the Minitab or SPSS outputs, but not the SAS output. If there is no association between success and exclusive territories, then we expect both firms with exclusive and lacking exclusive territories should have 27.6% unsuccessful firms. Because there are 28 firms that lack exclusive territories, we expect 27.6% × 28 $= \frac{47 \times 28}{170} = 7.74$ of these to be unsuccessful, which is the expected count in the second row and second column on the outputs.

9.7 The test statistic for a 5 × 3 table has $(r - 1)(c - 1) = (5 - 1)(3 - 1) = 8$ degrees of freedom associated with it.

9.9 a) The output below was produced by Minitab. The expected counts are printed below the observed counts.

	Men	Women	Total
Yes	63 48.70	27 41.30	90
No	233 247.30	224 209.70	457
Total	296	251	547

```
ChiSq =  4.198 +  4.950 +
         0.827 +  0.975 = 10.949
df = 1, p = 0.001
```

The reported value of $X^2 = 10.949$. The value of $z^2 = (3.30)^2 = 10.89$, so the two agree up to roundoff error in the z statistic.

b) The 0.001 critical value for a chi-square with df = 1 is 10.83. The 0.0005 critical value for the standard normal is 3.291, and $(3.291)^2 = 10.83$.

c) No relation between the variables says that whether or not a person is a label user has no relationship to gender, or equivalently the chance of label use is the same for men and women. This is the same as H_0: $p_1 = p_2$.

CHAPTER 9 EXERCISES

9.11 The column percents are included below the counts in the table below.

	Under 40	Over 40	Total
Partially/fully meets expectations	82 16.5%	230 30.3%	312 24.9%
Usually exceeds expectations	353 71.2%	496 65.4%	849 67.7%
Continually exceeds expectations	61 12.3%	32 4.2%	93 7.4%
Total	496	758	1254

```
ChiSq = 13.893 +   9.091 +
         0.880 +   0.576 +
        15.941 +  10.431 = 50.812
df = 2, p = 0.000
```

As shown by the chi-square analysis, the differences in percentages in the three performance categories are different for the younger and the older employees. The older employees (over 40) are almost twice as likely to fall into the lowest performance category but only 1/3 as likely to fall in the highest category.

9.13 Row percentages are printed below observed counts. There appears to be little difference in the response rates for the different industries.

	Response	No Response	Total
Metal Products	17 9.2%	168 90.8%	185 100%
Machinery	35 11.6%	266 88.4%	301 100%
Electrical equipment	75 13.6%	477 86.4%	552 100%
Transportation equipment	15 15.0%	85 85.0%	100 100%
Precision instrument	12 13.3%	78 86.7%	90 100%
Total	154	1074	1228

The chi-square analysis below was produced by Minitab. It shows no evidence the response rates vary by industry with a *P*-value of 0.513.

```
ChiSq =  1.657 +  0.238 +
         0.200 +  0.029 +
         0.482 +  0.069 +
         0.482 +  0.069 +
         0.045 +  0.006 = 3.277
df = 4, p = 0.513
```

9.15 Because $r = 5$ and $c = 2$, the degrees of freedom are $(5 - 1)(2 - 1) = 4$. According to the row corresponding to df = 4 in Table F, the value 5.39 corresponds to a tail probability of 0.25. So the *P*-value exceeds 0.25. Statistical software gives $P(\chi^2 > 3.955) = 0.4121$ as the *P*-value. There is no evidence of a difference in the income distribution of the customers at the two stores.

9.17 a) The counts, expected counts and the column proportions are listed in the table below. The phone call has a 68.2% response rate, a letter has a 43.7% response rate and no intervention has a 20.6% response rate.

	Intervention			
Response	Letter	Phone Call	None	Total
Yes	171 144.38 43.7%	146 79.02 68.2%	118 211.59 20.6%	435 36.9%
No	220 246.62 56.3%	68 134.98 31.8%	455 361.41 79.4%	743 63.1%
Total	391	214	573	1178

b) The hypotheses are H_0: There is no relationship between intervention and response rate, and H_a: There is a relationship.

c) The output below was produced by Minitab, The differences in response rates are significant. The key difference is that intervention seems to increase the response rate with a phone call being more effective than a letter.

```
ChiSq =    4.906 + 56.765 + 41.398 +
            2.872 + 33.234 + 24.237 = 163.413
df = 2, p = 0.000
```

9.19 The two surveys give information on nonresponse rate under various conditions. You can decide whether to use prenotification and which method to obtain an estimate of the response rate that you might expect in your survey. You should then determine the sample size necessary so that your statistical procedures have the required power and your estimates have the necessary precision. Use the information on nonresponse rate, and take a larger sample size than calculated to make sure that you have enough observations with nonresponse accounted for.

9.21 The two-way table is given below. From Exercise 9.20, X^2 = 38.411.

	Next year		
This year	Winner	Loser	Total
Winner	85	35	120
Loser	37	83	120
Total	122	118	240

The overall proportion of winners this year is $\hat{p} = \dfrac{X_1 + X_2}{n_1 + n_2} = \dfrac{85+35}{122+118} = 0.5000$. To evaluate the z statistic, first compute

$$SE_{Dp} = \sqrt{\hat{p}(1-\hat{p})\left(\frac{1}{n_1}+\frac{1}{n_2}\right)} = \sqrt{0.50(0.50)\left(\frac{1}{122}+\frac{1}{118}\right)} = 0.06456$$

and from this

$$z = \frac{\hat{p}_1 - \hat{p}_2}{SE_{Dp}} = \frac{\frac{85}{122} - \frac{35}{118}}{0.06456} = 6.1977$$

From the value of z, it is easily verified that $z^2 = X^2$.

9.23 The two-way table is given below, with the column percentages below the counts.

	Next year		
This year	Winner	Loser	Total
Winner	96	148	244
	39.8%	59.9%	50%
Loser	145	99	244
	60.2%	40.1%	50%
Total	241	247	488

```
ChiSq =   4.981 +  4.860 +
           4.981 +  4.860 = 19.683
df = 1, p = 0.000
```

As in Exercise 9.20, there is strong evidence of a relationship between winning or losing this year and winning or losing next year. The pattern is that 39.8% of the winners in the second year were winners in the first year, while 59.9% of the losers the second year had been winners the first year. However, in Exercise 9.20, 69.7% of the winners in the second year were winners in the first year, while 29.7% of the losers the second year had been winners the first year. In Exercise 9.20, good performance continued, while for the data in this exercise the opposite is true.

9.25 An association that holds for all of several groups can reverse direction when the data are combined to form a single group. This reversal is called Simpson's paradox.

a) For all flights combined, $X^2 = 13.572$ and $P < 0.001$, so the difference in percent on time for the two airlines is highly significant with America West being the winner.

b) For the five departure cities, we have Los Angeles ($X^2 = 3.241$ and $P = 0.072$), Phoenix ($X^2 = 2.346$ and $P = 0.126$), San Diego ($X^2 = 4.845$ and $P = 0.028$), San Francisco ($X^2 = 21.223$ and $P < 0.001$) and Seattle ($X^2 = 14.903$ and $P < 0.001$).

c) In all cases, Alaska airlines is the winner with the result being significant at the 5% level in three of the five cities, significant at the 10% level in one city and not significant at the 10% level in the last city. Not only are the results in opposite directions, but the effects that illustrate Simpson's paradox in this example do not appear to be due to chance alone.

9.27 The two-way table and chi-square results are given below. The row percentages are reported below the counts in the table.

	Low	Medium	High	Total
Agriculture	5	27	35	67
	7.5%	40.3%	52.2%	
Child development and family studies	1	32	54	87
	1.1%	36.8%	62.1%	
Engineering	12	129	94	235
	5.1%	54.9%	40.0%	
Liberal Arts and Education	7	77	129	213
	3.3%	36.2%	60.5%	
Management	3	44	28	75
	4.0%	58.7%	37.3%	
Science	7	29	24	60
	11.7%	48.3%	40.0%	
Technology	2	62	64	128
	1.6%	48.4%	50.0%	
Total	37	400	428	865

```
ChiSq =  1.589 +  0.512 +  0.103 +
         1.990 +  1.684 +  2.787 +
         0.377 +  3.803 +  4.268 +
         0.489 +  4.692 +  5.288 +
         0.013 +  2.503 +  2.236 +
         7.659 +  0.057 +  1.090 +
         2.206 +  0.133 +  0.007 = 43.487
df = 12, p = 0.000
```

The value of $X^2 = 43.487$ is highly significant, indicating a relationship between the PEOPLE score and field of study. Among the major differences, we note that science has an unusually large percentage of low-scoring students relative to other fields, while liberal arts and education has an unusually large percentage of high-scoring students relative to the other fields. These two table entries make the two largest contributions to the value of X^2.

9.29 Exercise 9.28 shows a strong increasing trend with a highly significant $X^2 = 39.677$ (P-value < 0.0005). The percentage of women students in pharmacy increased from around 20% to 60% during the period 1970 to 1986. The graph of the years 1987 to 2000 on the next page also shows what appears to be a fairly strong increasing trend, yet $X^2 = 9.969$ with 13 degrees of freedom (P-value = 0.696). Although there appears to be a strong increasing trend, the changes in the percents are quite small (from 60% to about 66%) and as we know from previous chapters, statistical significance is difficult to achieve when differences are small. In addition, the chi-square procedure is looking for any difference in percents between years. If we used a procedure that just looked for increasing or decreasing trends, it would be more sensitive to the pattern in these data. To summarize both data sets, women represented a very small minority in 1970 which increased quite rapidly until the mid '80s when it reached about 60%. The percentage continued to increase in the late 1980's but much more slowly, with some leveling off in the early '90s. The percentage started increasing again in the mid '90s but very slowly.

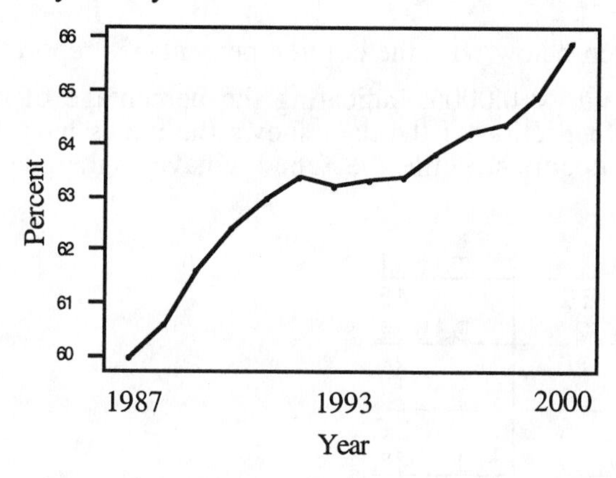

9.31 a) You should use column percents because we believe the "source" of the cat is the explanatory variable. The two-way table is given below with the column percentages reported below the counts in the table.

	Source			
	Private	Pet Store	Other	Total
Cases	124 36.2%	16 40.0%	76 27.2%	216 32.6%
Controls	219 63.8%	24 60.0%	203 72.8%	446 67.4%
Total	343	40	279	662

b) The chi-square analysis produced by Minitab is given below. At the 5% level of significance, we would conclude there is a relationship between the source of the cat and whether or not the cat is brought to an animal shelter. The percentage for "other" sources is lower for animals brought to a shelter.

```
ChiSq =   1.305 +   0.666 +   2.483 +
          0.632 +   0.323 +   1.202 = 6.611
df = 2, p = 0.037
```

9.33 The table below compares the sources of dogs and cats. The column percentages are reported below the counts. Dog or cat is the explanatory variable, so they are used as the column categories. $X^2 = 24.9$ and the P-value < 0.0005 indicating the source of dogs and cats are different. A much higher percentage of dogs than cats come from private sources, while a much higher percentage of cats than dogs come from other sources such as homes, the streets or shelters.

	Cat	Dog	Total
Private	219 49.1%	518 71.1%	737 62.8%
Pet Store	24 5.4%	68 9.3%	92 7.8%
Other	203 45.5%	142 19.5%	345 29.4%
Total	446	728	1174

```
ChiSq = 13.283 +  8.138 +
         3.431 +  2.102 +
        39.482 + 24.188 = 90.624
df = 2, p = 0.000
```

9.35 Exercise 9.12 is a test of independence based on a single sample; Exercise 9.13 is a comparison of several populations based on separate samples from each; Exercise 9.14 is a comparison of several populations based on separate samples from each; Exercise 9.16 is a test of independence based on a single sample.

9.37 The two-way table is given below with the column percentages reported below the counts in the table. $X^2 = 24.9$ and the *P*-value < 0.0005, indicating the percentage of males and females in the four categories are different. Inspection of the data shows the males have higher percentages in the two high social comparison categories, while the females have higher percentages in the two low social comparison categories.

	Female	Male	Total
HSC-HM	14 20.9%	31 46.2%	45 33.6%
HSC-LM	7 10.4%	18 26.9%	25 18.7%
LSC-HM	21 31.3%	5 7.5%	26 19.4%
LSC-LM	25 37.3%	13 19.4%	38 28.3%
Total	67	67	134

b) The direct comparison of males and females in the high vs. low social comparison categories in the 2 × 2 table obtained by summing over the mastery variable has a $X^2 = 23.45$ and a *P*-value < 0.0005. This is in agreement with what we found from the full 4 × 2 table.

c) The direct comparison of males and females in the high vs. low mastery categories in the 2 × 2 table obtained by summing over the social comparison variable has a $X^2 = 0.03$ and a *P*-value $= 0.863$, and shows no gender differences.

CHAPTER 10

INFERENCE FOR REGRESSION

SECTION 10.1

OVERVIEW

The statistical model for **simple linear regression** is

$$y_i = \beta_0 + \beta_1 x_i + \varepsilon_i$$

where $i = 1, 2, \ldots, n$. The deviations ε_i are assumed to be independent and normally distributed with mean 0 and standard deviation σ. The **parameters** of the model are the intercept β_0, the slope β_1, and σ. β_0 and β_1 are estimated by the slope b_0 and intercept b_1 of the **least-squares regression line.** Given n observations on an explanatory variable x and a response variable y,

$$(x_1, y_1), (x_2, y_2), \ldots, (x_n, y_n)$$

recall that the formula for the slope and intercept of the least-squares regression line are

$$b_1 = r\frac{s_y}{s_x}$$

and

$$b_0 = \bar{y} - b_1 \bar{x}$$

where r is the correlation between y and x, $\bar{y}$ is the mean of the y observations, s_y is the standard deviation of the y observations, $\bar{x}$ is the mean of the x observations, and s_x is the standard deviation of the x observations. The standard deviation σ is estimated by

$$s = \sqrt{\frac{\sum e_i^2}{n-2}}$$

where the e_i are the **residuals**

$$e_i = y_i - \hat{y}_i$$

and

$$\hat{y}_i = b_0 + b_1 x_i$$

b_0, b_1, and s are usually calculated using a calculator or statistical software.

A **level *C* confidence interval** for β_1 is

$$b_1 \pm t^* \mathrm{SE}_{b_1}$$

where t^* is the upper $(1 - C)/2$ critical value for the $t(n - 2)$ distribution and

$$SE_{b_1} = \frac{s}{\sqrt{\sum(x_i - \bar{x})^2}}$$

is the standard error of the slope b_1. SE_{b_1} is usually computed using a calculator or statistical software. (The formula above is actually given in Section 10.3 of the text, but reproduced here for continuity of exposition. If your class is not covering Section 10.3, you need not worry about the formula.) The **test of the hypothesis** $H_0{:}\beta_1 = 0$ is based on the t statistic

$$t = \frac{b_1}{SE_{b_1}}$$

with P-values computed from the $t(n - 2)$ distribution. There are similar formulas for confidence intervals and tests for β_0, but using the standard error of the intercept b_0

$$SE_{b_0} = s\sqrt{\frac{1}{n} + \frac{\bar{x}^2}{\sum(x_i - \bar{x})^2}}$$

in place of SE_{b_1}. As with SE_{b_1}, SE_{b_0} is usually computed using a calculator or statistical software. (The formula above is actually given in Section 10.3 of the text but reproduced here for continuity of exposition. If your class is not covering Section 10.3, you need not worry about the formula.) Note that inferences for the intercept are meaningful only in special cases.

When the variables y and x are jointly normal, the sample correlation is an estimate of the population correlation ρ. The test of H_0: $\rho = 0$ is based on the ***t* statistic**

$$t = \frac{r\sqrt{n-2}}{\sqrt{1-r^2}}$$

which has a $t(n - 2)$ distribution under H_0. This test statistic is numerically identical to the t statistic used to test H_0: $\beta_1 = 0$.

APPLY YOUR KNOWLEDGE

10.1 a) β_0 in this line is 4.7. It is the intercept, the value of the response variable when the explanatory variable is 0. It tells us that when U.S. returns are 0%, overseas returns are 4.7%.

b) β_1 in this line is 0.66. It is the slope, the amount the response variable will increase when the explanatory variable increases by 1 unit. It tells us that a 1% increase in U.S. returns is associated with a 0.66% increase in overseas returns. Because the slope is positive, it also tells us that U.S. and overseas returns are positively correlated.

c) The regression model is

$$\text{Mean Overseas Return} = 4.7 + 0.66 \times \text{U.S. Return} + \varepsilon$$

where the ε term in this model represents the deviations from the line and accounts for the variability in overseas returns when U.S. return remains the same.

10.3 a) A scatterplot with year on the horizontal axis and yield on the vertical axis will show the increase in yield over time. The plot is given below.

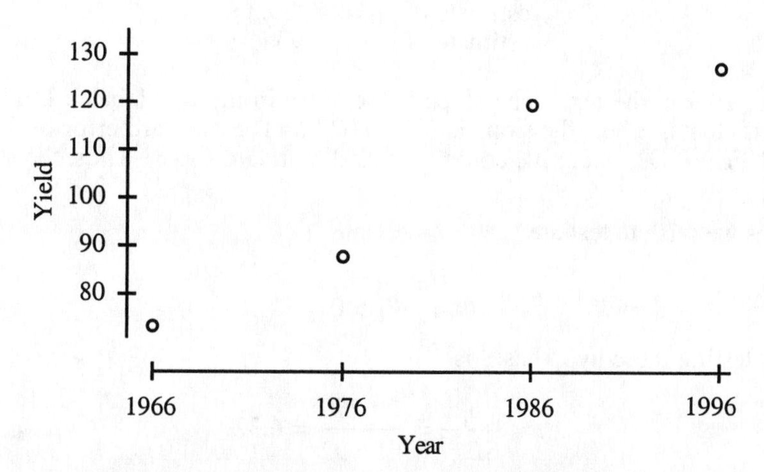

The plot does suggest a roughly linear relationship between yield and time.

b) Using statistical software, we find the equation of the least-squares line for predicting yield from year is

$$\text{Yield} = -3729.35 + 1.934 \times \text{Year}.$$

The scatterplot in part (a) with this line added is given below.

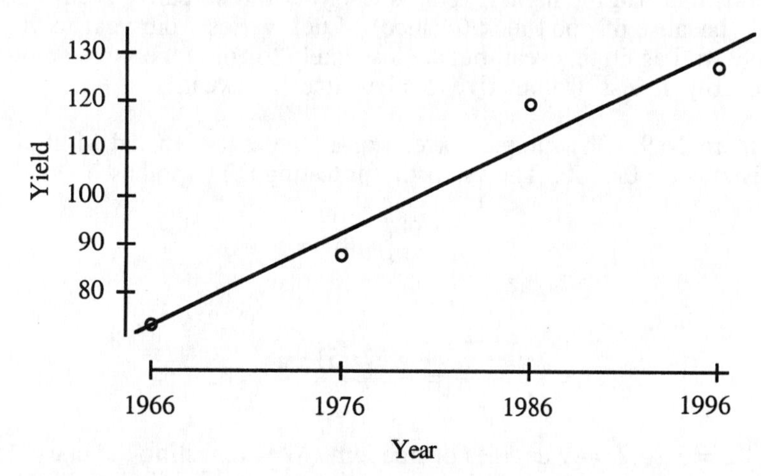

c) The residuals are observed values minus the value predicted by the least-squares regression line. Thus,

residual for 1966 = 73.1 – (–3729.35 + 1.934×1966) = 0.206
residual for 1976 = 88.0 – (–3729.35 + 1.934×1976) = –4.234
residual for 1986 = 119.4 – (–3729.35 + 1.934×1986) = 7.826
residual for 1996 = 127.1 – (–3729.35 + 1.934×1996) = –3.814

and the standard error s is

$$s = \sqrt{\frac{1}{n-2}\sum(y_i - \hat{y}_i)^2} = \sqrt{\frac{1}{n-2}\sum(e_i)^2} = \sqrt{\frac{1}{4-2}\left[(0.206)^2 + (-4.234)^2 + (7.826)^2 + (-3.814)^2\right]}$$
$$= 6.847$$

d) The regression model is

$$\text{Yield} = \beta_0 + \beta_1 \times \text{Year} + \varepsilon$$

and our estimates of the unknown parameters β_0 and β_1 in this model are

estimate of $\beta_0 = -3729.35$
estimate of $\beta_1 = 1.934$

10.5 Refer to Figure 10.9 in the text. The slope is the entry in the row labeled INFLATION under the column labeled Coefficients. Thus, the slope is $b_1 = 0.6270$. The standard error of the slope is the entry in the row labeled INFLATION under the column labeled Standard Error. Thus, the standard error of the slope is $SE_{b_1} = 0.0992$.

The hypotheses we wish to test are

$$H_0: \beta_1 = 0$$
$$H_a: \beta_1 > 0$$

and the t statistic for testing these hypotheses is

$$t = \frac{b_1}{SE_{b_1}} = \frac{0.6270}{0.0992} = 6.32$$

The t statistic has $n - 2 = 51 - 2 = 49$ degrees of freedom. We are testing the one-sided alternative H_a: $\beta_1 > 0$, so the P-value is the upper tail probability. There is no entry for 49 degrees of freedom, so we use the row labeled 40 df in Table D (the row closest to but less than 49). We see that a t of 6.32 is larger than the largest entry of 3.551 in the row. Thus, the P-value is smaller than the tail probability (column heading) of 0.0005 that corresponds to a t of 3.551. We conclude the P-value < 0.0005.

10.7 The mutual funds that had the highest returns last year did so partly because of they were a good investment and partly because of good luck (chance). Luck varies from year to year. These funds are likely to do reasonably well again this year, but are less likely to enjoy the same amount of good luck. As a result, they will probably do less well relative to other funds this year.

10.9 a) Refer to Figure 10.9. The sample correlation is the entry labeled Multiple R in the summary output. The value given is $r = 0.6700$. The t statistic for testing the hypotheses

$$H_0: \rho = 0$$
$$H_a: \rho > 0$$

is

$$t = \frac{r\sqrt{n-2}}{\sqrt{1-r^2}} = \frac{0.67\sqrt{51-2}}{\sqrt{1-0.67^2}} = 6.32.$$

The t statistic has $n - 2 = 51 - 2 = 49$ degrees of freedom. We are testing the one-sided alternative H_a: $\rho > 0$, so the P-value is the upper tail probability. There is no entry for 49 degrees of freedom, so we use the row labeled 40 df in Table D (the row closest to but less than 49). We see that a t of 6.32 is larger than the largest entry of 3.551 in the row. Thus, the P-value is smaller than the tail probability (column heading) of 0.0005 that corresponds to a t of 3.551. We conclude the P-value < 0.0005, and there is strong evidence that the population correlation ρ is > 0.

b) The t for the slope is given in the ANOVA table in the row labeled INFLATION (corresponding to the slope) and this entry is 6.3177. This is the same as the value we obtained in part (a) up to roundoff error.

SECTION 10.1 EXERCISES

10.11 a) Age can only take on integer (whole number) values. Each of the vertical stacks corresponds to one of these integer values. Fractional values (values between these integers) are not possible (because of how age was recorded) and hence the gaps between the vertical stacks.

b) Older men might earn more than younger men because of seniority (they have been with a company longer and have moved up to the higher end of the pay scale for their job) or because of their experience they have moved into managerial positions that pay higher salaries. Younger men might earn more than older men because they are better trained for certain types of high paying jobs (technology), have more education (more students go on for advanced degrees now than many years ago), or are more attractive to hire (companies tend to prefer younger workers than older workers).

In Figure 10.11, we see that the correlation (multiple R) is positive, so there is a positive association between age and income in the sample. In other words, higher incomes are associated with older workers. The relationship is very weak. The value of R^2 given in Figure 10.11 is 0.03525, so that age only explains about 3% of the variation in income. Also, it is nearly impossible to detect any obvious trend in the scatterplot.

c) The equation of the least-squares line for predicting income from age, according to the output displayed in Figure 10.11, is

$$\text{Income} = 24874.3745 + 892.1135 \times \text{Age}$$

The slope is 892.1135 and this tells us that for each additional year in age, the predicted income increases by $892.1135.

10.13 a) We see the skewness in the plot by looking at the vertical stacks of points. There are many points in the bottom (lower income) portion of each stack. At the upper (higher income) portion of each vertical stack the points are more dispersed, with only a very few at the highest incomes. Thus, each stack would correspond to a histogram with large bars at the lower incomes, with progressively smaller bars as income increasing, ending in a few, scattered very small bars at very high incomes. Such a histogram is right-skewed.

b) Regression inference is robust against a moderate lack of normality. As the text indicates (see the discussion preceding the box that gives inference for a regression slope), a special form of the central limit theorem tells us that the distributions of b_0 and b_1 will be approximately normal when we have a large sample. Here the sample size is 5712, which would be considered very large.

10.15 a) The intercept tells us what the T-bill percent will be when inflation is at 0%. We would expect β_0 to be greater than 0 simply because no one will purchase T-bills if they do not offer a rate greater than 0. One can achieve a return of 0 simply by putting one's money in a safe.

b) In Figure 10.9 we see that $b_0 = 2.6660$ and $SE_{b_0} = 0.5039$.

c) To determine if there is good evidence that $\beta_0 > 0$, we test the hypotheses

$$H_0\colon \beta_0 = 0$$
$$H_a\colon \beta_0 > 0$$

The t statistic for testing these hypotheses is

$$t = \frac{b_0}{SE_{b_0}} = \frac{2.6660}{0.5039} = 5.29$$

The t statistic has $n - 2 = 51 - 2 = 49$ degrees of freedom. We are testing the one-sided alternative $H_a\colon \beta_0 > 0$, so the P-value is the upper tail probability. There is no entry for 49 degrees of freedom, so we use the row labeled 40 df in Table D (the row closest to but less than 49). We see that a t of 5.29 is larger than the largest entry of 3.551 in the row. Thus, the P-value is smaller than the tail probability (column heading) of 0.0005 that corresponds to a t of 3.551. We conclude the P-value < 0.0005. This is strong evidence that $\beta_0 > 0$, provided we can assume that the 51 years of data can be regarded as a simple random sample from the population of all years. This may not be realistic because yearly data like these are often correlated.

d) The formula for a 95% confidence interval for β_0 is $b_0 \pm t^* SE_{b_0}$. The t distribution here has 49 degrees of freedom, so we use the table entry $t^* = 2.021$ for 40 degrees of freedom. The 95% confidence interval is therefore

$$b_0 \pm t^* SE_{b_0} = 2.6660 \pm (2.021)(0.5039) = 2.6660 \pm 1.0184 = (1.6476, 3.6844),$$

which agrees approximately with the output in Figure 10.9.

10.17 a) A plot of selling price versus number of square feet is given below.

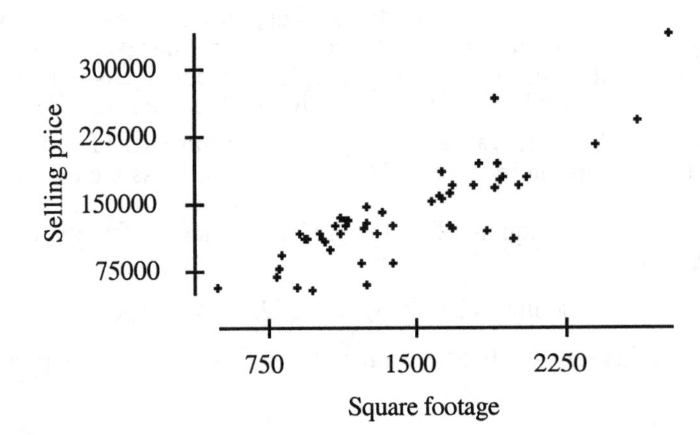

The pattern is roughly linear and the association is positive. From statistical software, $r^2 = 0.696$, so that the least-squares regression of selling price on square footage explains 69.6% of the variation in selling price. This indicates that square footage is helpful for predicting selling price.

b) Using statistical software, we find get the following.

R squared = 69.6%
s = 30349 with 50 – 2 = 48 degrees of freedom

Source	Sum of Squares	df	Mean Square	F-ratio
Regression	1.01422e+11	1	1.01422e+11	110
Residual	44211631690	48	921075660	

Variable	Coefficient	s.e. of Coeff	t-ratio	prob
Constant	4786.46	13455	0.356	0.7236
Square footage	92.8209	8.846	10.5	≤ 0.0001

From this, we find that the least-squares regression line is

$$\text{Selling price} = 4786.46 + 92.8209 \times \text{Square footage}$$

The test for significance of the slope β_1, namely the test of the hypotheses

$$H_0: \beta_1 = 0$$
$$H_a: \beta_1 \neq 0$$

has a P-value of less than 0.0001 and this tells us that there is a statistically significant straight line relationship between selling price and square footage.

10.19 The sample correlation is computed from statistical software and is found to be $r = 0.835$. The t statistic for testing the hypotheses

$$H_0: \rho = 0$$
$$H_a: \rho > 0$$

is thus

$$t = \frac{r\sqrt{n-2}}{\sqrt{1-r^2}} = \frac{0.835\sqrt{50-2}}{\sqrt{1-0.835^2}} = 10.51.$$

The t statistic has $n - 2 = 50 - 2 = 48$ degrees of freedom. We are testing the one-sided alternative H_a: $\rho > 0$, so the P-value is the upper tail probability. There is no entry for 48 degrees of freedom, so we use the row labeled 40 df in Table D (the row closest to but less than 49). We see that a t of 10.51 is larger than the largest entry of 3.551 in the row. Thus, the P-value is smaller than the tail probability (column heading) of 0.0005 that corresponds to a t of 3.551. We conclude the P-value < 0.0005, and there is strong evidence that the population correlation ρ between selling price and square footage is > 0.

10.21 a) The analysis with the one house removed is given below.

R squared = 71.8%
s = 27773 with 49 – 2 = 47 degrees of freedom

Source	Sum of Squares	df	Mean Square	F-ratio
Regression	92191830468	1	92191830468	120
Residual	36253872590	47	771358991	

Variable	Coefficient	s.e. of Coeff	t-ratio	prob
Constant	8039.26	12354	0.651	0.5184
Square footage	89.3029	8.169	10.9	≤ 0.0001

If this is compared to the analysis with all the houses (see the solution to Exercise 10.17), we see that r^2 increases from 69.6% to 71.8%, the intercept and slope for the least-squares regression line change from 4786.46 and 92.8209 to 8039.26 and 89.3029, respectively, and the *t* statistic for the slope changes from 10.5 to 10.9. The change in the intercept is the most dramatic, but the intercept is not particularly meaningful (no house has 0 square feet). r^2, the slope, and the *t* statistic for the slope have increased a little and the conclusion we reached in Exercise 10.17 is not changed (in fact the conclusion is stronger after removing the one house).

b) The analysis with the four houses removed is given below.

R squared = 59.1%
s = 24,722 with 46 – 2 = 44 degrees of freedom

Source	Sum of Squares	df	Mean Square	F-ratio
Regression	38863308076	1	38863308076	63.6
Residual	26891222162	44	611164140	

Variable	Coefficient	s.e. of Coeff	t-ratio	prob
Constant	27982.8	12996	2.15	0.0368
Square footage	73.3070	9.193	7.97	≤ 0.0001

If this is compared to the analysis with all the houses (see the solution to Exercise 10.17), we see that r^2 decreases from 69.6% to 59.1%, the intercept and slope for the least-squares regression line change from 4786.46 and 92.8209 to 27982 and 73.3070, respectively, and the *t* statistic for the slope changes from 10.5 to 7.97. The change in the intercept is the most dramatic, but the intercept is not particularly meaningful (no house has 0 square feet). r^2, the slope, and the *t* statistic for the slope have decreased a little, but the conclusion we reached in Exercise 10.17 is not changed.

10.23 a) A plot of the data is given below. We see that growth is much faster than linear

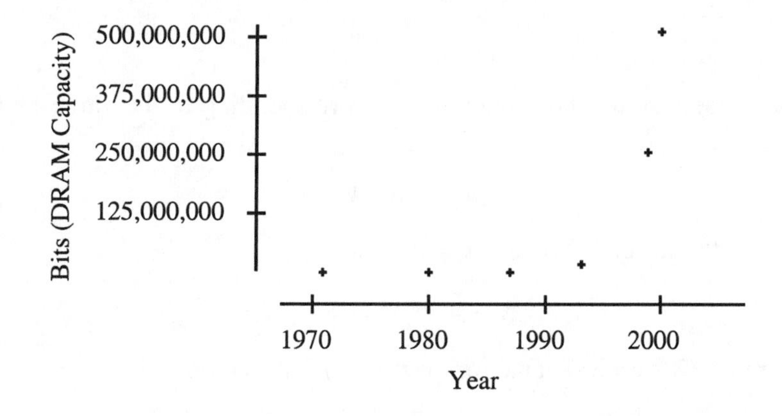

b) A plot of the logarithm of DRAM capacity against year is given below. The plot looks very linear.

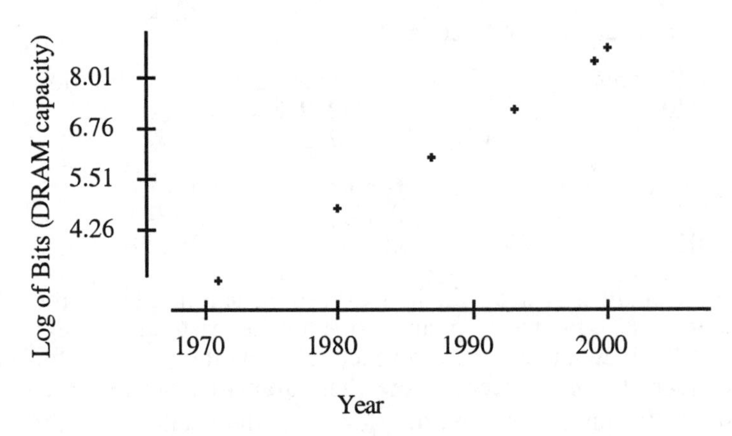

c) We use software to regress the logarithm of DRAM capacity against year. The results are given below.

R squared = 99.9%
s = 0.0755 with 6 – 2 = 4 degrees of freedom

Source	Sum of Squares	df	Mean Square	F-ratio
Regression	24.1787	1	24.1787	4239
Residual	0.022815	4	0.005704	

Variable	Coefficient	s.e. of Coeff	t-ratio	prob
Constant	–379.108	5.920	–64.0	≤0.0001
Year	0.193865	0.0030	65.1	≤0.0001

We want a 90% confidence interval. For this problem, we have 4 degrees of freedom, so we use $t^* = 2.132$ from the 4 df row of Table D corresponding to a 90% confidence level. From the results of our statistical software, we see that the slope $b_1 = 0.193865$ and standard error $SE_{b_1} = 0.0030$. Thus, the 90% confidence interval for the slope is

$$b_1 \pm t^* SE_{b_1} = 0.193865 \pm (2.132)(0.0030) = 0.193865 \pm 0.006396 = (0.187469, 0.200261)$$

SECTION 10.2

OVERVIEW

The **estimated mean response** for the subpopulation corresponding to the value x^* of the explanatory variable is

$$\hat{\mu}_y = b_0 + b_1 x^*$$

A **level *C* confidence interval for the mean response** is

$$\hat{\mu}_y \pm t^* SE_{\hat{\mu}}$$

where t^* is the upper $(1 - C)/2$ critical value for the $t(n - 2)$ distribution and

$$SE_{\hat{\mu}} = s\sqrt{\frac{1}{n} + \frac{(x^* - \bar{x}^2)}{\sum(x_i - \bar{x}^2)}}$$

$SE_{\hat{\mu}}$ is usually computed using a calculator or statistical software. (The formula above is actually given in Section 10.3 of the text, but reproduced here for continuity of exposition. If your class is not covering Section 10.3, you need not worry about the formula.)

The **estimated value of the response variable** y for a future observation from the subpopulation corresponding to the value x^* of the explanatory variable is

$$\hat{y} = b_0 + b_1 x^*$$

A **level C prediction interval** for the estimated response is

$$\hat{y} \pm t^* SE_{\hat{y}}$$

where t^* is the upper $(1 - C)/2$ critical value for the $t(n - 2)$ distribution and

$$SE_{\hat{y}} = s\sqrt{1 + \frac{1}{n} + \frac{(x^* - \bar{x})^2}{\sum(x_i - \bar{x})^2}}$$

$SE_{\hat{y}}$ is usually computed using a calculator or statistical software. (The formula above is actually given in Section 10.3 of the text but reproduced here for continuity of exposition. If your class is not covering Section 10.3, you need not worry about the formula.)

APPLY YOUR KNOWLEDGE

10.25 a) Here, $\hat{y} = 423.2$, $SE_{\hat{\mu}} = 15.6$, and the sample size is 59. Thus, for a 95% confidence interval we need the t^* for $59 - 2 = 57$ degrees of freedom. There is no entry for 57 degrees of freedom, so we use the row labeled 50 df in Table D (the row closest to but less than 57). This gives $t^* = 2.009$ for 95% confidence. Thus, a 95% confidence interval for the mean earnings of workers with 125 months' experience is

$$\hat{y} \pm t^* SE_{\hat{\mu}} = 423.2 \pm (2.009)(15.6) = 423.2 \pm 31.34 = (391.86, 454.54)$$

This agrees (up to roundoff error) with the Minitab results.

b) For a 90% confidence interval, we need the t^* for $59 - 2 = 57$ degrees of freedom. There is no entry for 57 degrees of freedom, so we use the row labeled 50 df in Table D (the row closest to but less than 57). This gives $t^* = 1.676$ for 90% confidence. Thus, a 90% confidence interval for the mean earnings of workers with 125 months' experience is

$$\hat{y} \pm t^* SE_{\hat{\mu}} = 423.2 \pm (1.676)(15.6) = 423.2 \pm 26.15 = (397.05, 449.35)$$

SECTION 10.2 EXERCISES

10.27 a) From Figure 10.11, we see that the least-squares regression line is

$$\text{Income} = 24874.3745 + 892.1135 \times \text{Age}$$

For Age = 30, the fitted value for income from the least-squares regression line is

$$\hat{y} = 24874.3745 + 892.1135 \times \text{Age} = 24874.3745 + 892.1135 \times 30 = 51637.78.$$

This agrees with the Minitab output.

b) We can read the 95% confidence interval for the mean income of all 30-year old men from the Minitab output. Under the column labeled 95.0% CI, we see in the row corresponding to a fit of 51638 (the first row) that the desired interval is (49,780, 53,496).

c) We can read the 95% prediction interval for the predicted income of a 30-year-old man from the Minitab output. Under the column labeled 95.0% PI we see in the row corresponding to a fit of 51638 (the first row) that the desired interval is (–41735, 145010). Awide interval such as this is not very useful.

10.29 From the Minitab output we see that the predicted income for age = 40 is $\hat{y}$ = 60559 and the standard error for estimating a mean response is $SE_{\hat{\mu}} = 637$. For a 99% confidence interval, we need the t^* for 5712 – 2 = 5710 degrees of freedom. There is no entry for 5710 degrees of freedom, but the closest row smaller than 5710 is the row labeled 1000 in Table D. This gives $t^* = 2.581$ for 99% confidence. Thus, a 99% confidence interval for the mean income of all 40-year old-men is

$$\hat{y} \pm t^* SE_{\hat{\mu}} = 60559 \pm (2.581)(637) = 60559 \pm 1644 = (58915, 62203)$$

10.31 a) For a 95% confidence interval using the one-sample t procedure based on 195 observations, we need the t^* for 195 – 1 = 194 degrees of freedom. There is no entry for 194 degrees of freedom, but the closest row smaller than 194 is the row labeled 100 in Table D. This gives $t^* = 1.984$ for 95% confidence. Thus, using the one-sample t procedure, a 95% confidence interval for the mean income μ_y of 30-year-old men is

$$\bar{y} \pm t^* \frac{s_y}{\sqrt{n}} = 49880 \pm (1.984)\frac{38250}{\sqrt{195}} = 49880 \pm 5434 = (44446, 55314)$$

b) This interval is wider than the 95% confidence interval for μ_y of 30-year-old men based on the regression output. This is not surprising. The regression output uses all 5712 men in the data, while the one-sample t procedure uses only the 195 men of age 30. We would expect the method that uses the much larger sample size to produce a narrower confidence interval.

10.33 From Minitab we obtain the following output for the predicted BAC for someone who has drunk five beers.

Fit	Stdev.Fit	95% C.I.	95% P.I.
0.07712	0.00513	(0.06611, 0.08812)	(0.03191, 0.12233)

Because we want to predict the BAC of an individual (Steve) we want a 90% prediction interval for the BAC of someone who drinks five beers. Minitab give a 95% prediction interval. For 95% confidence with $n = 16$, we use the value of t^* for 16 – 2 = 14 degrees of freedom. From Table D, we see that $t^* = 2.145$. Using the values in the output, we see the interval must be of the form

$$\hat{y} \pm t^* SE_{\hat{y}} = 0.07712 \pm (2.145)\, SE_{\hat{y}} = (0.03191, 0.12233).$$

Thus, $0.07712 + (1.761) SE_{\hat{y}} = 0.12233$, or $SE_{\hat{y}} = (0.12233 - 0.07712)/(2.145) = 0.0211$. For 90% confidence with $n = 16$, we use the value of t^* for 16 – 2 = 14 degrees of freedom. From Table D, we see that $t^* = 1.761$. Thus, our 90% prediction interval for the BAC of someone who drinks five beers is

$$\hat{y} \pm t^* SE_{\hat{y}} = 0.07712 \pm (1.761)(0.0211) = 0.07712 \pm 0.03716 = (0.03996, 0.11428)$$

This interval includes values larger than 0.08, so Steve can't be confident that he won't be arrested if he drives and is stopped.

SECTION 10.3

OVERVIEW

The **analysis of variance equation** for simple linear regression expresses the total variation in the responses as the sum of two sources: the linear relationship of y with x and the residual variation in responses for the same x. The equations is expressed in terms of the sums of squares.

$$\text{Total variation in } y = \text{Total SS} = \sum(y_i - \bar{y})^2$$

$$\text{Variation along the line} = \text{Regression SS} = \sum(\hat{y}_i - \bar{y})^2$$

$$\text{Variation about the line} = \text{Residual SS} = \sum(y_i - \hat{y}_i)^2$$

The degrees of freedom for these sums of squares are divided in the same way with

$$\text{Total df} = n - 1$$

$$\text{Regression df} = 1$$

$$\text{Residual df} = n - 2$$

The mean sum of squares are defined by the relation

$$\text{MS} = \text{mean sum of squares} = \frac{\text{sums of squares}}{\text{degrees of freedom}}$$

The residual mean square is the square of the regression standard error.

The ANOVA table gives the degrees of freedom, sums of squares, and mean squares for total, regression, and residual variation. The **ANOVA F statistic** is the ratio

$$F = (\text{Regression MS})/(\text{Residual MS})$$

It is the square of the t statistic for testing H_0: $\beta_1 = 0$ versus the two-sided alternative.

The **square of the sample correlation** can be written as

$$r^2 = \frac{\text{Regression SS}}{\text{Total SS}}$$

and is interpreted as the proportion of the variability in the response variable y that is explained by the explanatory variable x in the simple linear regression.

APPLY YOUR KNOWLEDGE

10.35 From Figure 10.9, we know that the regression standard error is $s = 2.1801$. From the data in Table 10.2, we calculate the variance of the inflation percentages to be

$$\frac{1}{n-1}\sum(x_i - \bar{x})^2 = 9.61227$$

and because $n = 51$, we see that

$$\sum(x_i - \bar{x})^2 = 50(9.61227) = 480.6135$$

Thus, we see that

$$\text{SE}_{b_1} = \frac{s}{\sqrt{\sum(x_i - \bar{x})^2}} = \frac{2.1801}{480.6135} = 0.09944$$

This agrees (to roundoff error) with the Excel output, which gives $SE_{b_1} = 0.0992$.

10.37 The hypotheses being tested are

$$H_0: \beta_1 = 0$$
$$H_a: \beta_1 \neq 0$$

The two test statistics given in the output are $F = 39.914$ (for the ANOVA test) and $t = 6.3177$ (for the t test for the slope). We notice that

$$\sqrt{F} = \sqrt{39.914} = 6.31775$$

which up to roundoff error equals $t = 6.3177$. The common P-value is 7.563E–08.

10.39 a) $r^2 = \dfrac{\text{Regression SS}}{\text{Total SS}}$. The output gives Regression SS = 189.705 and Total SS = 422.596, hence

$$r^2 = 189.705/422.596 = 0.4489$$

b) The regression standard error $s = \sqrt{\text{Residual MS}}$. The output gives Residual MS = 4.753, so

$$s = \sqrt{4.753} = 2.1801$$

SECTION 10.3 EXERCISES

10.41 We know that Total SS = Regression SS + Residual SS and that Total df = Regression df + Residual df. From the information given in the output this implies 13598.3 = 3445.9 + Residual SS and 29 = 1 + Residual df. Solving these equations, we find Residual SS = 13598.3 – 3445.9 = 10152.4 and Residual df = 29 – 1 = 28. Thus,

$$s = \sqrt{\text{Residual MS}} = \sqrt{(\text{Residual SS})/(\text{Residual df})} = \sqrt{10152.4/28} = 19.0417$$

Also,

$$r^2 = \frac{\text{Regression SS}}{\text{Total SS}} = 3445.9/13598.3 = 0.2534$$

10.43 a) To determine if reputation helps explain profitability, we test the hypotheses

$$H_0: \beta_1 = 0$$
$$H_a: \beta_1 \neq 0$$

The t statistic for testing these hypotheses can be read from the output (see the column labeled T for H_0: Parameter = 0). The value is

$$t = 6.041$$

and from the output we see that the P-value for the test is 0.0001.

We conclude that there is strong evidence against the null hypothesis and therefore strong evidence that the slope of the population regression line of profitability on reputation is nonzero (thus, good evidence that reputation helps explain profitability).

b) r^2 is the proportion of the variation in profitability among these companies that is explained by regression on reputation. From the output, we see that $r^2 = 0.1936$, and converting to percents, that 19.36% of the variation in profitability among these companies that is explained by regression on reputation.

c) Although the results of part (a) tell us that the test of whether the slope of the regression line of profitability on reputation is 0 is statistically significant, the value of r^2 in part (b) tells us that only 19.36% of the variation in profitability among these companies is explained by regression on reputation.

This is not a large percentage and indicates that reputation may not give very good predictions of profitability. In other words, while there is statistically significant evidence that reputation helps explain profitability, it may not be very useful for making practical predictions of profitability.

10.45 For companies with a reputation score of $x = 7$, using the output we see that the least-squares regression line predicts profitability to be

$$\text{profitability} = -0.147573 + 0.039111(7) = 0.126204$$

A 95% confidence interval for the mean profitability for all companies with reputation score $x = 7$ is

$$0.126204 \pm t^* \mathrm{SE}_{\hat{\mu}}$$

To complete our computations, we need t^* and $\mathrm{SE}_{\hat{\mu}}$. The sample size here is $n = 154$. For a 95% confidence interval based on 154 observations, we need the t^* for $154 - 2 = 152$ degrees of freedom. There is no entry for 152 degrees of freedom, but the closest row smaller than 152 is the row labeled 100 in Table D. This gives $t^* = 1.984$ for 95% confidence.

To compute $\mathrm{SE}_{\hat{\mu}}$, we begin by noting that

$$\mathrm{SE}_{\hat{\mu}} = s\sqrt{\frac{1}{n} + \frac{(x^* - \bar{x})^2}{\sum (x_i - \bar{x})^2}}$$

From the output we have s = Root MSE = 0.07208. We know that $n = 154$. $x^* = 7$ because we want a confidence interval for the mean profitability for all companies with reputation score of 7. We are told that

$$0.8101 = s_x^2 = \frac{\sum (x_i - \bar{x})^2}{n-1}$$

and thus

$$\sum (x_i - \bar{x})^2 = (n - 1)s_x^2 = 153 \times 0.8101 = 123.94$$

The last quantity we need is $\bar{x}$. The output tells us that $\bar{y}$ = Dep Mean = 0.1000. Because the regression line passes through the point $(\bar{x}, \bar{y})$, we know $0.1000 = \bar{y} = -0.147573 + 0.039111\bar{x}$ and hence

$$\bar{x} = (0.1000 + 0.147573)/0.039111 = 6.33$$

Putting all this information together, we have

$$\mathrm{SE}_{\hat{\mu}} = s\sqrt{\frac{1}{n} + \frac{(x^* - \bar{x})^2}{\sum (x_i - \bar{x})^2}} = 0.07208\sqrt{\frac{1}{154} + \frac{(7 - 6.33)^2}{123.94}} = 0.007249$$

and so our 95% confidence interval for the mean profitability for all companies with reputation score $x =$ 7 is

$$0.126204 \pm t^* \mathrm{SE}_{\hat{\mu}} = 0.126204 \pm 1.984(0.007249) = 0.126204 \pm 0.014382 = (0.111822, 0.140586)$$

10.47 The ANOVA F statistic (36.492 in the output) is the square of the t statistic (6.041 in the output) for the slope. The P-values should be equal. We notice that $6.041^2 = 36.494$, which agrees with the value of the F statistic in the output up to roundoff error, and from the output that P-value for F statistic = 0.0001 = P-value for t statistic for the slope.

10.49 $r^2 = \dfrac{\text{Regression SS}}{\text{Total SS}} = 0.18957/0.97920 = 0.1936$, which is equal to the value of R-square in the output.

CHAPTER 10 REVIEW EXERCISES

10.51 The plot indicates that there is a negative association between age and selling price, and that the relationship between the two looks roughly linear. The association appears to be moderately strong.

r^2 = 46.5%, so that 46.5% of the variation in selling price is explained by the least-squares regression line of selling price on age.

The least-squares regression line is

$$\hat{y} = 189226 - 1334.5 \times \text{age}$$

The slope is negative, as we would expect from the negative association we see in the plot. The slope tells us the 1-year increase in the age of the house corresponds to a decrease of –$1334.5, on average, in the 2000 selling price. The intercept tells us that, on average, a new house (age = 0) sells for $189,226.

10.53 a) A house built in 1990 is 10 years old, so age = 10. From the output (see the first entry in the Fit column) we see that, on average, these houses sell for $175,881 in 2000. This can be obtained from the equation of the least-squares regression line, namely

$$\hat{y} = 189226 - 1334.5 \times \text{age} = 189226 - 1334.5 \times 10 = 175881$$

From the output (see the first entry under the column labeled 95.0% CI) a 95% confidence interval for the mean selling price in 2000 of houses built in 1990 is ($159,569, $192,193).

b) From the output (see the first entry under the column labeled 95.0% PI), a 95% interval for the predicted selling price in 2000 of houses built in 1990 is ($93,236, $258,526).

c) Individual houses built in the same year can vary considerably in characteristics that affect the selling price (location, size, quality of construction) and this makes it difficult to accurately predict the value of one house based on age alone. Predicting the average price of all houses of a given age is more reliable. The average "balances out" the effects of other factors that affect selling price.

10.55 A 90% confidence interval for the mean price of houses in Ames that were 30 years old is

$$\hat{y} \pm t^* SE_{\hat{\mu}}$$

where $\hat{y}$ = 149191 (obtained from the Fit column) and $SE_{\hat{\mu}}$ = 5930 (obtained from the StDev Fit column). For a 90% confidence interval based on 50 observations, we need the t^* for 50 – 2 = 48 degrees of freedom. There is no entry for 48 degrees of freedom, but the closest row smaller than 48 is the row labeled 40 in Table D. This gives t^* = 1.684 for 90% confidence. Thus, our 90% confidence interval is

$$\hat{y} \pm t^* SE_{\hat{\mu}} = 149191 \pm (1.684)(5930) = 149191 \pm 9986 = (139205, 159177)$$

10.57 The scatterplot follows. The plotting symbol x is used for the 34 women who work in large banks and the symbol o for the 25 who work in small banks.

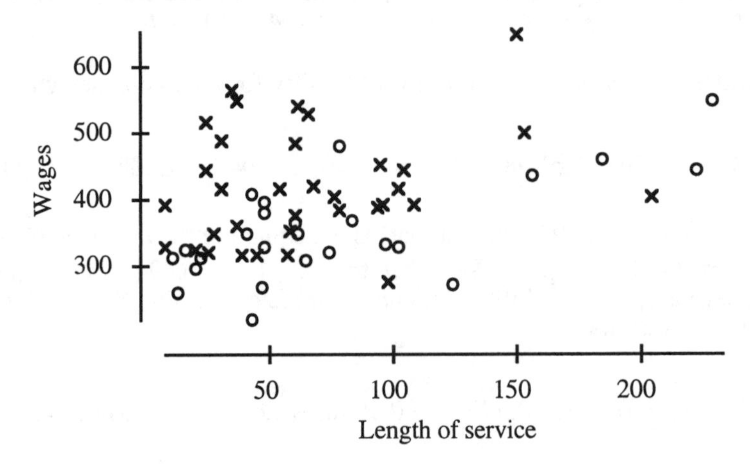

For a given length of service, the women who work in large banks appear to have generally higher wages. The points plotted as o appear to be more closely clustered around a line than the points plotted as x. Thus, it appears that regressing wages on length of service will do a better job of explaining wages for women who work in small banks than for women who work in large banks.

10.59 The analysis for women who work for large banks is

R squared = 4.8% R squared (adjusted) = 1.8%
s = 85.18 with 34 – 2 = 32 degrees of freedom

Source	Sum of Squares	df	Mean Square	F-ratio
Regression	11691.5	1	11691.5	1.61
Residual	232170	32	7255.32	

Variable	Coefficient	s.e. of Coeff	t-ratio	prob
Constant	390.236	26.34	14.8	≤ 0.0001
Length of service	0.424959	0.3348	1.27	0.2134

We see that the P-value for the test of the hypothesis that the slope is 0 is 0.2134. Thus, for women who work in large banks there is not strong evidence of a straight line relationship (linear association) between wages and length of service.

The analysis for women who work in small banks is

R squared = 46.7% R squared (adjusted) = 44.4%
s = 56.52 with 25 – 2 = 23 degrees of freedom

Source	Sum of Squares	df	Mean Square	F-ratio
Regression	64406.5	1	64406.5	20.2
Residual	73469.5	23	3194.33	

Variable	Coefficient	s.e. of Coeff	t-ratio	prob
Constant	289.185	18.37	15.7	≤ 0.0001
Length of service	0.840855	0.1873	4.49	0.0002

We see that the P-value for the test of the hypothesis that the slope is 0 is 0.0002. Thus, for women who work in small banks there is strong evidence of a straight line relationship (linear association) between wages and length of service. The value of r^2 is 44.4%, so the least-squares regression line of wages on length of service explains 44.4% of the variation in wages.

From these analyses, we might use length of service to predict wages for women who work in small banks, but not for women who work in large banks (at least in Indiana, assuming these data are a random sample of all women who work in banks in Indiana).

10.61 a) Using software, we find the equation of the least-squares regression line of lean on year is

$$\hat{y} = -61.1209 + 9.31868 \times \text{year}$$

From this equation, we would predict the lean in 1918 (year = 18) to be

$$\hat{y} = -61.1209 + 9.31868 \times \text{year} = -61.1209 + 9.31868 \times 18 = 106.6$$

or 2.91066 meters.

b) A plot of the data follows.

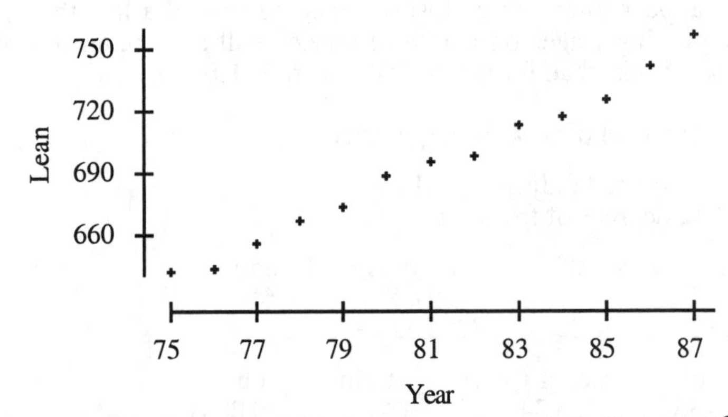

These data closely follow a straight line. In addition, using software we find that $r^2 = 99.8$, which tells us that for the years 1975 to 1987, the least-squares regression line of lean on year explains 99.8% of the variation in lean. Both the plot and r^2 are evidence that the least-squares regression line gives excellent fit to the data for 1975 to 1987. Using statistical software, we find the standard error of regression is s = 4.181. However the absolute difference between the observed value in 1918 (coded value of 71) and the value predicted by the least-squares regression line (coded value 106.6) is 35.6. This is a multiple of more than 8 times s, and would be considered an extreme outlier from the pattern of the data from 1975 to 1987. As we have seen before, extrapolation is dangerous.

10.63 The test statistic is

$$t = \frac{b_1}{\mathrm{SE}_{b_1}} = \frac{0.82}{0.38} = 2.16$$

The t statistic has $n - 2 = 55 - 2 = 53$ degrees of freedom. We are testing the two-sided alternative that the slope differs from 0, so the P-value is twice the upper tail probability. There is no entry for 53 degrees of freedom, so we use the row labeled 50 df in Table D (the row closest to but less than 53). We see that a t of 2.16 is between the entries of 2.109 and 2.403 corresponding to upper tail probabilities of 0.02 and 0.01, respectively. Thus, the P-value is between $2\times 0.01 = 0.02$ and $2 \times 0.02 = 0.04$, i.e., $0.02 < P\text{-value} < 0.04$. We conclude that there is reasonably strong evidence against the hypothesis of no linear relationship between the pretest score and the score on the final exam.

CHAPTER 11

MULTIPLE REGRESSION

SECTION 11.1

OVERVIEW

Multiple linear regression extends the techniques of simple linear regression to situations involving $p > 1$ explanatory variables $x_1, x_2, \ldots, x_p$. The data consists of the values of the response y and the p explanatory variables for n individuals or cases. Data analysis begins by examining the distribution of the variables individually and then drawing scatterplots to explore the relationships between the variables.

The **multiple regression equation** predicts the value of the response y as a linear function of the explanatory variables

$$\hat{y}_i = b_0 + b_1 x_{i1} + b_2 x_{i2} + \ldots + b_p x_{ip}$$

where the coefficients b_i are estimated using the method of least squares. The variability of the responses about the multiple regression equation is measured in terms of the regression standard error s,

$$s = \sqrt{\text{MSE}} = \frac{\Sigma e_i^2}{n-p-1}$$

where the e_i are the **residuals**

$$e_i = y_i - \hat{y}_i$$

The **distribution of the residuals** should be examined and the residuals should be plotted against each of the p explanatory variables.

APPLY YOUR KNOWLEDGE

11.1 a) The response variable is bank assets.
b) The explanatory variables are the number of banks and deposits.
c) Because there are two explanatory variables, $p = 2$.
d) The sample size is $n = 54$, the 50 states plus Guam, Puerto Rico, the Virgin Islands and the District of Columbia.

11.3 The distribution of sales is skewed to the right, but the skewness is not as extreme as for assets. There are two high outliers. The two companies that were high outliers in assets were General Electric and Citigroup, while the two companies that were high outliers in sales were General Motors and Wal-Mart. It is not surprising that Wal-Mart has high sales relative to its assets because its primary business function is the distribuion of products to final users. This would require less in the form of assets and increases the amount of sales.

11.5 The correlation between log(profits) and log(sales) is 0.526 as compared to 0.538 on the original scale. The correlation between log(profits) and log(assets) is 0.569 as compared to 0.533 on the original scale. The correlation between the explanatory variables log(assets) and log(sales) is 0.643 as compared to 0.455 on the original scale. Except for the two explanatory variables, there is little change in the correlations. The three scatterplots are given on the next page. The linear association between log(assets) and log(sales) appears much stronger in the scatterplots. The high outliers were eliminated from the other two plots, but there is a new outlier on the log scale corresponding to International Paper. International Paper had very low profits and on the log scale the point for International Paper is further from the other companies than it was on the original scale.

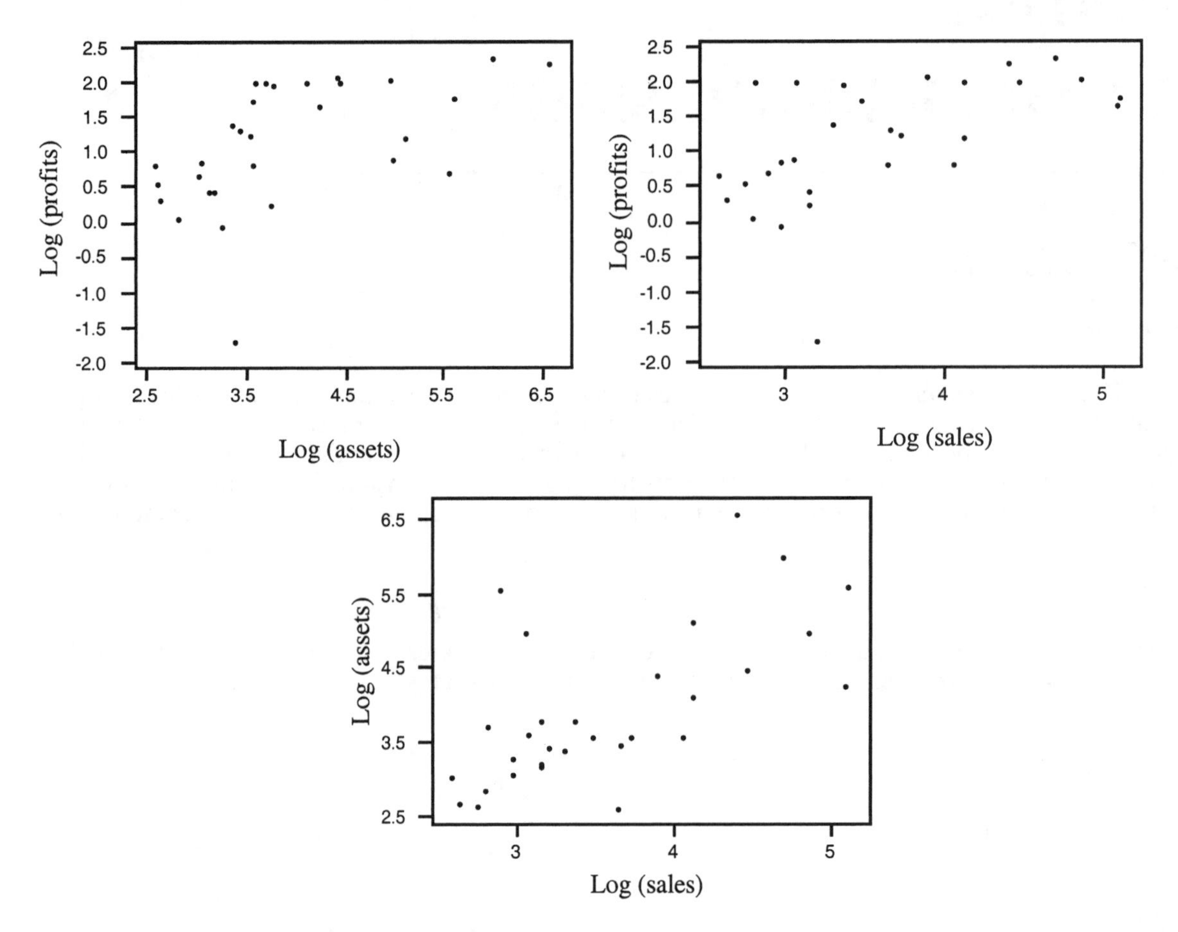

11.7 The data for the three log variables was input into Minitab. On the log scale, the regression equation is

$$\log(\text{profits}) = -1.50 + 0.238 \log(\text{assets}) + 0.478 \log(\text{sales})$$

11.9 General Motors and Wal-Mart have large outliers in sales, which is one of the two explanatory variables. The residuals for both of these companies are negative and they may be pulling the regression function down. The regression equations with and without these two companies are given below. The coefficients of both assets and sales have changed considerably without the influence of these two companies. The coefficient of sales has more than doubled and the coefficient of assets is much smaller.

All companies:	profits = 2.34 + 0.00741 assets + 0.0261 sales
General Motors and Wal-Mart deleted:	profits = 1.55 + 0.00496 assets + 0.0553 sales

11.11 For Excel, the unrounded values are $s = 2.449581635$ and $s^2 = 6.000450185$. The name given to s in the output is Standard Error. For Minitab, the unrounded values are $s = 2.450$ and $s^2 = 6.000$. The name given to s in the output is s. For SPSS, the unrounded values are $s = 2.44958$ and $s^2 = 6.000$. The name given to s in the output is Std. Error of the Estimate. For SAS, the unrounded values are $s = 2.44958$ and $s^2 = 6.00045$. The name given to s in the output is Root MSE. Many software packages print far more significant digits than necessary as part of the output. Unfortunately, the terminology used for s in the four packages is a bit different. For most of the other quantities reported, the terminology is more standardized.

SECTION 11.1 EXERCISES

11.13 a) The sample sizes, means, standard deviations, and the five-number summaries are reported for the three variables measured on the brokerages.

Variable	N	Mean	StDev	Median	Min	Max	Q1	Q3
Share	10	8.96	7.74	8.85	1.30	27.50	2.80	11.60
Accounts	10	794	886	509	125	2500	134	909
Assets	10	48.9	76.2	15.35	1.3	219.0	5.9	38.8

b) The three stemplots are given below. For 10 observations, this is the most appropriate graphical display of the distribution.

Market Share (leaves are 1's)
```
0|1223
0|89
1|012
1|
2|
2|7
```

Accounts (leaves are 100's)
```
0|11124
0|569
1|
1|
2|3
2|5
```

Assets (leaves are 10's)
```
0|00001123
0|
1|
1|6
2|1
```

c) All three distributions appear to be skewed to the right with high outliers. Charles Schwab is a high outlier in all three variables. Fidelity is an outlier in both accounts and assets with values of these two variables close to the values of Schwab, but it is not an outlier in market share (as measured by number of trades per day).

11.15 The three variables were input to Minitab giving the regression equation

$$\text{Share} = 5.16 - 0.00031 \text{ Accounts} + 0.0828 \text{ Assets}$$

The value of s reported in the output is 5.488.

11.17 The sample sizes, means, standard deviations, and the five-number summaries are recomputed for the three variables measured on the brokerages, with Charles Schwab and Fidelity eliminated. All four means are considerably smaller. The smallest effect is on shares because Fidelity is not an outlier in this variable. There is an enormous decrease in the standard deviations as the high outliers had a large effect on this measure of variability. The minimums stay the same and there is little effect on the first quartiles, although the medians and third quartiles are much smaller for the variables accounts and assets. The maximum become smaller as the previous maximums were the high outliers.

Variable	N	Mean	StDev	Median	Min	Max	Q1	Q3
Share	8	6.60	4.63	6.00	1.30	12.90	2.50	10.80
Accounts	8	392	293	316.5	125	909	132	602.5
Assets	8	13.76	12.27	9.00	1.30	38.80	5.70	20.30

The three stemplots are reproduced below and the appearance of all three is quite different than when the outliers were present. The two things to notice are there are more splits in the stems of the market share stemplot than there were before, and the units of the other two stemplots have changed. This allows us to see features of the distribution of the other eight companies that were obscured when the outliers were present. For example, the eight remaining brokerages fall into two fairly separated groups in terms of market share, a fact that was not as obvious in the original stemplot.

Market Share (leaves are 1's)
```
0|1
0|223
0|
0|
0|8
1|01
1|2
```

Accounts (leaves are 10's)
```
1|233
2|0
1|
4|2
5|9
6|1
7|
8|
9|0
```

Assets (leaves are 1's)
```
0|1
0|556
1|1
1|9
2|1
2|
3|
3|8
```

11.19 With Charles Schwab and Fidelity eliminated, the new regression equation is

$$\text{Share} = 1.85 + 0.00663 \text{ Accounts} + 0.157 \text{ Assets}$$

and the new value of s is 3.501. The coefficients have changed and the value of s is now much smaller.

11.21 a) For a small data set such as this, a stemplot is the best graphical description. All four distributions are right-skewed and gross sales, cash items, and credit items all have high outliers. The table below gives the some numerical summaries for the four variables. Because of the right skewness and outliers, the means are higher than the medians.

```
Gross sales
0|9
1|1679
2|02334456889
3|049
4|128
5|9
6|4
7|
8|9
```

```
Cash Items
0|5889
1|001
1|5556899
2|0134
2|677
3|3
3|6
4|3
4|
5|
5|5
```

```
Credit Items
0|3
0|5688
1|0334444
1|5778
2|13
2|68
3|12
3|8
4|
4|
5|
5|67
```

```
Check Items
0|000012233344
0|55679
1|0124
1|6
2|1
2|68
```

Variable	N	Mean	Median	StDev
gross	25	320.3	263.3	180.1
cash	25	20.52	19.00	11.80
credit	25	20.04	15.00	14.07
check	25	7.68	5.00	7.98

b) The graphic on the next page contains a plot of each pair of variables. It was produced in Minitab under the option "Matrix Plot" and the use of this option avoids having to ask for each plot individually. Although it is a little more difficult to read the scales than on a larger plot, the general patterns are fairly clear. The pairwise correlations between all pairs of variables are given as well.

The first row of the matrix plot gives the plots of gross sales versus the variables cash items, credit items, and check items. The positive association and linear trend are apparent in all three plots, although there are a few outliers. The correlation between gross sales and check items is the lowest and the low correlation is due to the outlier present in the upper left-hand corner. The three plots in the first column are the same except that gross sales is on the horizontal axis and the other variables are on the vertical axis. The remaining plots show the relationships between the three explanatory variables. The associations between the pairs of explanatory variables are not as strong.

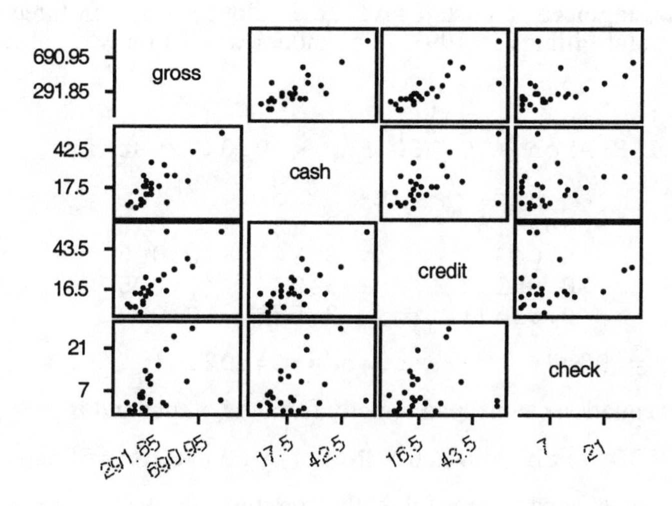

Correlations

	gross	cash	credit
cash	0.817		
credit	0.821	0.516	
check	0.458	0.353	0.176

11.23 The model using cash items, check items, and credit card items to predict gross sales was fit and the residuals were obtained. The plots of the residuals versus each of the three explanatory variables and the normal quantile plot are given below. You need to be careful not to overinterpret plots with only 25 points. One of the points with a large residual (underpredicted by the model) seems to have very large values for both cash items and credit card items.

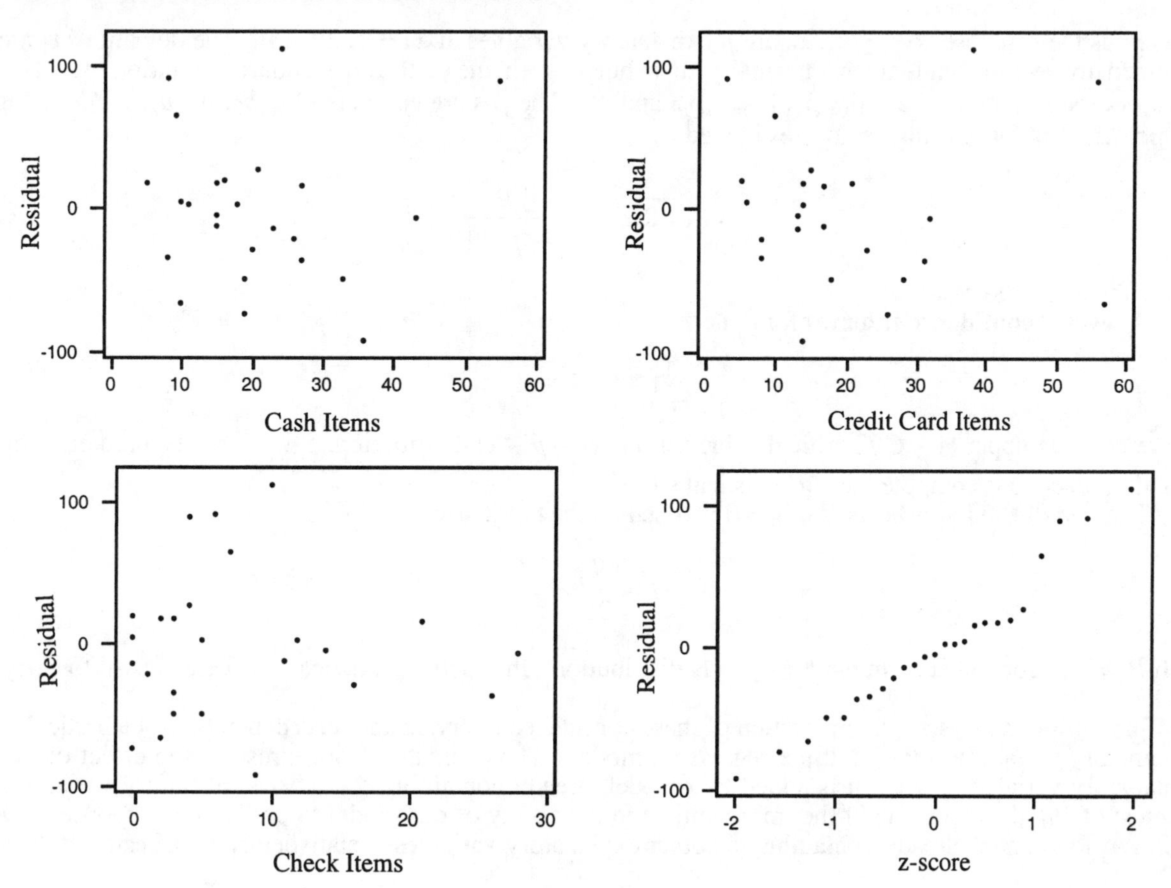

11.25 The variables architects, engineers, and staff give the employee counts in three categories. These are to be used to predict the total billings in 1998. The model was fit on Minitab with the following output.

```
The regression equation is
TBill98 = 0.883 + 0.138 Arch + 0.160 Eng + 0.0478 Staff

Predictor        Coef       Stdev     t-ratio       p
Constant       0.8832      0.4165        2.12   0.046
Arch          0.13788     0.04369        3.16   0.005
Eng           0.16009     0.05114        3.13   0.005
Staff         0.04784     0.01341        3.57   0.002

s = 1.162        R-sq = 93.5%      R-sq(adj) = 92.6%
```

a) The fitted regression equation can be read directly from the output and is

```
TBill98 = 0.883 + 0.138 Arch + 0.160 Eng + 0.0478 Staff
```

b) The regression standard error can be read directly from the output and is $s = 1.162$.

SECTION 11.2

OVERVIEW

The statistical model for **multiple linear regression** is

$$y_i = \beta_0 + \beta_1 x_{i1} + \beta_2 x_{i2} + \ldots + \beta_p x_{ip} + \varepsilon_i$$

where y is the response, $x_1, x_2, \ldots, x_p$ are p explanatory variables and $i = 1, 2, \ldots, n$. The deviations ε_i are assumed to be independent and normally distributed with mean 0 and standard deviation σ. The **parameters** of the model are $\beta_0, \beta_1, \beta_2, \ldots, \beta_p$, and σ. The β's are estimated by $b_0, b_1, b_2, \ldots, b_p$ using the **principle of least-squares.** σ is estimated by

$$s = \sqrt{\text{MSE}} = \frac{\sum e_i^2}{n - p - 1}$$

where the e_i are the **residuals.**

A **level *C* confidence interval** for β_j is

$$b_j \pm t^* \text{SE}_{b_j}$$

where t^* is the upper $(1 - C)/2$ critical value for the $t(n - p - 1)$ distribution. SE_{b_j} is the standard error of b_j and in practice is computed using statistical software.

The **test of the hypothesis** H_0: $\beta_j = 0$ is based on the ***t* statistic**

$$t = \frac{b_j}{\text{SE}_{b_j}}$$

with *P*-values computed from the $t(n - p - 1)$ distribution. In practice statistical software is used to carry out these tests.

In multiple regression, interpretation of these confidence intervals and tests depends on the particular explanatory variables in the multiple regression model. The estimate of β_j represents the effect of the explanatory variable x_j when it is added to a model already containing the other explanatory variables. The test of H_0: $\beta_j = 0$ tells us if the improvement in the ability of our model to predict the response y by adding x_j to a model already containing the other explanatory variables is statistically significant. It does not tell us if x_j would be useful for predicting the response in multiple regression models with a different collection of explanatory variables.

The **ANOVA table** for a multiple regression is analogous to that in simple linear regression. It gives the sum of squares, the mean squares and degrees of freedom for regression and residual sources of variation. The ANOVA F is the regression mean square divided by the residual mean square and is used to test the hypothesis H_0: $\beta_1 = \beta_2 = \ldots = \beta_p = 0$. Under H_0, this statistic has an $F(p, n-p-1)$ distribution.

The **squared multiple correlation** can be written as

$$R^2 = \text{SSM/SST}$$

and is interpreted as the proportion of the variability in the response variable y that is explained by the explanatory variables $x_1, x_2, \ldots, x_p$ in the multiple regression.

It is also possible to test the null hypothesis that a **collection of q explanatory variables** all have coefficients equal to zero. Use

$$F = \left(\frac{n-p-1}{q}\right)\left(\frac{R_1^2 - R_2^2}{1 - R_1^2}\right)$$

where R_1^2 is the squared multiple correlation with all variables in the model and R_2^2 is the squared multiple correlation with the q variables deleted. Under the null hypothesis, this statistic has an F distribution with q degrees of freedom in the numerator and $n - p - 1$ degrees of freedom in the denominator.

APPLY YOUR KNOWLEDGE

11.27 The table below compares the output for the variable high school science (HSS) given by the four packages. The only differences are in the number of significant digits reported.

Package	Coefficient	Standard error	t statistic	P-value
Excel	0.034315568	0.03755888	0.913647251	0.361902429
Minitab	0.03432	0.03756	0.91	0.362
SPSS	3.432E-02	.038	.914	.362
SAS	0.03432	0.03756	0.91	0.3619

HSS does not help us much to predict GPA, given that math and English grades are available to use for prediction.

11.29 With HSS eliminated, the new regression equation and a portion of the regression output is reported below. The coefficients of both HSM and HSE change as well as their t values. HSE, which appeared to be of little use in predicting GPA when both math and science were available to use for prediction, now appears to be of more use when only math is available. However, HSM still seems to be the most important variable in predicting GPA. The value of s is almost exactly the same as when all three variables were in the model.

```
GPA = 0.624 + 0.183 HSM + 0.0607 HSE

Predictor        Coef        Stdev     t-ratio        p
Constant       0.6242       0.2917        2.14    0.033
HSM           0.18265      0.03196        5.72    0.000
HSE           0.06067      0.03473        1.75    0.082

s = 0.6996        R-sq = 20.2%       R-sq(adj) = 19.4%
```

11.31 Although the two models are different, based on Exercise 11.29 we would not be surprised if the predictions of college GPA from HSM and HSE alone is about the same as the predictions based on HSM, HSS, and HSE. For a straight-C student, the table below gives the fit (predicted value) followed by the 95% confidence interval for the mean GPA and a 95% prediction interval. The first line gives the results with all three predictors and the second line has HSS dropped from the model. As expected, the answers are quite close.

```
                   Fit         95.0% C.I.              95.0% P.I.
HSM,HSS,HSE      1.5975     (1.2927,   1.9024)     (0.1852,   3.0098)
HSM,HSE          1.5818     (1.2749,   1.8887)     (0.1685,   2.9951)
```

11.33 The table below gives the values of R^2 for models using different sets of explanatory variables. It is clear that HSM is necessary in the model as without it the value of R^2 is quite low. However, once HSM is in the model, addition of other variable seems to be of limited use. The conclusion is that almost all of the information for predicting GPA is contained in HSM.

Model	R^2
a) hsm, hss, hse	20.5%
b) hsm, hse	20.2%
c) hsm, hss	20.0%
d) hss, hse	12.3%
e) hsm	19.1%

11.35 We want to find out if the three high school grade variables help explain GPA in a model that contains the two SAT variables. To answer this question, you need to run a model with all five predictors and a model with only SAT scores and obtain the R^2 values from these two models. These R^2 values can be used to compute an F test of the hypotheses that the coefficients of the $q = 3$ grade variables are zero. Following the general notation for the F test, in a model with all $p = 5$ explanatory variables, $R_1^2 =$ 21.15%, and after removing the $q = 3$ grade variables and rerunning the regression we have $R_2^2 = 6.34\%$. The F test for a collection of regression coefficients equal to zero has test statistic

$$F = \left(\frac{n-p-1}{q}\right)\left(\frac{R_1^2 - R_2^2}{1-R_1^2}\right) = \left(\frac{224-5-1}{3}\right)\left(\frac{0.2115-0.0634}{1-0.2115}\right) = 13.65$$

and degrees of freedom $q = 3$ and $n - p - 1 = 218$. Software gives $P < 0.0001$. High school grades contribute significantly to explaining GPA when SAT scores are already in the model.

SECTION 11.2 EXERCISES

11.37 a) You need to carry out the test that the regression coefficients for the three explanatory variables are all zero, which is the analysis of variance F test. We are told that SSR = 90, SSE = 1000, $n = 104$, and there are $p = 3$ explanatory variables in the model. MSR = SSR / p = 90 / 3 = 30, and MSE = SSE / $n - p - 1$ = 1000 / 100 = 10. This gives F = MSR /MSE = 30 / 10 = 3, which has the $F(p, n - p - 1)$ = $F(3, 100)$ distribution. Software gives $P = 0.0342$.

b) The value of R^2 = SSR / SST = 90 / (1000 + 90) = 8.2%. Very little of the variability in the response is explained by this set of three explanatory variables.

11.39 a) The hypotheses tested by an individual t statistic about the jth explanatory variable is

$$H_0: \beta_j = 0 \text{ and } H_a: \beta_j \neq 0$$

It is a test that the jth explanatory variable is useful for prediction, given that all the other variables are available to us for prediction. The degrees of freedom for the t statistics are $n - p - 1 = 2229 - 13 - 1 =$ 2215, because there are 2229 loans and 13 explanatory variables. At the 5% level, values of t that are less than –1.96 or greater than 1.96 will lead to rejection of the null hypothesis.

b) The explanatory variables whose coefficients are significantly different from zero are loan size, length of loan, percent down payment, cosigner, unsecured loan, total income, bad credit report, young borrower, own home and years at current address. If a t statistic is not significant, we conclude that the explanatory variable is not useful in predicting the interest rate charge when all other variables were available to use for prediction.

c) Loan size — negative means the interest rate is lower for larger loans; length of loan — negative means the interest rate is lower for longer length loans; percent down payment — negative means the interest rate is lower for a higher percent down payment; cosigner — negative means the interest rate is lower when there is a cosigner; unsecured loan — positive means the interest rate is higher for an unsecured loan; total income — negative means the interest rate is lower for those with higher total income; bad credit report — positive means the interest rate is higher when there is a bad credit report; young borrower — positive means the interest rate is higher when there is a young borrower; own home — negative means the interest rate is lower when the borrower owns a home; years at current address — negative means the interest rate is lower when the years at current address is higher.

11.41 a) The hypotheses tested by an individual t statistic about the jth explanatory variable is

$$H_0: \beta_j = 0 \text{ and } H_a: \beta_j \neq 0$$

It is a test that the jth explanatory variable is useful for prediction, given that all the other variables are available to us for prediction. The degrees of freedom for the t statistics are $n - p - 1 = 5664 - 13 - 1 = 5650$, because there are 5664 loans and 13 explanatory variables. At the 5% level, values of t that are less than –1.96 or greater than 1.96 will lead to rejection of the null hypothesis.

b) The explanatory variables whose coefficients are significantly different from zero are loan size, length of loan, percent down payment and unsecured loan.

c) Loan size — negative means the interest rate is lower for larger loans; length of loan — negative means the interest rate is lower for longer length loans; percent down payment — negative means the interest rate is lower for a higher percent down payment; unsecured loan — positive means the interest rate is higher for an unsecured loan.

11.43 a) The model says that y varies normally with a mean

$$\mu_{\text{GPA}} = \beta_0 + 9\beta_1 + 8\beta_2 + 7\beta_3$$

b) The estimate for the mean of the subpopulation of students with an A– in math, B+ in science, and B in English is

$$\widehat{\text{GPA}} = 0.590 + 9(0.169) + 8(0.0343) + 7(0.0451) = 2.70$$

The GPA of students with an A– in math, B+ in science, and B in English has a normal distribution with an estimated mean of 2.70.

11.45 a) The multiple regression model is

$$y_i = \beta_0 + \beta_1 x_{i1} + \beta_2 x_{i2} + \varepsilon_i$$

where y is the 1999 salary, x_1 is the 1996 salary and x_2 is years in rank.

b) The parameter of the model are β_0, β_1, β_2, and σ.

c) Using Minitab to fit the model results in the output

```
The regression equation is
1999 = 7499 + 1.10 1996 - 268 years

Predictor        Coef        Stdev     t-ratio        p
Constant         7499         9290        0.81    0.434
1996           1.1022       0.1268        8.69    0.000
years          -267.6        210.9       -1.27    0.227

s = 4123       R-sq = 94.5%      R-sq(adj) = 93.6%

Analysis of Variance

SOURCE        DF           SS          MS          F        p
Regression     2   3790817024  1895408512     111.50    0.000
Error         13    220984848    16998834
Total         15   4011801856
```

Reading from the output, the parameter estimates are $b_0 = 7499$, $b_1 = 1.1022$, $b_2 = -267.6$, and $s = 4123$.

d) The test statistic is the ANOVA F test, F = MSR / MSE = 11.50 from the output. The degrees of freedom are $p = 2$ and $n - p - 1 = 13$, and the P-value from the analysis of variance output is 0.000. We conclude years in rank and 1996 salary contain information that can be used to predict 1999 salary.

e) R^2 = SSR / SST = 94.5% from the output.

f) From the output, the test that the coefficient of salary is zero has $t = 8.69$, df = 13 and P-value = 0.000. In a model containing years in rank, the 1996 salary has information for predicting the 1999 salary. The test that the coefficient of years in rank is zero has $t = -1.27$, df = 13 and P-value = 0.227. In a model containing the 1996 salary, years in rank has little additional information for predicting the 1999 salary.

11.47 The model with only years in rank was fit in Minitab and produced the following partial output

```
The regression equation is
1999 = 86973 + 1308 years

Predictor        Coef        Stdev     t-ratio        p
Constant        86973         4145       20.98    0.000
years          1308.0        270.9        4.83    0.000

s = 10370        R-sq = 62.5%      R-sq(adj) = 59.8%
```

a) The fitted equation is $\text{Salary}_{1999} = 86973 + 1308$ years

b) The test that the coefficient of years in rank is zero is $t = 4.83$, df = 14, and P-value = 0.000. By itself, years in rank is useful for predicting 1999 salary.

c) The test in Exercise 11.45 is a test of the usefulness of years in rank for predicting 1999 salary in a model that already contains 1996 salary. The correlation between years in rank and 1996 salary is 0.860, so there is little additional information in the variable years in rank once we have already included 1996 salary in the model. However, if we only have years in rank available, this exercise shows that it is useful by itself for prediction. Regardless, of the two variables, years in rank and 1996 salary, the 1996 salary is a better single predictor of 1999 salary as it is more highly correlated with 1999 salary.

SECTION 11.3

OVERVIEW

Model building in multiple linear regression begins with an examination of the distribution of the variables and the relationships among them. **Categorical explanatory variables** can be included in the model. If any of the categorical variables have very few cases, they should be combined with similar categories or else examined separately.

If there are **curved relationships,** then variables can be transformed or depending on the nature of the curvilinear relationship, inclusion of quadratic terms in the model may help the fit. In some instances, the effect of one explanatory variable depends on the value of another explanatory variable. In this case, the variables **interact** and terms that measure interaction can be included in the model. Many software packages provide **variable selection methods** as part of the regression analysis and they can be used to find the models of each subset size with the highest R^2.

APPLY YOUR KNOWLEDGE

11.49 We drop the last two digits, use the hundreds and thousands digits as leaves, and the ten thousand and hundred thousand digits as stems. This leads to the stemplot on the next page. The seven homes excluded (those that cost more than \$150,000 or have more than 1800 square feet) are underlined in the stemplot. Six of these are the six most expensive homes. Three of these are clearly outliers (the three most expensive) for this location. They cost far more than the remaining homes.

```
05| 20, 29
06| 29, 49, 50, 99
07| 25, 29, 39, 39, 50, 69, 69
08| 00, 15, 19, 29, 45, 49, 49, 49, 70, 79, 90, 90, 96, 99
09| 28, 39, 49, 49, 60, 99
10| 49
11| 49, 49, 99
12| 49, 49, 90, 90
13|
14|
15|
16|
17| 39, 99
18|
19| 95
```

11.51 For 1000-square-foot homes,

Predicted price = 45,298 + 34.32362(1000) = 45,298 + 34,323.62 = \$79,621.62

For 1500-square-foot homes,

Predicted price = 45,298 + 34.32362(1500) = 45,298 + 51,485.43 = \$96,783.43

11.53 For 1000-square-foot homes,

Predicted price $= 81{,}273 - 30.13753(1000) + 0.02710(1000)^2 = 81{,}273 - 30{,}137.53 + 27{,}100 = \$72{,}853.47$

For 1500-square-foot homes,

Predicted price $= 81{,}273 - 30.13753(1500) + 0.02710(1500)^2 = 81{,}273 - 45{,}206.30 + 60{,}975 = \$97{,}041.71$

The predictions in Exercise 11.51 based on the simple linear regression model were for 1000 square feet, 79,621.62 (slightly larger than the prediction for the quadratic model) and for 1500 square feet, 96,783.43 (slightly smaller than the prediction for the quadratic regression model). Overall, the two sets of predictions are fairly similar.

11.55 From Example 11.16, we know that for homes with two or fewer bedrooms the mean price is $\bar{x}_1 =$ 75,700, and for homes with three or more bedrooms the mean price is $\bar{x}_2 = 90{,}846$. From the output in Figure 11.15 we see that Root MSE = 15,297, and this is the pooled estimate of the standard deviation, s_p, needed for the pooled two-sample t test discussed in Chapter 7. If we eliminate the seven homes corresponding to those that either cost more than 150,000 or have more than 1800 square feet, we see in Table 11.6 that there are $n_1 = 13$ homes with two or fewer bedrooms and $n_2 = 24$ homes with three or more bedrooms. The pooled t test statistic is thus,

$$t = \frac{\bar{x}_1 - \bar{x}_2}{s_p\sqrt{\frac{1}{n_1} + \frac{1}{n_2}}} = \frac{90846 - 75700}{15297\sqrt{\frac{1}{13} + \frac{1}{24}}} = \frac{15146}{5267.807} = 2.875.$$

The test statistics has $n_1 + n_2 - 2 = 13 + 24 - 2 = 35$ degrees of freedom. 35 df is not given in Table D, so we look at the next closest (and smaller) line of Table D, which is 30 df. A value of 2.875 lies between the t critical values with upper tail probabilities 0.005 and 0.0025, respectively. Because we have a two-sided test, the P-value satisfies $2\times0.0025 < P\text{-value} < 2\times0.005$ or

$$0.005 < P\text{-value} < 0.01$$

These results agree (up to roundoff) with the values of the *t* statistic, the degrees of freedom, and the *P*-value for the coefficient of Bed3 in Example 11.16.

11.57 A plot of price versus number of garage spaces follows.

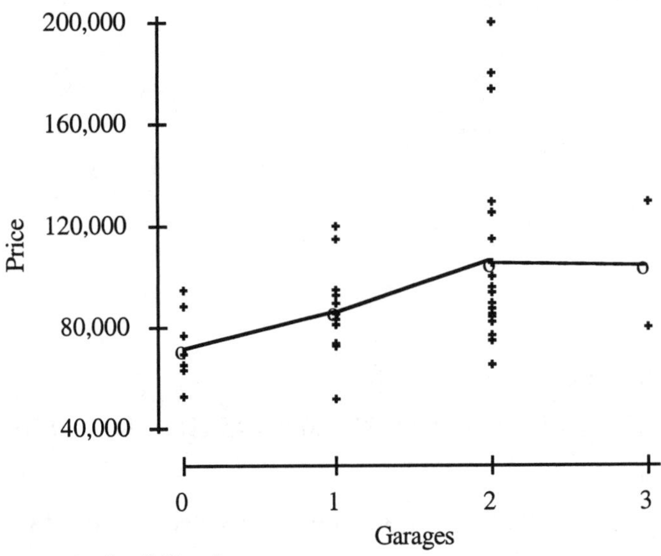

The mean price for each group is the following.

0 garage spaces: mean price = 72,900
1 garage space: mean price = 85,128.6
2 garage spaces: mean price = 107,833
3 garage spaces: mean price = 105,500

These are added to the plot (as o's) and are connected by lines. The trend is increasing and roughly linear as one goes from 0 to 2 garages and then levels off.

11.59 From the regression equation in Example 11.22, we see that the predicted price for a 140- square-foot home with an extra half bath is

Predicted price = 63,375 + 15.15(1400) – 58,800(1) + 52.81(1400)(1)
= 63,375 + 21,210 – 58,800 + 7393 = 99,719

The predicted price for a 1400 square foot home without an extra half bath is

Predicted price = 63,375 + 15.15(1400) – 58,800(0) + 52.81(1400)(0) = 63,375 + 21,210 – 0 + 0
= 84,585

The difference in price is

Predicted price with extra half bath – Predicted price without extra half bath = 99,719 – 84,585 = 15,134

11.61 No home in the data with under 1000 square feet has an extra half bath. Because we have no data on smaller homes with an extra half bath, our regression equation is probably not trustworthy for predicting the effect of an extra half bath on the price of such homes. It is usually unreliable to extrapolate outside the range of our data.

SECTION 11.3 EXERCISES

11.63 a) A sketch of $\mu_y = 5 + 2x + 3x^2$ is

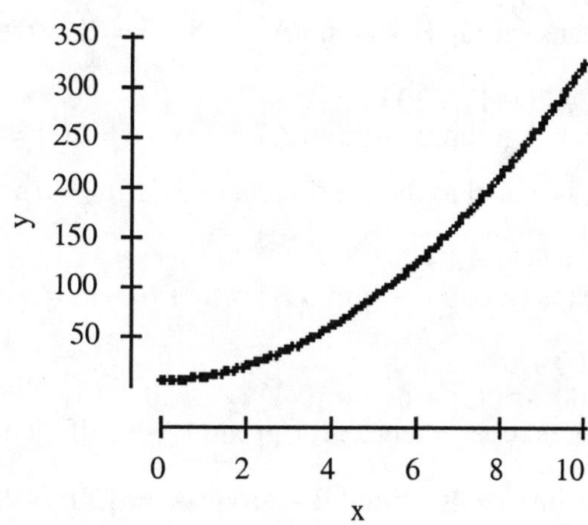

The relationship between μ_y and x is curved and increasing.

b) A sketch of $\mu_y = 7 - 30x + 3x^2$ is

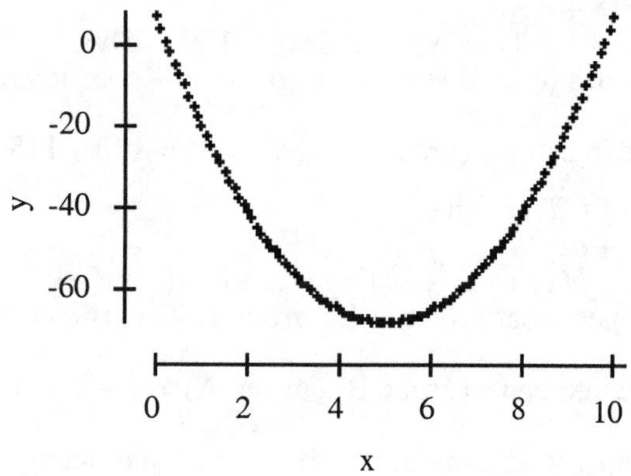

The relationship between μ_y and x is curved (u shaped), first decreasing and then increasing.

c) A sketch of $\mu_y = 600 - 30x - 3x^2$ is

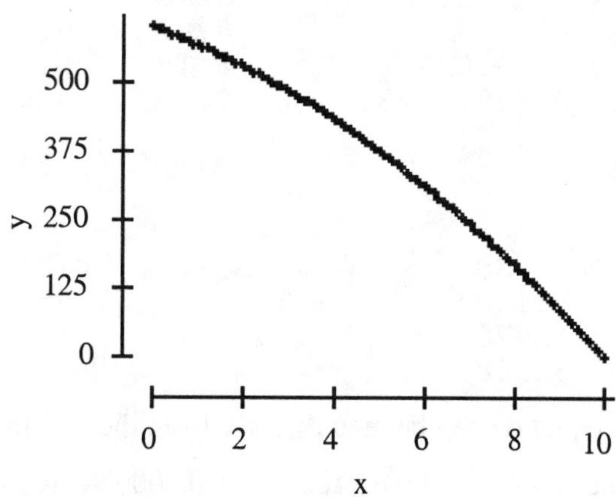

The relationship between μ_y and x is curved and decreasing.

11.65 a) For group A: $\mu_y = 5$
For group B: $\mu_y = 5 + 2(1) = 7$
The difference in means (group B – group A) = 7 – 5 = 2 = the coefficient of x.

b) For group A: $\mu_y = 5$
For group B: $\mu_y = 5 + 20(1) = 25$
The difference in means (group B – group A) = 25 – 5 = 20 = the coefficient of x.

c) For group A: $\mu_y = 5$
For group B: $\mu_y = 5 + 200(1) = 205$
The difference in means (group B – group A) = 205 – 5 = 200 = the coefficient of x.

The difference will, in general, be equal to the coefficient of x. If $\mu_y = a + bx$, then
For group A: $\mu_y = a$
For group B: $\mu_y = a + b(1) = a + b$
The difference in means (group B – group A) = $(a + b) - a = b$ = the coefficient of x.

11.67 a) For group A: $\mu_y = 100 + 6x_2$
For group B: $\mu_y = 100 + 50(1) + 6x_2 + 10(1)x_2 = 150 + 16x_2$
The difference in slopes (coefficient x_2 for group B – coefficient x_2 for group A) = 16 – 6 = 10 = the coefficient of x_1x_2.
The difference in the intercepts (group B – group A) = 150 – 100 = 50 = coefficient of x_1.

b) For group A: $\mu_y = 100 + 6x_2$
For group B: $\mu_y = 100 - 10(1) + 6x_2 - 6(1)x_2 = 90 + 0x_2$
The difference in slopes (coefficient x_2 for group B – coefficient x_2 for group A) = 0 – 6 = –6 = the coefficient of x_1x_2.
The difference in the intercepts (group B – group A) = 90 – 100 = –10 = coefficient of x_1.

c) For group A: $\mu_y = 118 + 6x_2$
For group B: $\mu_y = 118 + 2(1) + 6x_2 - 16(1)x_2 = 120 - 10x_2$
The difference in slopes (coefficient x_2 for group B – coefficient x_2 for group A) = –10 – 6 = –16 = the coefficient of x_1x_2.
The difference in the intercepts (group B – group A) = 120 – 118 = 2 = coefficient of x_1.

In general, if $\mu_y = a + bx_1 + cx_2 + dx_1x_2$, then
For group A: $\mu_y = a + cx_2$
For group B: $\mu_y = a + b(1) + cx_2 + d(1)x_2 = a + b + (c + d)x_2$
The difference in slopes (coefficient x_2 for group B – coefficient x_2 for group A) = $(c + d) - c$ = d = the coefficient of x_1x_2.
The difference in the intercepts (group B – group A) = $a + b - a = b$ = coefficient of x_1.

11.69 The Minitab output from fitting a model with account and accounts squared to predict assets is given below.

```
The regression equation is
Assets = 7.61 - 0.0046 Accounts +0.000034 Account2

Predictor         Coef        Stdev     t-ratio        p
Constant         7.608        8.503        0.89    0.401
Accounts      -0.00457      0.02378       -0.19    0.853
Account2    0.00003361   0.00000893        3.76    0.007

s = 12.41       R-sq = 97.9%     R-sq(adj) = 97.3%

Analysis of Variance

SOURCE       DF         SS          MS          F        p
Regression    2      51130       25565     165.95    0.000
Error         7       1078         154
Total         9      52208
```

a) The fitted regression equation can be read directly from the output and is

Assets = 7.61 – 0.0046 Account + 0.000034 Account2

b) The confidence interval for a regression coefficient is $b_j \pm t^* \text{SE}_{b_j}$ where $t^* = 2.365$ as we are working with a t distribution with 7 degrees of freedom. The values of b_j and SE_{b_j} can be read from the output under Coef and Stdev, respectively. The 95% confidence interval for the coefficient of the squared term is

Account2: $b_j \pm t^* \text{SE}_{b_j} = 0.000034 \pm (2.365)(0.00000893) = 0.000034 \pm 0.000021$

c) Reading from the output we have $t = 3.76$, df = 7 and $P = 0.007$. The quadratic term is useful for predicting assets in a model that already contains the linear term.

d) The fitted model with only accounts is Assets = –17.1 + 0.0832 Account. The variables account and the square of account are highly correlated and the meaning of the linear term in a model with the square term is different than the meaning of the linear term when the square term is absent.

11.71 a) Following the general notation for the F test, the model which includes years in rank and the square of years in rank has $p = 2$ explanatory variables, $R_1^2 = 65.38\%$. After removing the square term or one variable so $q = 1$ and rerunning the regression, we have $R_2^2 = 62.48\%$.

b) The F test for a collection of regression coefficients equal to zero has test statistic

$$F = \left(\frac{n-p-1}{q}\right)\left(\frac{R_1^2 - R_2^2}{1-R_1^2}\right) = \left(\frac{16-2-1}{1}\right)\left(\frac{0.6538-0.6248}{1-0.6538}\right) = 1.09$$

and degrees of freedom $q = 1$ and $n - p - 1 = 13$. Software gives $P = 0.3155$. The square term does not contribute significantly to explaining salary.

c) From the previous exercise, $t = -1.05$, and $t^2 = 1.10$, which agrees with F up to rounding error.

CHAPTER 11 REVIEW EXERCISES

11.73 a) With 100 degrees of freedom, $t^* = 1.98$ and the prediction interval is

$$\$2.136 \pm (1.98)(\$0.013) = (\$2.110, \$2.162)$$

b) Yes. The actual price of $2.13 is consistent with the prediction interval.

11.75 The amounts of three vitamins are measured in loaves of bread over a seven-day period. The vitamin content is regressed against time to study the effect of storage time on each vitamin. The coefficient of Days is negative in all three cases as expected, because additional vitamins should not be produced with storage. The deterioration of vitamin E is minimal, if at all. Vitamin A content deteriorates significantly over the seven-day period, while vitamin C is highly significant. Because amounts of each vitamin in the loaves are quite different, the slopes cannot be compared directly.

Vitamin A:	Vit A = 3.34 – 0.0388 Days;	$t = -2.84$;	$P = 0.022$
Vitamin C:	Vit C = 46.0 – 6.05 Days;	$t = -10.62$;	$P = 0.000$
Vitamin E:	Vit E = 95.4 – 0.074 Days;	$t = -0.34$;	$P = 0.744$

11.77 a) The partial Minitab output is given below.

```
The regression equation is
Wages = 354 + 63.9 Size

Predictor        Coef        Stdev     t-ratio        p
Constant       354.20        16.37       21.64    0.000
Size            63.86        21.56        2.96    0.004

s = 81.84        R-sq = 13.3%      R-sq(adj) = 11.8%
```

Reading from the output, $t = 2.96$ with $n - 2 = 59 - 2 = 57$ degrees of freedom and $P = 0.004$. Although size is statistically significant, R^2 is only 11.8% so the model with only size will probably not be adequate for prediction purposes.

b) The Minitab output for the t statistic assuming equal variances (pooled t) is given below and agrees with the regression output in terms of the t statistic, degrees of freedom and P value. The reason is the regression model with a 0 or 1 for size corresponds to

$$\mu_{\text{small}} = \beta_0 + \beta_1 x = \beta_0 \text{ and } \mu_{\text{large}} = \beta_0 + \beta_1 x = \beta_0 + \beta_1 .$$

The parameter β_1 in the regression corresponds to the difference in the means of the two groups, so a test that $\beta_1 = 0$ is the same as a test that the two group means are equal.

```
Two-sample T for Wages
Size    N         Mean         StDev     SE Mean
1      34        418.1          86.0          15
0      25        354.2          75.8          15

T-Test mu 1 = mu 0 (vs not =): T = 2.96   P = 0.0045   DF = 57
Both use Pooled StDev = 81.8
```

c) The plot of the residuals versus LOS show an increasing trend suggesting that adding LOS to the model that contains size would be useful for predicting wages. This turns out to be the case as LOS is highly significant in the two-variable model including size and LOS.

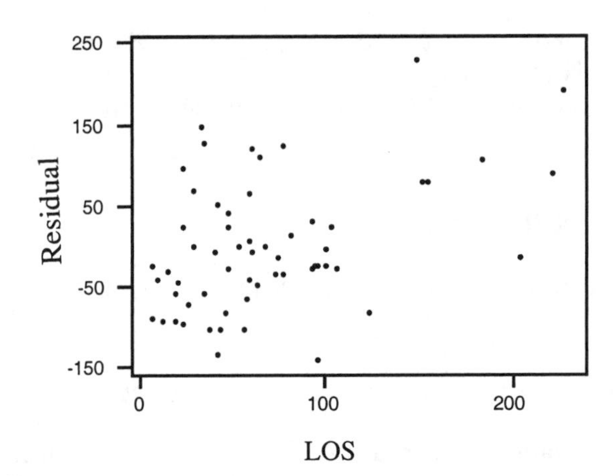

11.79 a) The Minitab output from fitting a model using year to predict corn is given below. A plot of year versus corn on the next page shows that yield has tended to go up during the 40 years the data was collected, and the general trend is fairly linear. It is not surprising that the effect of year is highly significant, $t = 13.06$ with 38 degrees of freedom and $P = 0.000$. The value of R^2 is 81.8% suggesting that much of the variability in corn yield is explained by the linear effect of year.

```
The regression equation is
Corn = -3545 + 1.84 Year

Predictor        Coef        Stdev     t-ratio        p
Constant      -3544.6        278.4      -12.73    0.000
Year           1.8396       0.1408       13.06    0.000

s = 10.28        R-sq = 81.8%     R-sq(adj) = 81.3%

Analysis of Variance

SOURCE        DF          SS           MS         F        p
Regression     1       18038        18038    170.60    0.000
Error         38        4018          106
Total         39       22055
```

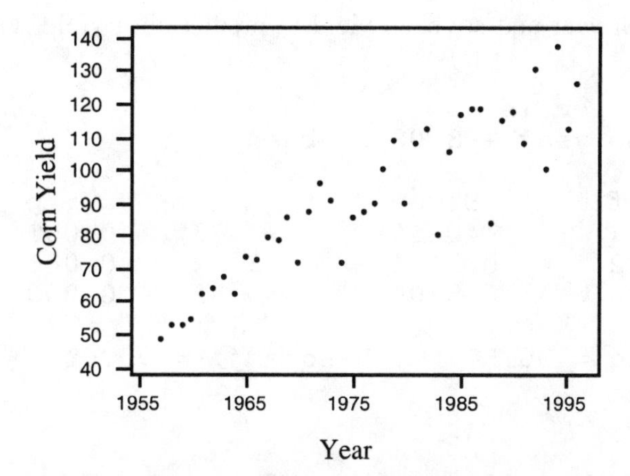

b) The normal quantile plot below shows no strong departures from normality. There are about three low outliers and these are also apparent from the plot of corn versus year. These years had yields that were below the general increasing linear trend.

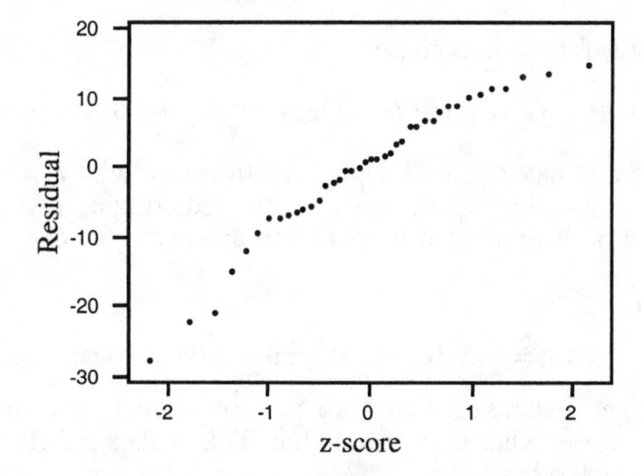

c) The plot of residuals versus soybean yield shows an increasing trend. This suggests that soybean yield might be useful in addition to year for predicting corn yield. This is not surprising as any departures that occur from the increasing linear trend in year for corn may occur for soybeans as well if these are generally poor years for crops. In this case, the soybean yield will have additional information. It can be seen from the plot of residuals versus soybean yield that the years in which there were large negative residuals (corn yield was below the general linear trend), soybean yields tended to be low as well.

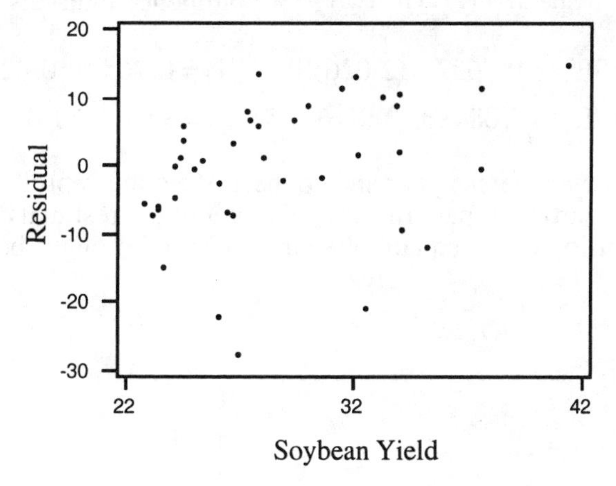

11.81 The model using both year and soybean yield to predict corn yield was fit on Minitab with the results given below.

```
The regression equation is
corn = -1510 + 0.765 year + 3.08 soybean

Predictor        Coef        Stdev     t-ratio        p
Constant      -1510.3        404.6       -3.73    0.001
year           0.7652       0.2114        3.62    0.001
soybean        3.0848       0.5300        5.82    0.000

s = 7.529        R-sq = 90.5%       R-sq(adj) = 90.0%

Analysis of Variance

SOURCE        DF          SS           MS         F        p
Regression     2     19957.9       9978.9    176.05    0.000
Error         37      2097.3         56.7
Total         39     22055.2
```

a) The ANOVA F test is for the hypotheses

$$H_0:\beta_1 = \beta_2 = 0 \text{ and } H_a\text{:at least one of the } \beta_j \text{ is not 0}$$

From the output, F = 176.05 and has 2 and 37 degrees of freedom, with P = 0.000. We conclude that at least one of the variables year or soybean yield is useful for predicting corn yield .

b) 90.5% of the variation is explained by both variables compared with 87.1% for soybean yield alone or 81.8% for year alone.

c) The fitted model is

$$\text{corn} = -1510 + 0.765 \text{ year} + 3.08 \text{ soybean}$$

The coefficients have changed because when both variables are in the model we are assessing the contribution of each given that the other is in the model. This makes a difference when the explanatory variables are correlated with each other.

d) The t statistics for year and soybean yield are 3.62 and 5.82, respectively. Both are highly significant indicating each variable makes an additional contribution for predicting corn yield with the other variable already in the model.

e) The confidence interval for a regression coefficient is $b_j \pm t^* \text{SE}_{b_j}$ where t^*= 2.026 as we are working with a t distribution with 37 degrees of freedom. The values of b_j and SE_{b_j} can be read from the output under Coef and Stdev, respectively. The two 95% confidence intervals are

Year: $b_j \pm t^* \text{SE}_{b_j} = 0.7652 \pm (2.026)(0.2114) = 0.7652 \pm 0.4283$

Soybean Yield: $b_j \pm t^* \text{SE}_{b_j} = 3.0848 \pm (2.026)(0.5300) = 3.0848 \pm 1.0738$

f) The plots of the residuals versus year and soybean yield are reproduced on the next page. The plot of the residuals shows no particular pattern, while the plot of the residuals versus year shows a curved pattern that suggests the inclusion of the square of year in the model might be helpful in explaining corn yield.

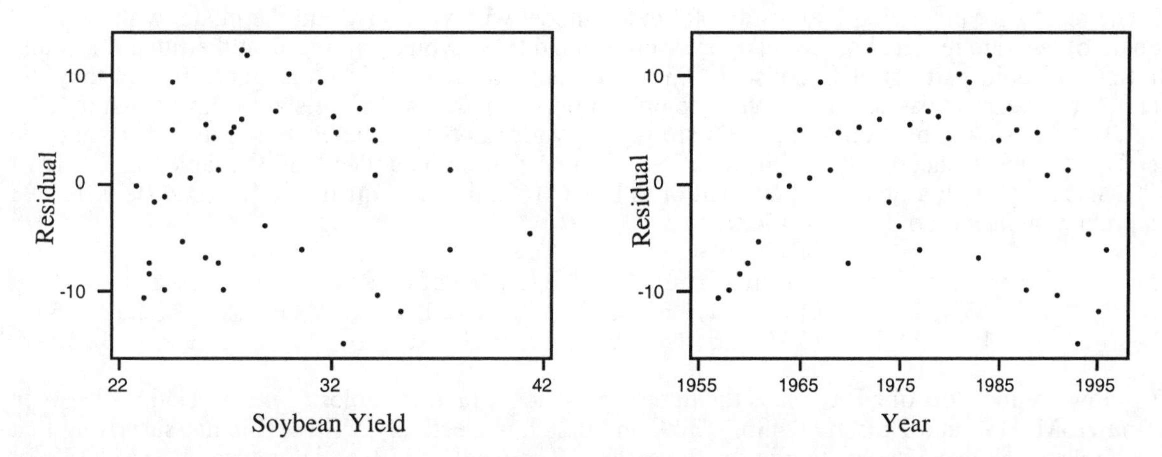

11.83 a) The partial output for the model with year and year2 = (year – 1978.5) was fit on Minitab with the results below. For year2, $t = -1.48$ with 37 degrees of freedom and $P = 0.148$. The result is that with year in the model, the square of year is not that useful for prediction. This is a different result than the one obtained in the previous exercise.

```
The regression equation is
corn = -3385 + 1.76 year - 0.0198 year2

Predictor        Coef         Stdev     t-ratio        p
Constant      -3385.0         294.7      -11.49    0.000
year           1.7602        0.1488       11.83    0.000
year2        -0.01985       0.01345       -1.48    0.148

s = 10.13          R-sq = 82.8%      R-sq(adj) = 81.9%
```

b) The plot in Exercise 11.82 suggests that adding the square of year to a model that contained year and soybean yield would be helpful in predicting corn yield. This turned out to be correct, as when the square of year was added to the model that contained year and soybean yield in Exercise 11.82, the coefficient of year2 was statistically significant. However, because the explanatory variables are correlated, the contribution of year2 in a model that contains only year will not necessarily be the same as its contribution in a model that contains both year and soybean yield.

c) The fitted line and the fitted quadratic are both given in the plot below. In general, the fitted values are quite similar, which is why the quadratic term was not significant. The larger differences tend to occur in the earlier and later years where the predicted values from the linear fit are higher than that for the quadratic.

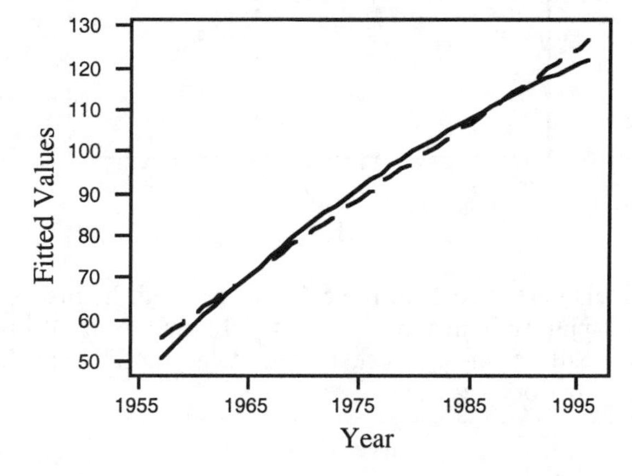

11.85 The results for predicting 2001 and 2002 in the model with year only and the model with year and the square of year are reported below. To better understand these, you should look at the fitted linear and quadratic models in part (c) of Exercise 11.83. The quadratic fit is below the linear fit and they get further apart as year increases. This is why the predictions from the two models are further apart in 2002 than in 2001. Also, the prediction intervals are getting wider as the predicted year gets further into the future. There is more uncertainty in this situation. To make sure you understand the table, the linear fit for 2002 is 138.24 with a prediction interval of (115.95, 160.54). The quadratic fit for 2002 is 127.98 with a prediction interval of (101.88, 154.09).

Year	Linear Fit	Linear P.I.	Quadratic Fit	Quadratic P.I.
2001	136.41	(114.20,158.61)	127.14	(101.82,152.46)
2002	138.24	(115.95,160.54)	127.98	(101.88,154.09)

11.87 Below are the plots of GPA versus the three high school grade variables. The correlations between GPA and HSM, HSS and HSE are 0.436, 0.329, and 0.289, respectively. Thus, it is not surprising that there is only a weak linear association in the three plots. There are no influential values of HSM, HSS, or HSE and there are no obvious outliers. The trend in all the plots is that as the high school grade variable goes up, there is a general tendency for the GPA to go up as well.

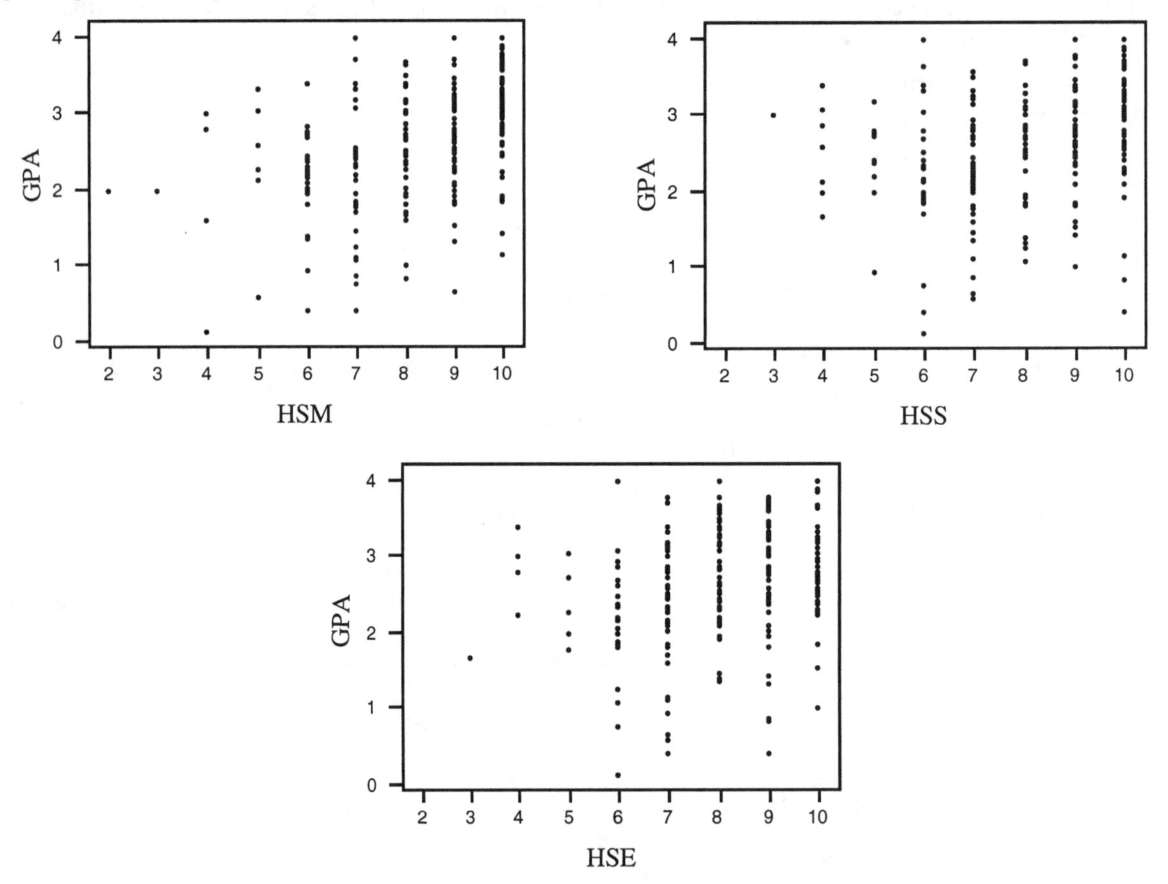

11.89 The plots of the residuals versus each of the SAT scores and the predicted values are given on the next page. There is no obvious pattern in any of the plots. The one thing of interest is that there are more large negative residuals than positive ones, suggesting that the model may be overpredicting GPA for certain combinations of the explanatory variables.

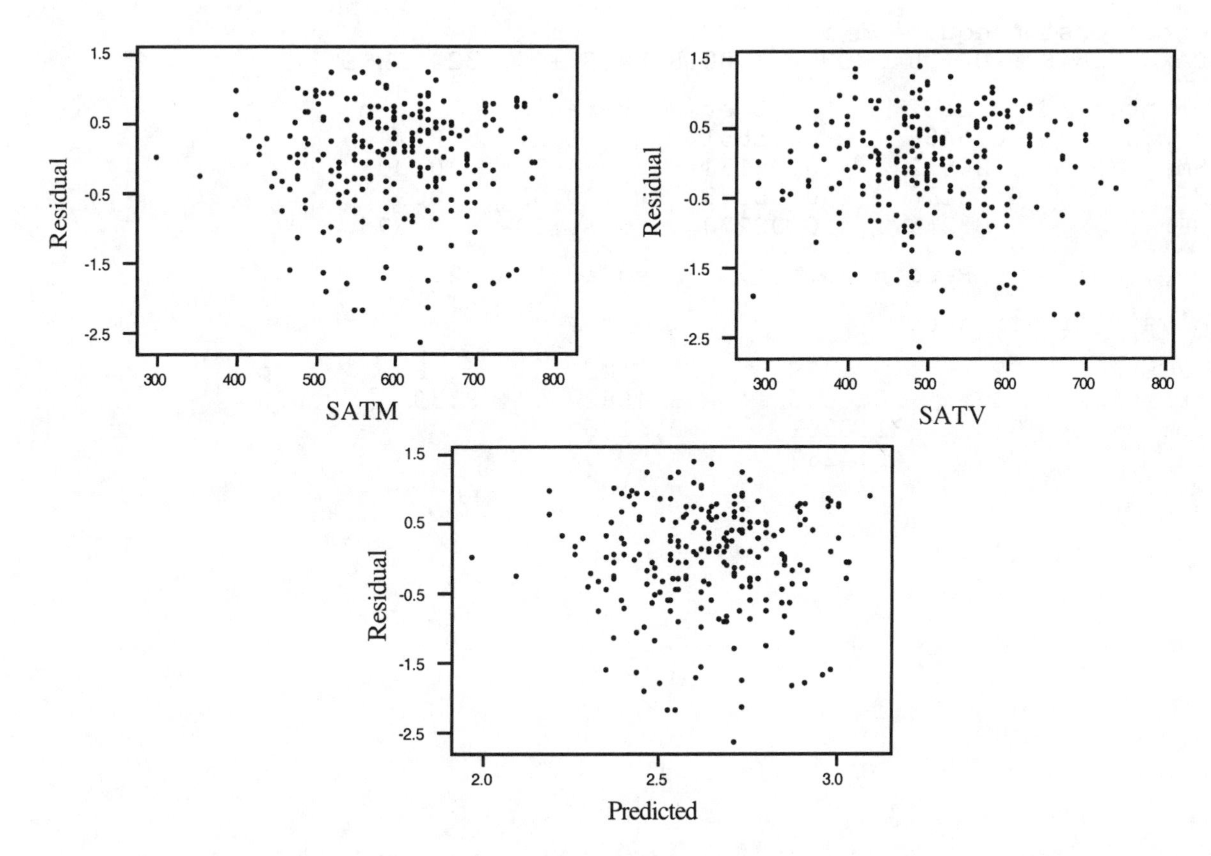

11.91 The Minitab output from fitting a model to predict GPA from HSE and SATV is given below. When looking at the *t* ratios, we see that SATV is of little use in predicting GPA in a model that already contains HSE. This is similar to the situation for the math predictors in that SATM was of little use in predicting GPA in a model that already contained HSM. The big difference in the models using the math variable versus those using the verbal variables is in the value of R^2, which measures how much of the variability in GPA is explained by the model. The model with the math variables explains close to 20% of the variability, while the model with the two verbal variables explains only 8.6%.

```
The regression equation is
GPA = 1.28 + 0.143 HSE +0.000394 SATV

Predictor        Coef         Stdev     t-ratio       p
Constant       1.2750        0.3474        3.67   0.000
HSE           0.14348       0.03428        4.19   0.000
SATV        0.0003942     0.0005582        0.71   0.481

s = 0.7487        R-sq = 8.6%       R-sq(adj) = 7.7%

Analysis of Variance

SOURCE        DF          SS          MS         F        p
Regression     2     11.5936      5.7968     10.34    0.000
Error        221    123.8692      0.5605
Total        223    135.4628
```

11.93 The model for females only was fitted on Minitab with the results given on the next page. The general results are quite similar to those found for males. Neither the science grade nor the English grade is of much use in predicting GPA in a model that contains the math grade variable. However, the use of high school grade variables as predictors explains more of the variability in GPA for females than for males, with an R^2 of 25.1% for females versus an R^2 of 18.4% for males.

```
The regression equation is
FGPA = 0.648 + 0.205 FHSM + 0.0018 FHSS + 0.0324 FHSE

Predictor        Coef       Stdev    t-ratio        p
Constant       0.6484      0.5551       1.17    0.247
FHSM          0.20512     0.06134       3.34    0.001
FHSS          0.00178     0.05873       0.03    0.976
FHSE          0.03243     0.08270       0.39    0.696

s = 0.6431      R-sq = 25.1%     R-sq(adj) = 22.1%

Analysis of Variance

SOURCE       DF          SS          MS          F        p
Regression    3     10.4046      3.4682       8.39    0.000
Error        75     31.0201      0.4136
Total        78     41.4247
```

CHAPTER 12

STATISTICS FOR QUALITY: CONTROL AND CAPABILITY

SECTION 12.1

OVERVIEW

In practice, work is often organized into a chain of activities that lead to some result. Such a chain of activities that turns inputs into outputs is called a **process**. A process can be described by a **flow chart,** which is a picture of the stages of a process. A **cause-and-effect diagram,** which displays the logical relationships between the inputs and output of a process, is also useful for describing and understanding a process.

All processes have variation. If the pattern of variation is stable over time, the process is said to be in statistical control. In this case, the sources of variation are called **common causes.** If the pattern is disrupted by some unusual event, **special cause** variation is added to the common cause variation. **Control charts** are statistical plots intended to warn when a process is disrupted or **out of control.**

Standard **3σ control charts** plot the values of some statistic Q for regular samples from the process against the time order in which the samples were collected. The **center line** of the chart is at the mean of Q. The **control limits** lie three standard deviations of Q above (the **upper control limit**) and below (the **lower control limit**) the center line. A point outside the control limits is an **out-of-control signal.** For **process monitoring** of a process that has been in control, the mean and standard deviations used to establish the center line and control limits are based on past data and are updated regularly.

When we measure some quantitative characteristic of a process, we use $\bar{x}$ and *s* **charts** for process control. The $\bar{x}$ chart plots the sample means of samples of size n from the process and the s chart the sample standard deviations. The s chart monitors variation within individual samples from the process. If the s chart is in control, the $\bar{x}$ chart monitors variation from sample to sample. To interpret charts, always look first at the s chart.

For a process that is in control with mean μ and standard deviation σ, the 3σ $\bar{x}$ chart based on samples of size n has center line and control limits

$$\text{CL} = \mu,\ \text{UCL} = \mu + 3\frac{\sigma}{\sqrt{n}},\ \text{LCL} = \mu - 3\frac{\sigma}{\sqrt{n}}$$

The 3σ s chart has control limits

$$\text{UCL} = (c_4 + 2c_5)\sigma = B_6\sigma,\ \text{LCL} = c_4 - 2c_5)\sigma = B_5\sigma$$

and the values of c_4, c_5, B_5, and B_6 can be found in Table 12.3 in the text for values of n from 2 to 10.

APPLY YOUR KNOWLEDGE

12.1 For this exercise, it is important to choose a process that you know well so that you can describe it carefully and recognize those factors that affect the process. We take as our example the process of making a cup of coffee. A possible flow chart and cause-and-effect diagram of the process follow.

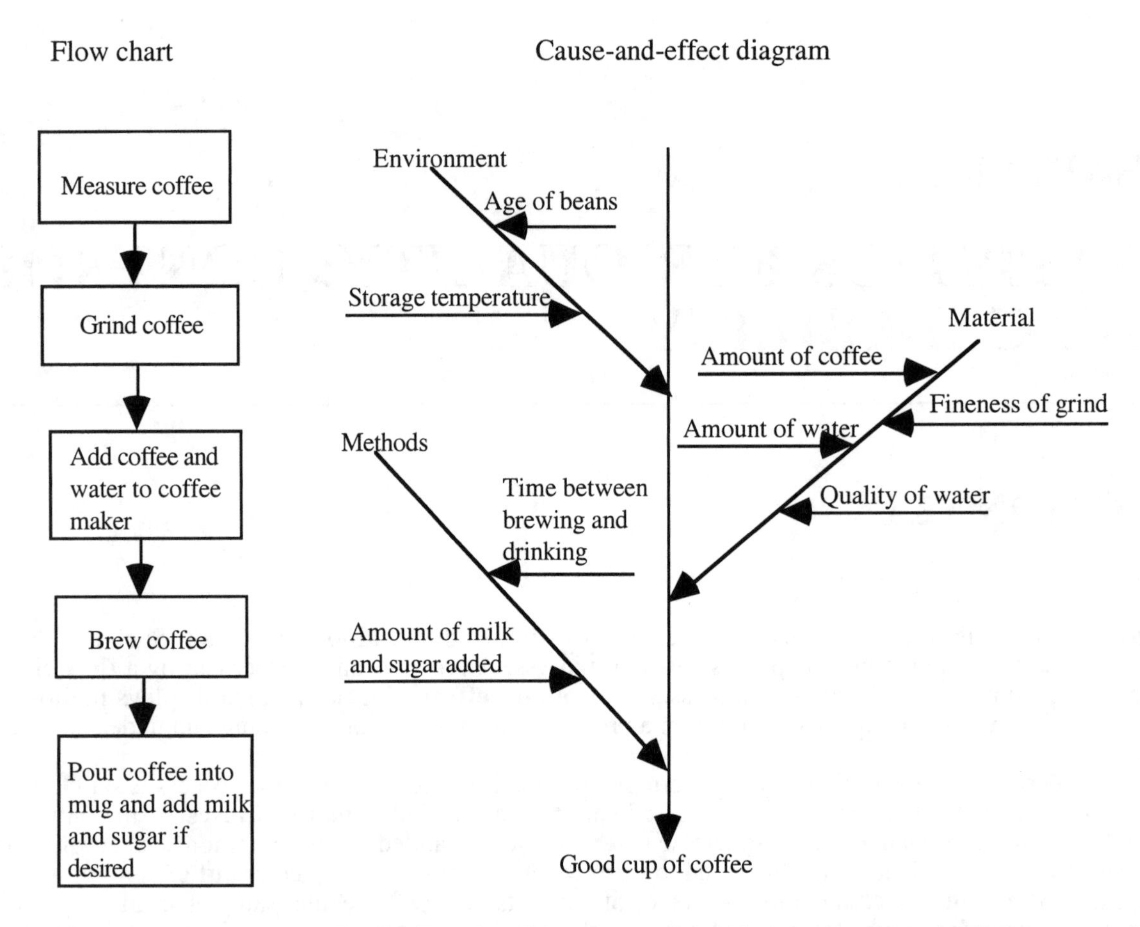

12.3 Adding the percents listed, one finds that the percent of total losses that these 9 DRGs account for is 80.5%. A Pareto chart of losses by DRG follows.

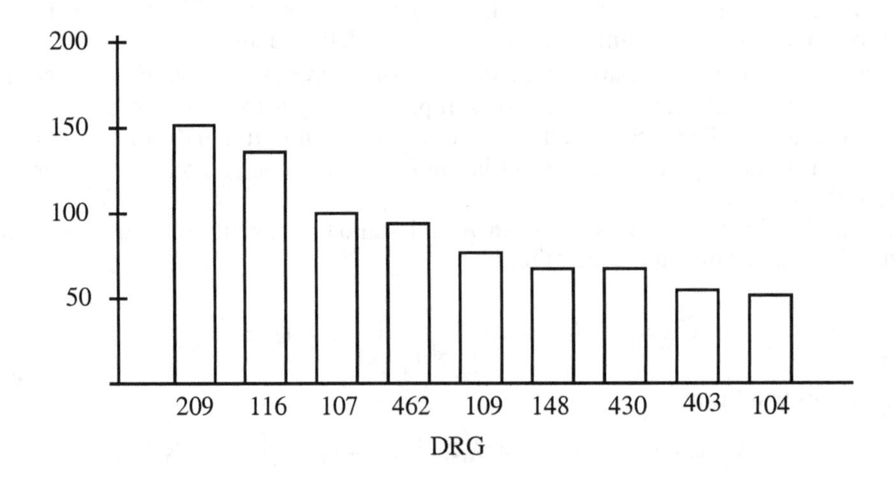

The hospital ought to study DRGs 209 and 116 first in attempting to reduce its losses. These are the two DRGs with the largest percent losses and combined account for nearly 30% of all losses.

12.5 In Exercise 12.1, we described the process of making a good cup of coffee. Some sources of common-cause variation are variation in how long the coffee has been stored and the conditions under which it has been stored, variation in the measured amount of coffee used, variation in finely ground the coffee is, variation in the amount of water added to the coffee maker, variation in the length of time the coffee sits between when it has finished brewing and when it is drunk, and variation in the amount of milk and/or sugar added.

Some special causes that might at times drive the process out of control would be a bad batch of coffee beans, a serious mismeasurement of the amount of coffee used or the amount of water used, a malfunction of the coffee maker or a power outage, interruptions that result in the coffee sitting a long time before it is drunk, and the use of milk that has gone bad.

12.7 a) The center line will be at height μ, which in this case we are told has target value $\mu = 11.5$. The control limits are (using $\sigma = 0.2$ and $n = 4$)

$$\text{UCL} = \text{upper control limit} = \mu + 3\frac{\sigma}{\sqrt{n}} = 11.5 + 3\frac{0.2}{\sqrt{4}} = 11.5 + 3(0.1) = 11.8$$

$$\text{LCL} = \text{lower control limit} = \mu - 3\frac{\sigma}{\sqrt{n}} = 11.5 - 3\frac{0.2}{\sqrt{4}} = 11.5 - 3(0.1) = 11.2$$

b) The $\bar{x}$ charts are as follows (we used Minitab to create the charts). Points outside the control limits are circled.

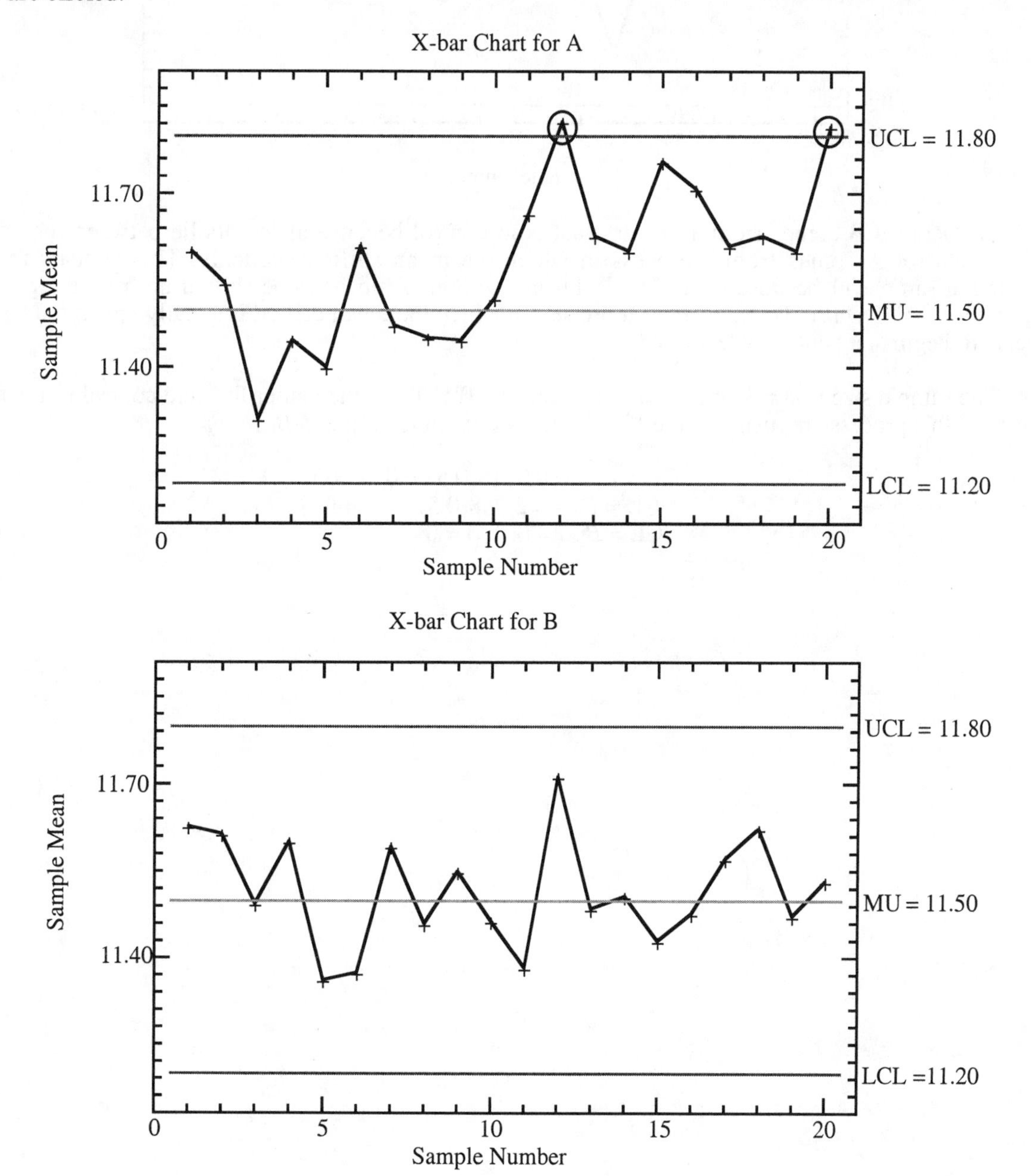

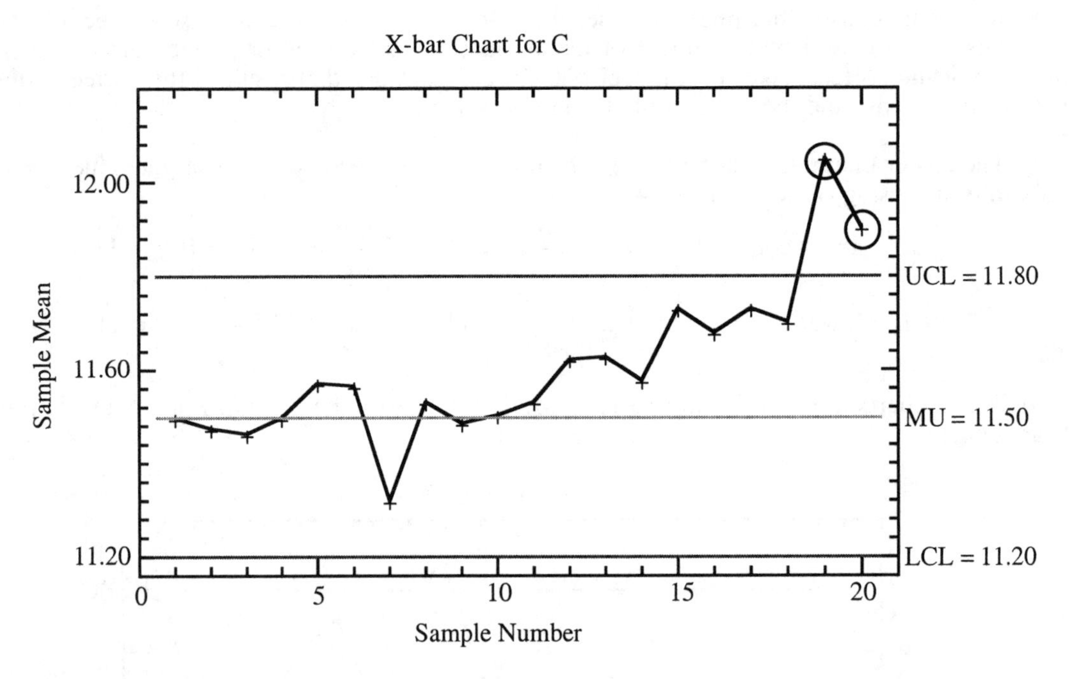

c) Data set B comes from a process that is in control because all points lie between the control limits. Data set A comes from a process in which the mean shifted suddenly. This appears to have occurred at sample 11 because from sample 11 on, the points appear to be shifted up from the previous values observed. Data set C comes from a process in which the mean drifted gradually upward. This drift appears to begin at a bout sample 11 or 12.

12.9 The sample size is $n = 4$, and we are told that $\sigma = 0.5$. Thus, the center line and control limits for an s chart for this process are (using Table 12.3 for the values of c_4, B_5, and B_6)

$$\begin{aligned} CL &= c_4\sigma = 0.9213(0.5) = 0.46065 \\ UCL &= B_6\sigma = 2.088(0.5) = 1.044 \\ LCL &= B_5\sigma = 0(0.5) = 0 \end{aligned}$$

SECTION 12.1 EXERCISES

12.11 Here is my flowchart for the daily process of getting to work on time.

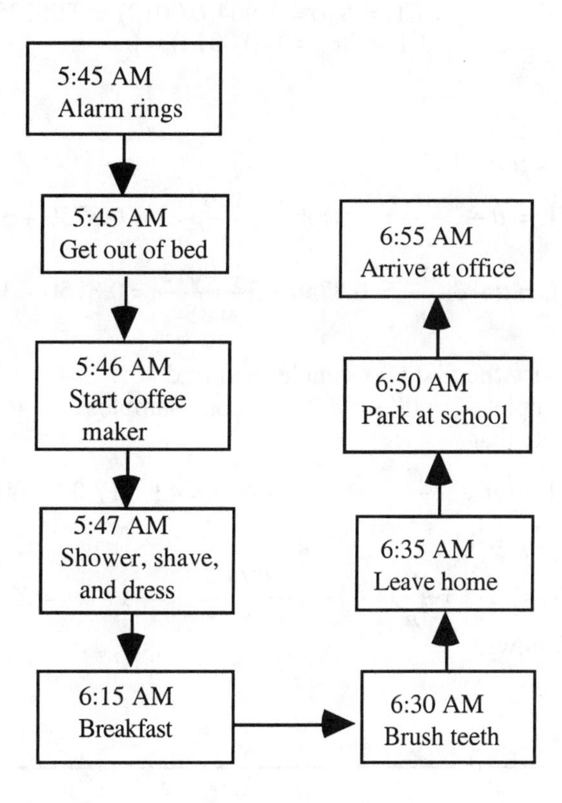

12.13 My main reasons for late arrivals at work are "don't get out of bed when the alarm rings" (responsible for about 40% of late arrivals), "too long eating breakfast" (responsible for about 25% of late arrivals), "too long showering, shaving, and dressing" (responsible for about 20% of late arrivals), and "slow traffic on the way to work" (responsible for about 15% of late arrivals). A Pareto chart that reflects this follows.

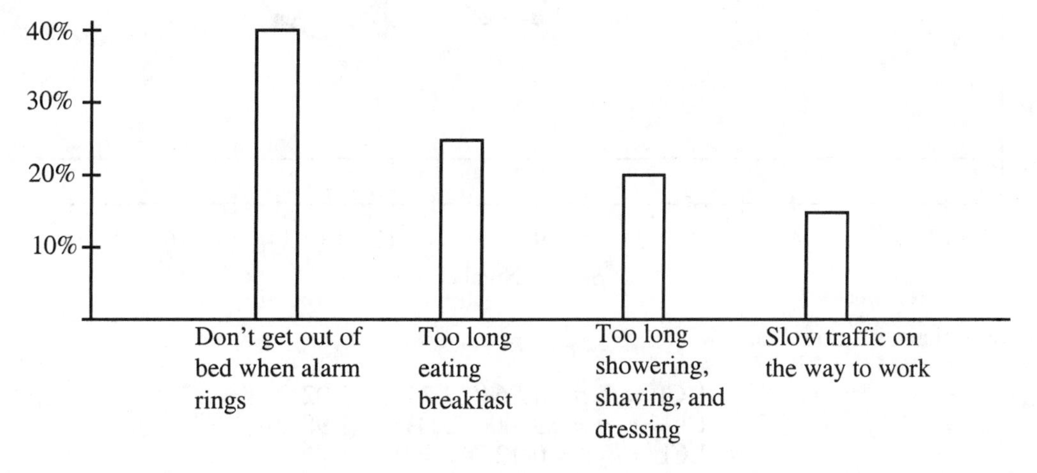

12.15 We are told that $n = 5$, $\mu = 0.8750$, and $\sigma = 0.0012$. Thus, for the s chart, using Table 12.3 for the values of c_4, B_5, and B_6)

$$\begin{aligned}
CL &= c_4\sigma = 0.9400(0.0012) = 0.001128 \\
UCL &= B_6\sigma = 1.964(0.0012) = 0.0023568 \\
LCL &= B_5\sigma = 0(0.0012) = 0
\end{aligned}$$

For the $\bar{x}$ chart,

$$CL = \mu = 0.8750$$

$$UCL = \mu + 3\frac{\sigma}{\sqrt{n}} = 0.8750 + 3\frac{0.0012}{\sqrt{5}} = 0.8750 + 3(0.00054) = 0.87662$$

$$LCL = \mu - 3\frac{\sigma}{\sqrt{n}} = 0.8750 - 3\frac{0.0012}{\sqrt{5}} = 0.8750 - 3(0.00054) = 0.87338$$

12.17 We compute $\bar{x}$ and s for the first two samples and find

First sample: $\bar{x} = 48$, $s = 8.94$; Second sample: $\bar{x} = 46$, $s = 13.03$

To make the $\bar{x}$ chart, we note that

$$UCL = \mu + 3\frac{\sigma}{\sqrt{n}} = 43 + 3\frac{12.74}{\sqrt{5}} = 43 + 17.09 = 60.09$$

$$CL = \mu = 43$$

$$LCL = \mu - 3\frac{\sigma}{\sqrt{n}} = 43 - 3\frac{12.74}{\sqrt{5}} = 43 - 17.09 = 25.91$$

resulting in the chart that follows.

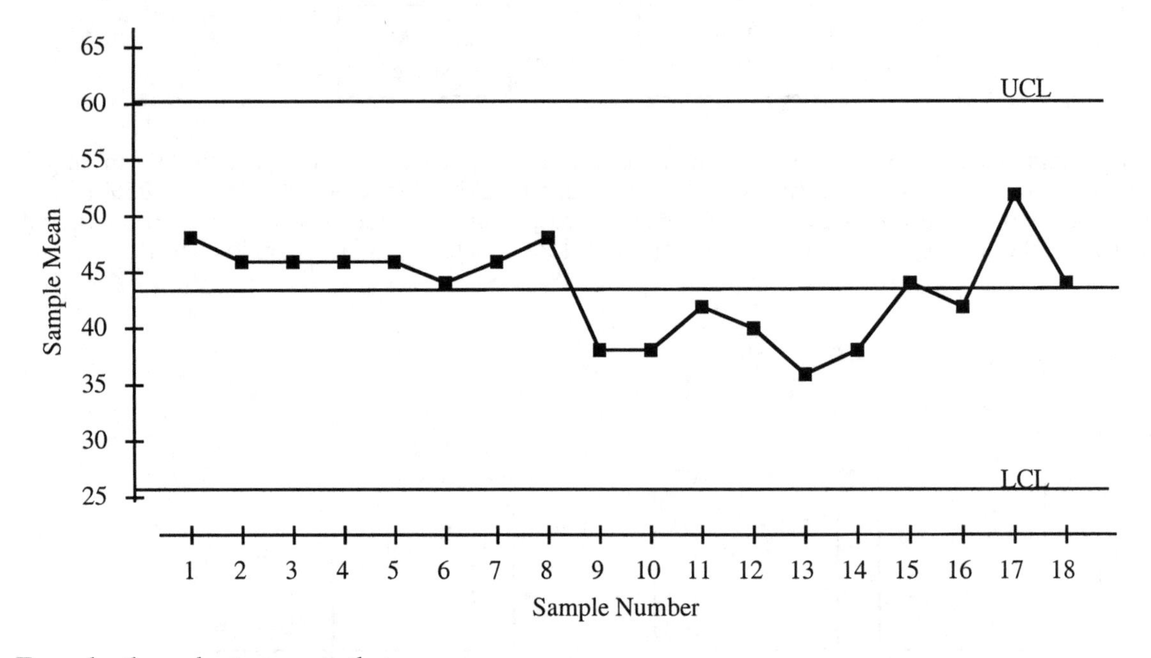

To make the s chart, we note that

$$\begin{aligned}
UCL &= B_6\sigma = 1.964(12.74) = 25.02 \\
CL &= c_4\sigma = 0.9400(12.74)) = 11.98 \\
LCL &= B_5\sigma = 0(12.74)) = 0
\end{aligned}$$

resulting in the chart that is given on the next page.

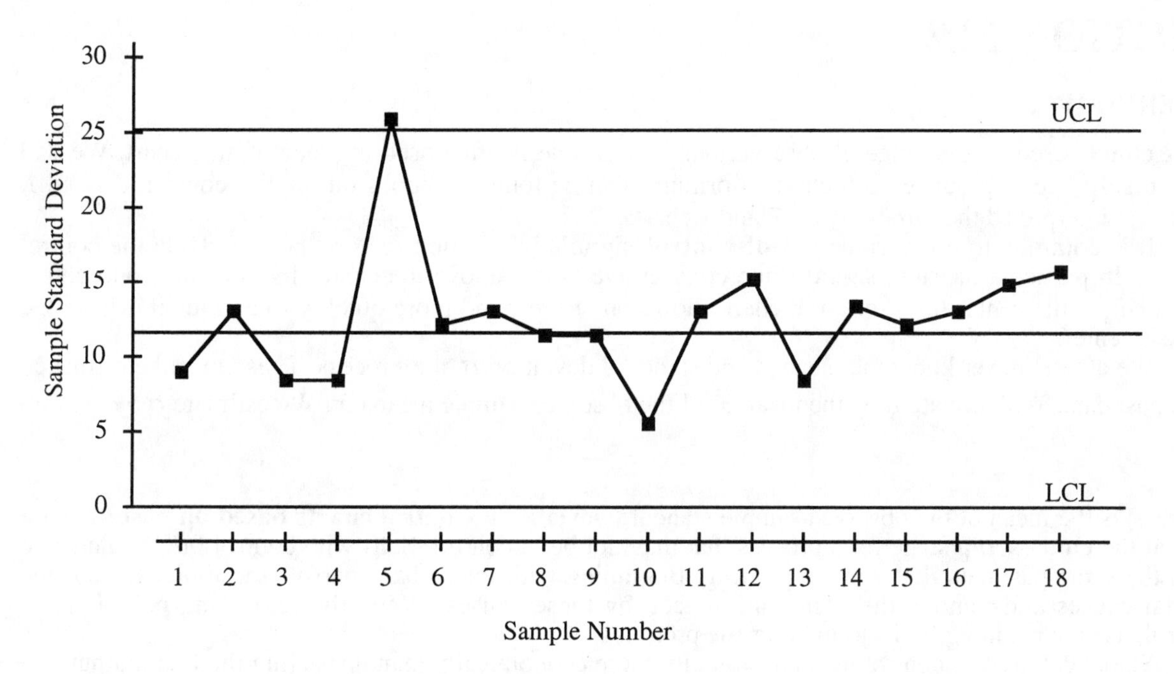

The s chart shows a lack of control at sample point 5 (8:15 3/8), but otherwise neither chart shows a lack of control. We would want to find out what happened at sample 5 to cause a lack of control in the s chart.

12.19 We assume that the observed data follow a normal distribution. After the disruption, observations will be normally distributed with mean $\mu = 693$ and standard deviation $\sigma = 12$. Thus, the mean $\bar{x}$ of a random sample of $n = 4$ observations after the disruption will have a

$$N(\mu, \frac{\sigma}{\sqrt{n}}) = N(693, \frac{12}{\sqrt{4}}) = N(693, 6)$$

distribution. We want to compute the probability that $\bar{x}$ will be beyond the control limits, i.e., will either be larger than UCL = 713 or smaller than LCL = 687. The desired probability is

$$\begin{aligned} P(\bar{x} > 713 \text{ or } \bar{x} < 687) &= P(\bar{x} > 713) + P(\bar{x} < 687) \\ &= P(\frac{\bar{x}-693}{6} > \frac{713-693}{6}) + P(\frac{\bar{x}-693}{6} < \frac{687-693}{6}) \\ &= P(Z > 3.33) + P(Z < -1) \\ &= 0.0004 + 0.1587 \\ &= 0.1591 \end{aligned}$$

using Table A.

12.21 a) If the process mean is μ and the standard deviation σ, then for samples of size n the $\bar{x}$ the 2σ control limits are based on $\mu_{\bar{x}} = \mu$ and $\sigma_{\bar{x}} = \frac{\sigma}{\sqrt{n}}$ and are

$$\text{UCL} = \mu_{\bar{x}} + 2\sigma_{\bar{x}} = \mu + 2\frac{\sigma}{\sqrt{n}}, \text{ LCL} = \mu_{\bar{x}} - 2\sigma_{\bar{x}} = \mu - 2\frac{\sigma}{\sqrt{n}}$$

b) For an s chart, we have seen in the text that $\mu_s = c_4\sigma$ and $\sigma_s = c_5\sigma$, so the 2σ control limits are

$$\text{UCL} = \mu_s + 2\sigma_s = c_4\sigma + 2c_5\sigma = (c_4 + 2c_5)\sigma, \text{ LCL} = \mu_s - 2\sigma_s = c_4\sigma - 2c_5\sigma = (c_4 - 2c_5)\sigma$$

SECTION 12.2

OVERVIEW

An ***R* chart** based on the range of observations in a sample is often used in place of an s chart. We will rely on software to produce such charts. Formulas can be found in books on quality control. $\bar{x}$ and R charts are interpreted the same way as $\bar{x}$ and s charts.

It is common to use various **out-of-control signals** in addition to "one point outside the control limits." In particular, a **runs signal** (nine consecutive points above the center line or nine consecutive points below the center line) for an $\bar{x}$ chart allows one to respond more quickly to a gradual drift in the process center.

We almost never know the mean μ and standard deviation σ of a process. These must be estimated from past data. We estimate μ by the mean $\bar{\bar{x}}$ of the observed sample means $\bar{x}$. We estimate σ by

$$\hat{\sigma} = \frac{\bar{s}}{c_4}$$

where $\bar{s}$ is the mean of the observed sample standard deviations. **Control charts based on past data** are used at the **chart setup** stage for a process that may not be in control. Start with control limits calculated from the same past data that you are plotting. Beginning with the s chart, narrow the limits as you find special causes and remove the points influenced by these causes. When the remaining points are in control, use the resulting limits to monitor the process.

Statistical process control maintains quality more economically than inspecting the final output of a process. Samples that are **rational subgroups** (subgroups that capture the features of the process in which we are interested) are important to effective control charts. A process in control is stable, so that we can predict its behavior. If individual measurements have a Normal distribution, we can give the **natural tolerances.**

A process is **capable** if it can meet or exceed the requirements placed on it. Control (stability over time) does not in itself improve capability. Remember that control describes the internal state of the process, whereas capability relates the state of the process to external specifications.

APPLY YOUR KNOWLEDGE

12.23 The samples will be either the mean of times from the experienced clerk or the mean of times from the inexperienced clerk. Means of times from the experienced clerk will be relatively small and vary little from sample to sample. Means of times from the inexperienced clerk will be relatively large and vary little from sample to sample. Presumably, a given sample is equally likely to be from the experienced clerk or the inexperienced clerk, and which clerk a sample comes from will be random. Thus, the $\bar{x}$ chart should display two types of points: those that have relatively small values and those with relatively large values. Both types should occur about equally often in the chart, and the pattern of large and small values should appear random. An $\bar{x}$ chart pattern that corresponds to this is given below. We have not included control limits on the chart because the purpose is merely to illustrate the pattern of points that we might expect to see.

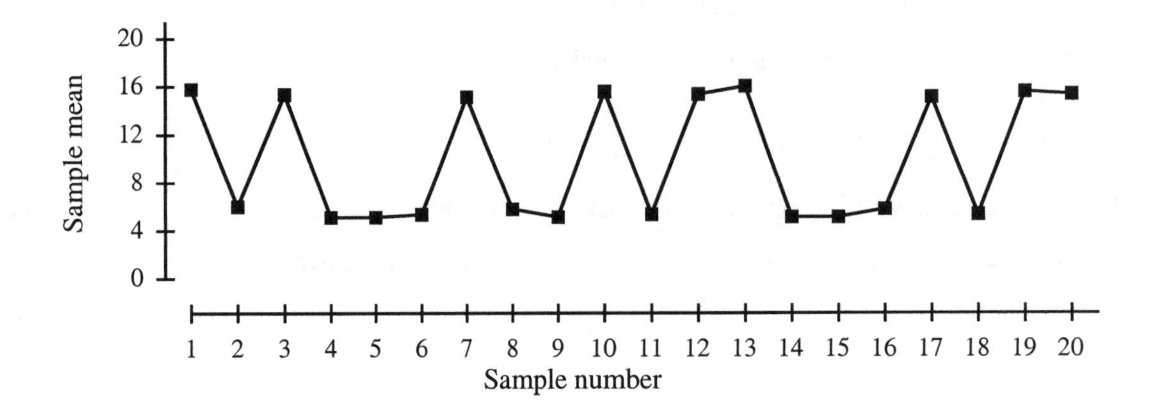

12.25 a) From the values of $\bar{x}$ and s in Table 12.1, we compute (using software)

$$\bar{\bar{x}} = \text{mean of the 20 values of } \bar{x} = 275.065$$
$$\bar{s} = \text{mean of the 20 values of } s = 34.55$$

hence we estimate μ to be

$$\hat{\mu} = \bar{\bar{x}} = 275.065$$

and we estimate σ to be (using the fact that the samples are each of size $n = 4$ and according to Table 12.3, $c_4 = 0.9213$)

$$\hat{\sigma} = \frac{\bar{s}}{c_4} = \frac{34.55}{0.9213} = 37.5$$

b) If we look at the s chart in Figure 12.7 we see that most of the points lie below 40 (and more than half of those below 40 lie well below 40), while of the points above 40, all but one (sample 12) are only slightly larger than 40. The s chart suggests that typical values of s are below 40, which is consistent with the estimate of σ in part (a).

12.27 By practicing statistical process control, the manufacturer of the monitors must sample and measure important characteristics of the monitors they produce. If these characteristics are out of control, the process must be adjusted to bring it back to being in control. The manufacturer is, in essence, inspecting samples of the monitors they produce and fixing any problems that arise. The control charts they create are a record of this inspection process. Incoming inspection is thus redundant and no longer necessary.

12.29 A normal quantile plot of the 120 individual patients in Table 12.6 is given below.

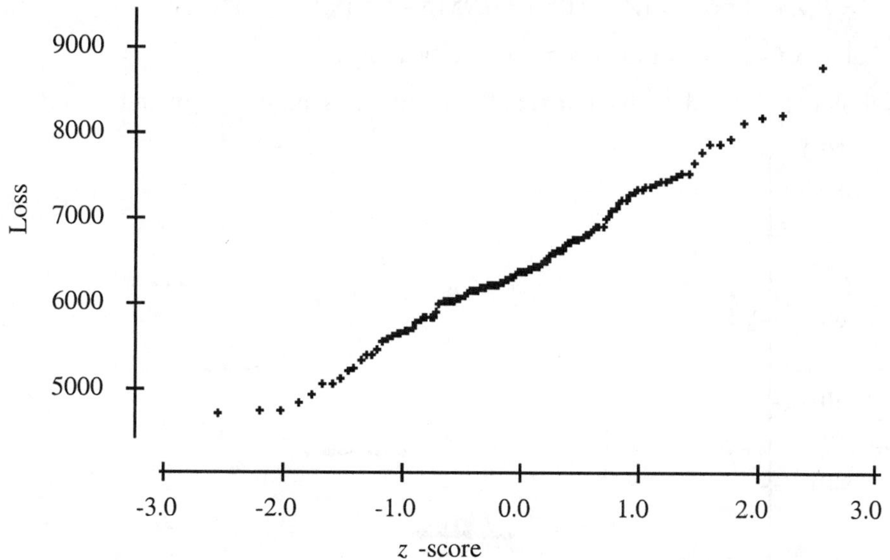

The plot shows no serious departures from normality. For normal data, the natural tolerances are trustworthy, so based on our normal quantile plot the natural tolerances we found in the previous exercise are trustworthy.

12.31 The new specifications are that the mesh tension should be between 150 and 350 mV. According to Problem 12.30, the standard deviation of the process is 38.4 mV. If we adjust the process mean to be 250 mV, then (assuming mesh tension is normally distributed) mesh tensions will follow an $N(\mu = 250, \sigma = 38.4)$ distribution. The probability a monitor has a mesh tension x that is within the new specifications is

$$\begin{aligned} P(150 < x < 350) &= P\left(\frac{150-250}{38.4} < \frac{x-250}{38.4} < \frac{350-250}{38.4}\right) = P(-2.60 < Z < 2.60) \\ &= P(Z < 2.60) - P(Z < -2.60) = 0.9953 - 0.0047 = 0.9906 \end{aligned}$$

where we have used Table A to compute $P(Z < 2.60)$ and $P(Z < -2.60)$. Thus, 99.06% of monitors will meet the new specifications if the process is adjusted to have center 250 mV.

SECTION 12.2 EXERCISES

12.33 The natural tolerances are $\mu \pm 3\sigma$. We do not know μ and σ, so we must estimate them from the data. We remove sample 5 from the data. Based on the remaining 17 samples, we find

$$\bar{\bar{x}} = \text{mean of the 17 values of } \bar{x} = 43.41$$
$$\bar{s} = \text{mean of the 17 values of } s = 11.65$$

hence we estimate μ to be

$$\hat{\mu} = \bar{\bar{x}} = 43.41$$

and we estimate σ to be (using the fact that the samples are each of size $n = 5$ and according to Table 12.3, $c_4 = 0.9400$)

$$\hat{\sigma} = \frac{\bar{s}}{c_4} = \frac{11.65}{0.9400} = 12.39$$

Based on these estimates, the natural tolerances for the distance between the holes are

$$\hat{\mu} \pm 3\hat{\sigma} = 43.41 \pm 3(12.39) = 43.41 \pm 37.17 \text{ or } 6.24 \text{ to } 80.58.$$

12.35 Based on the 17 samples that were in control, we saw in Exercise 12.33 that estimates of μ and σ are $\hat{\mu} = 43.41$ and $\hat{\sigma} = 12.39$. We therefore assume that distances between holes vary from meter to meter according to an $N(43.41, 12.39)$ distribution. The probability that the distance x between holes in a randomly selected meter is between 54 ± 10 (i.e., between 44 and 64) is thus

$$P(44 < x < 64) = P\left(\frac{44-43.41}{12.39} < \frac{x-43.41}{12.39} < \frac{64-43.41}{12.39}\right) = P(0.05 < Z < 1.66)$$
$$= P(Z < 1.66) - P(Z < 0.05) = 0.9515 - 0.5199 = 0.4316$$

We conclude that about 43.16% of meters meet specifications.

12.37 A normal plot of all the data that remain after removing sample 5 is given below.

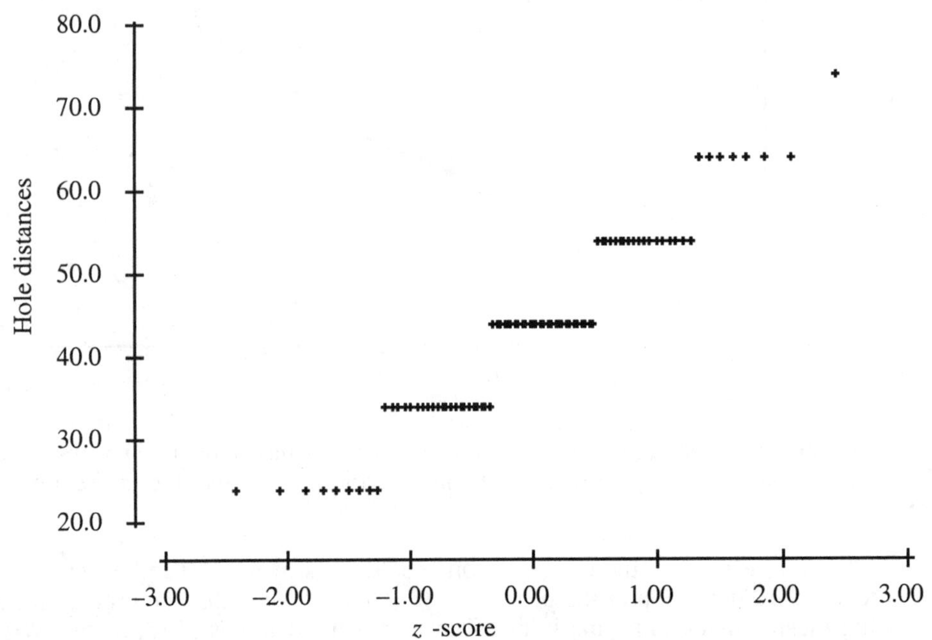

The large-scale pattern looks normal in that the centers of the horizontal bands of points appear to lie along a straight line. The most striking feature is the horizontal bands of points. These horizontal bands correspond to observations having the same value, resulting from the limited precision (round off). With greater precision, these observations would undoubtedly differ, but the differences are lost when we round off. The plot suggests that if we had more precise measurements the data would appear to come from a normal distribution. However, because of the lack of precision, we should probably view these data as only approximately normal and thus the calculations in Exercise 12.35 should only be considered approximate.

12.39 a) Presumably, the measurement of interest is the time to complete the checks. If all pilots suddenly adopt the policy of "work to rule" (and are now more careful and thorough about doing all the checks), one would expect to see a sudden increase in the time to complete the checks, but probably not much of a change in the variability of the time. Thus, we would expect to see a sudden change in level on the $\bar{x}$ chart.

b) Presumably, "accuracy" means less variable measurements. Thus, samples of measurements made by the laser system will have a much smaller standard deviation than samples of measurements made by hand. We would expect to see this reflected in a sudden change (decrease) in the level of an s or R chart.

c) Presumably, the measurement of interest is either the temperature in the office or the number of invoices prepared. If it is the temperature of the office, then as time passes, the temperature will gradually increase as outdoor temperature increases. This would be reflected in a gradual shift up on the $\bar{x}$ chart. If the measurement of interest is the number of invoices prepared, as the temperature in the office gradually increases, the workers gradually become less comfortable. They may take more breaks for water or become tired more quickly and thus accomplish less. We might expect to see a gradual decrease in the number of invoices prepared and hence a gradual shift down in the $\bar{x}$ chart.

12.41 The winning time for the Boston marathon has been gradually decreasing over time. The process has not become stable, but instead has steadily drifted downward. Thus two sources of variation are in the data. One is the process variation that we would observe if times were stable (random variation that we might observe over short periods), and the other is variation due the downward trend in times (the large differences between times from the early 1950s compared to recent times). In using the standard deviation s of all the times between 1950 and 2002, we include both of the sources of variation. However, to determine if times in the next few years are unusual, we should consider only the process variation (variability in recent times where the process is relatively stable). Using s overestimates the process variation, resulting in control limits that are too wide to effectively signal unusually fast or slow times.

SECTION 12.3

OVERVIEW

The lower and upper specification limits (LSL and USL) for a process define the acceptable set of values of the output of a process. **Capability indexes** compare process variability σ to the process specifications. The capability index C_p is defined to be

$$C_p = \frac{\text{USL} - \text{LSL}}{6\sigma}$$

and if the process mean is μ, the capability index C_{pk} is defined to be

$$C_{pk} = \frac{|\mu - \text{nearer spec limit}|}{3\sigma}$$

We set C_{pk} to be 0 if the process mean lies outside the specification limits. Larger values of these indexes indicate higher capability. We usually do not know μ and σ and so we use estimates $\hat{\mu}$ and $\hat{\sigma}$ in the formulas for C_p and C_{pk} leading to the estimates

$$\hat{C}_p = \frac{\text{USL} - \text{LSL}}{6\hat{\sigma}}$$

$$\hat{C}_{pk} = \frac{|\hat{\mu} - \text{nearer spec limit}|}{3\hat{\sigma}}$$

Interpretation of C_p and C_{pk} requires that measurements on the process output have a roughly Normal distribution. These indexes are not meaningful unless the process is in control so that its center and variability are stable.

Estimates of C_p and C_{pk} can be quite inaccurate when based on small numbers of observations, due to sampling variability. You should mistrust estimates not based on at least 100 measurements.

APPLY YOUR KNOWLEDGE

12.43 LSL and USL are the lower and upper specification limits and define the acceptable set of values for *individual* measurements on the output of a process. These limits represent the *desired performance* of the process. LCL and UCL for $\bar{x}$ are the lower and upper control limits for the *means of samples* of several individual measurements on the output of a process. They represent the *actual performance* of the process when it is in control (after special causes have been removed).

12.45 a) We are given that estimates of estimate μ and σ are $\hat{\mu}$ = 275.065 and $\hat{\sigma}$ = 38.38. LSL = 100 and USL = 400, so we estimate C_p and C_{pk} to be

$$\hat{C}_p = \frac{\text{USL} - \text{LSL}}{6\hat{\sigma}} = \frac{400 - 100}{6(38.38)} = 1.303$$

$$\hat{C}_{pk} = \frac{|\hat{\mu} - \text{nearer spec limit}|}{3\hat{\sigma}} = \frac{|275.065 - 400|}{3(38.38)} = 1.085$$

b) Specifications are now LSL = 150 and USL = 350, so now

$$\hat{C}_p = \frac{\text{USL} - \text{LSL}}{6\hat{\sigma}} = \frac{350 - 150}{6(38.38)} = 0.869$$

$$\hat{C}_{pk} = \frac{|\hat{\mu} - \text{nearer spec limit}|}{3\hat{\sigma}} = \frac{|275.065 - 350|}{3(38.38)} = 0.651$$

12.47 a) A sketch of the density curve with LSL and USL marked follows.

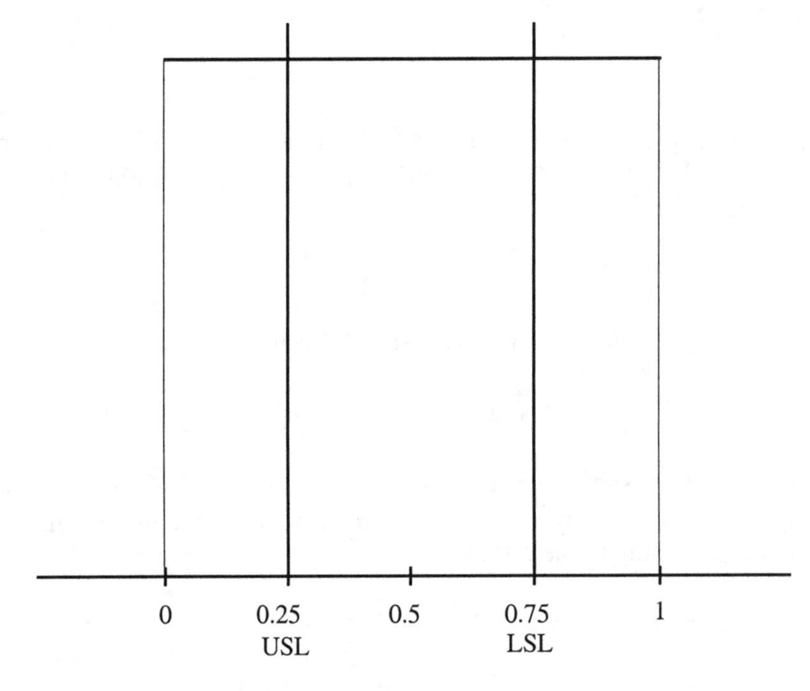

We are to ld that $\mu = 0.5$ and $\sigma = 1/12$, so

$$C_{pk} = \frac{|\mu - \text{nearer spec limit}|}{3\sigma} = \frac{|0.5 - 0.25|}{3(1/12)} = 1.$$

From the skethc of the density curve, we see that the area between USL and LSL is 0.5, so 50% of the output meets specifications.

b) Output follows an $N(0.5, 1/12)$ distribution. Let X denote a $N(0.5, 1/12$ random variable. Then the proportionof the output meeting specifications is

$$P(\text{LSL} < X < \text{USL}) = P(0.25 < X < 0.75) = P\left(\frac{0.25-0.5}{1/12} < z < \frac{0.75-0.5}{1/12}\right) = P(-3 < z < 3) = 0.9974.$$

Thus, 99.74% of the output meets specifications.

c) Interpretation of C_{pk} requires that measurements on the process output have roughly a Normal distribution.

SECTION 12.3 EXERCISES

12.49 a) Based on the remaining 17 samples, we find

$$\bar{\bar{x}} = \text{mean of the 17 values of } \bar{x} = 43.41$$
$$\bar{s} = \text{mean of the 17 values of } s = 11.65$$

hence we estimate μ to be

$$\hat{\mu} = \bar{\bar{x}} = 43.41$$

and we estimate σ to be (using the fact that the samples are each of size $n = 5$ and according to Table 12.3, $c_4 = 0.9400$)

$$\hat{\sigma} = \frac{\bar{s}}{c_4} = \frac{11.65}{0.9400} = 12.39$$

A sketch of an $N(43.41, 12.39)$ distribution showing the specification limits 54 ± 10 (or LSL = 44 and USL = 64) is given below.

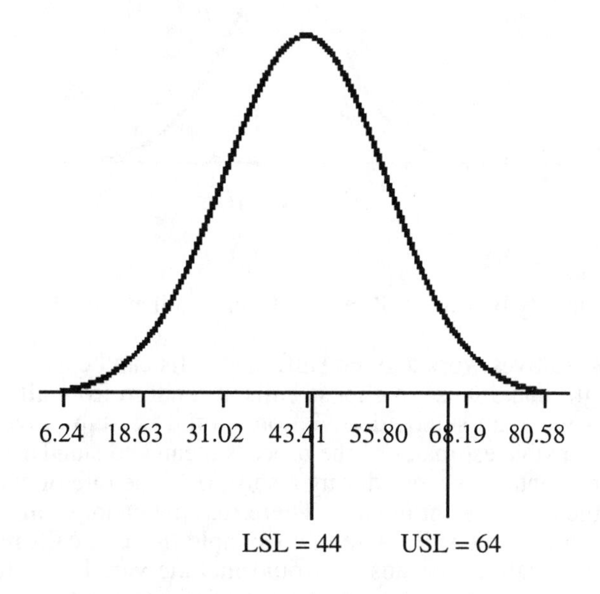

b) We estimate

$$\hat{C}_p = \frac{\text{USL} - \text{LSL}}{6\hat{\sigma}} = \frac{64-44}{6(12.39)} = 0.27$$

$$\hat{C}_{pk} = \frac{|\hat{\mu} - \text{nearer spec limit}|}{3\hat{\sigma}} = \frac{|43.41-44|}{3(12.39)} = 0.02$$

The capability is poor (both indices are very small). The reasons are that the process is not centered (we estimate the process mean to be 43.41 but the midpoint of the specification limits is 54) and the process variability is large (we saw in Exercise 12.33 that the natural tolerances for the process are $\hat{\mu} \pm 3\hat{\sigma} = 43.41 \pm 3(12.39) = 43.41 \pm 37.17$, which is much wider than the specification limits of 54 ± 10).

12.51 a) We would estimate the process mean and standard deviation to be $\hat{\mu} = \bar{\bar{x}} = \bar{x} = 14.99$ and $\hat{\sigma} = s = 0.2239$. In its current state, therefore, measurements on the process will vary (approximately) according to an $N(14.99, 0.2239)$ distribution. The probability that a measurement x lies within the specifications 15 ± 0.5 , i.e., is between 14.5 and 15.5, is

$$P(14.5 < x < 15.5) = P\left(\frac{14.50-14.99}{0.2239} < \frac{x-14.99}{0.2239} < \frac{15.50-14.99}{0.2239}\right) = P(-2.18 < Z < 2.28)$$
$$= P(Z < 2.28) - P(Z < -2.18) = 0.9887 - 0.0146 = 0.9741$$

Thus, about 97.41% of clip openings will meet specifications if the process remains in its current state.

b) We estimate C_{pk} to be

$$\hat{C}_{pk} = \frac{|\hat{\mu} - \text{nearer spec limit}|}{3\hat{\sigma}} = \frac{|14.99-14.50|}{3(0.2239)} = 0.73$$

12.53 $C_p = \dfrac{\text{USL} - \text{LSL}}{6\sigma}$ so if $C_p \geq 2$,we must have USL – LSL $\geq 2(6\sigma) = 12\sigma$. If the process is properly centered, then its mean μ will be halfway between USL and LSL, and hence USL and LSL will be at least 6σ from μ. A sketch of a normal distribution in which USL and LSL are 6σ from μ is given below.

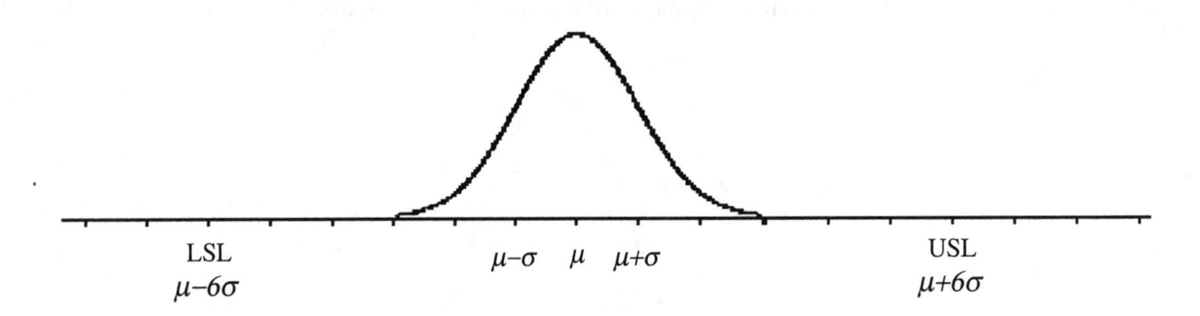

This is called six sigma quality because LSL and USL are both at least 6σ from the process mean μ.

12.55 If the same representatives work a given shift, and calls can be expected to arrive at random during a given shift, then it might make sense to choose calls at random from all calls received during the shift. In this case, the process mean ad standard deviation should be stable over the entire shift and random selection should lead to sensible estimates of the process mean and standard deviation..

If the different representatives work during a shift, or if the rate of calls varies over a shift, then the process may not be stable over the entire shift. There may be changes in either the process mean or the process standard deviation over the shift. A random sample from all calls received during the shift would overestimate the process variability because it would include variability due to special causes. Stability would more likely be seen over only much shorter time periods and thus it would be more reasonable to time six consecutive calls.

12.57 From Table 12.9, we see that the three outliers are a call of 276 seconds (occurring in Sample 28) resulting in a value of $s = 107.20$, a call of 244 seconds (occurring in Sample 42 resulting in a value of $s = 93.68$, and a call of 333 seconds (occurring in Sample 46) resulting in a value of $s = 119.53$. When the outliers are omitted, the values of s become

Sample 28: $s = 9.28$ Sample 42: $s = 6.71$ Sample 46: $s = 31.01$

SECTION 12.4

OVERVIEW

There are control charts for several different types of process measurements. One important type is the p **chart** which is a control chart based on plotting sample proportions $\hat{p}$ from regular samples from a process against the order in which the samples were taken. We estimate the process proportion p of "successes" by

$$\bar{p} = \frac{\text{total number of successes in past samples}}{\text{total number of opportunities in these samples}}$$

and then the control limits for a p chart for future samples of size n are

$$\text{UCL} = \bar{p} + 3\sqrt{\frac{\bar{p}(1-\bar{p})}{n}},\ \text{CL} = \bar{p},\ \text{LCL} = \bar{p} - 3\sqrt{\frac{\bar{p}(1-\bar{p})}{n}}$$

The interpretation of p charts is very similar to that of $\bar{x}$ charts. The out-of-control signals used are also the same as for $\bar{x}$ charts.

SECTION 12.4 EXERCISES

12.59 a) The average number of invoices over the 10-month period is 2875 per month. Because each invoice has the potential to be unpaid, the total number of invoices in the 10-month period is the total number of opportunities for unpaid invoices. Thus,

$$2875 = \frac{\text{total number of opportunities for unpaid invoices}}{10}$$

so that

$$\text{total number of opportunities for unpaid invoices} = 10(2875) = 28{,}750$$

960 of these invoices were unpaid after 30 days (these would be our "successes" in past samples). so

$$\bar{p} = \frac{\text{total number of successes in past samples}}{\text{total number of opportunities in these samples}} = \frac{960}{28750} = 0.0334$$

b) We expect 2875 invoices per month, so our p chart will be based on samples of size $n = 2875$. The control limits for a p chart for future samples of size $n = 2875$ are

$$\text{UCL} = \bar{p} + 3\sqrt{\frac{\bar{p}(1-\bar{p})}{n}} = 0.0334 + 3\sqrt{\frac{0.0334(1-0.0334)}{2875}} = 0.0334 + 0.0101 = 0.0435$$

$$\text{CL} = \bar{p} = 0.0334$$

$$\text{LCL} = \bar{p} - 3\sqrt{\frac{\bar{p}(1-\bar{p})}{n}} = 0.0334 - 3\sqrt{\frac{0.0334(1-0.0334)}{2875}} = 0.0334 - 0.0101 = 0.0233$$

12.61 The total number of opportunities for missing or deformed rivets is just the total number of 34700 rivets, because each rivet has the possibility of being missing or deformed. The number of "successes" in past samples is just the 208 missing or deformed rivets in the recent data. We therefore estimate the process proportion p of "successes" from the recent data by

$$\bar{p} = \frac{\text{total number of successes in past samples}}{\text{total number of opportunities in these samples}} = \frac{208}{34700} = 0.00599$$

The next wing contains $n = 1070$ rivets, and the control limits for a p chart for future samples of size $n = 1070$ are

$$\text{UCL} = \bar{p} + 3\sqrt{\frac{\bar{p}(1-\bar{p})}{n}} = 0.00599 + 3\sqrt{\frac{0.00599(1-0.00599)}{1070}} = 0.00599 + 0.00708 = 0.01307$$

$$\text{CL} = \bar{p} = 0.00599$$

$$\text{LCL} = \bar{p} - 3\sqrt{\frac{\bar{p}(1-\bar{p})}{n}} = 0.00599 - 3\sqrt{\frac{0.00599(1-0.00599)}{1070}} = 0.00599 - 0.00708 = 0$$

Note that in the LCL, we set negative values to 0 because a proportion can never be less than 0.

12.63 a) The total number of opportunities for a student to have three or more unexcused absences from school in a month is just the total of the numbers of students each month. This is 911 + 947 + 939 + 942 + 918 + 920 + 931 + 925 + 902 + 883 = 9218. The total number of "successes" is just the total of the numbers of students that had three or more unexcused absences from school in a month. This is 291 + 349 + 364 + 335 + 301 + 322 + 344 + 324 + 303 + 344 = 3277. Thus,

$$\bar{p} = \frac{\text{total number of successes in past samples}}{\text{total number of opportunities in these samples}} = \frac{3277}{9218} = 0.356$$

We also find the average number of students per month for this 10-month period is

$$\bar{n} = \frac{9218}{10} = 921.8$$

b) Assuming $n = \bar{n} = 921.8$ each month, the control limits for a p chart for future samples of size $n = 921.8$ are

$$\text{UCL} = \bar{p} + 3\sqrt{\frac{\bar{p}(1-\bar{p})}{n}} = 0.356 + 3\sqrt{\frac{0.356(1-0.356)}{921.8}} = 0.356 + 0.0473 = 0.4033$$

$$\text{CL} = \bar{p} = 0.356$$

$$\text{LCL} = \bar{p} - 3\sqrt{\frac{\bar{p}(1-\bar{p})}{n}} = 0.356 - 3\sqrt{\frac{0.356(1-0.356)}{921.8}} = 0.356 - 0.0473 = 0.3087$$

A p chart based on these control limits is

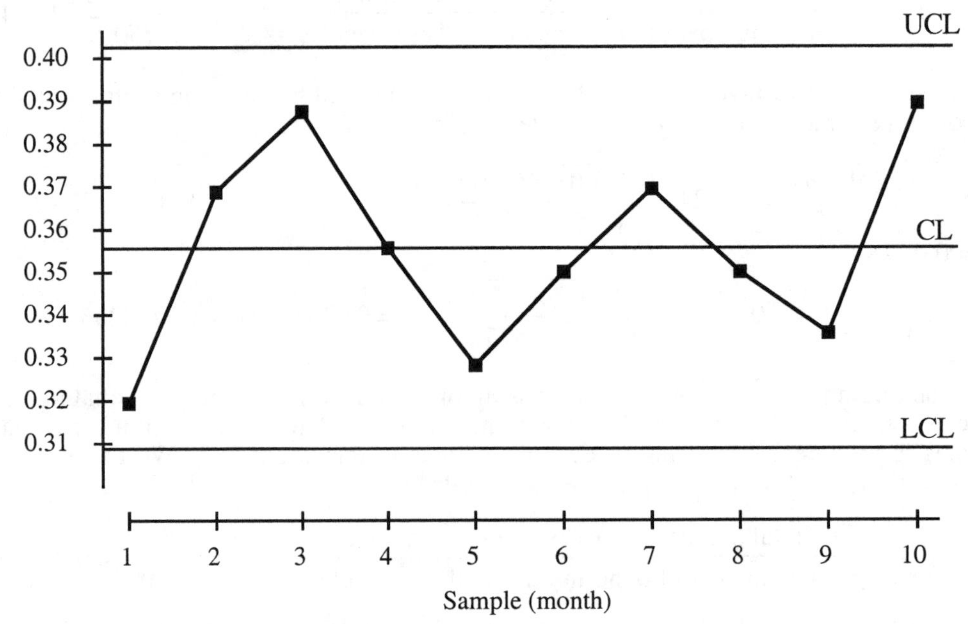

The process (proportion of students per month with three or more unexcused absences) appears to be in control, i.e., there are no months with unusually high or low proportions of absences, and no obvious trends.

c) For October,

$$\text{UCL} = \bar{p} + 3\sqrt{\frac{\bar{p}(1-\bar{p})}{n}} = 0.356 + 3\sqrt{\frac{0.356(1-0.356)}{947}} = 0.356 + 0.0467 = 0.4027$$

$$\text{LCL} = \bar{p} - 3\sqrt{\frac{\bar{p}(1-\bar{p})}{n}} = 0.356 - 3\sqrt{\frac{0.356(1-0.356)}{947}} = 0.356 - 0.0467 = 0.3093$$

For June,

$$\text{UCL} = \bar{p} + 3\sqrt{\frac{\bar{p}(1-\bar{p})}{n}} = 0.356 + 3\sqrt{\frac{0.356(1-0.356)}{883}} = 0.356 + 0.0483 = 0.4043$$

$$\text{LCL} = \bar{p} - 3\sqrt{\frac{\bar{p}(1-\bar{p})}{n}} = 0.356 - 3\sqrt{\frac{0.356(1-0.356)}{883}} = 0.356 - 0.0483 = 0.3077$$

These limits are added to the p chart, reproduced below.

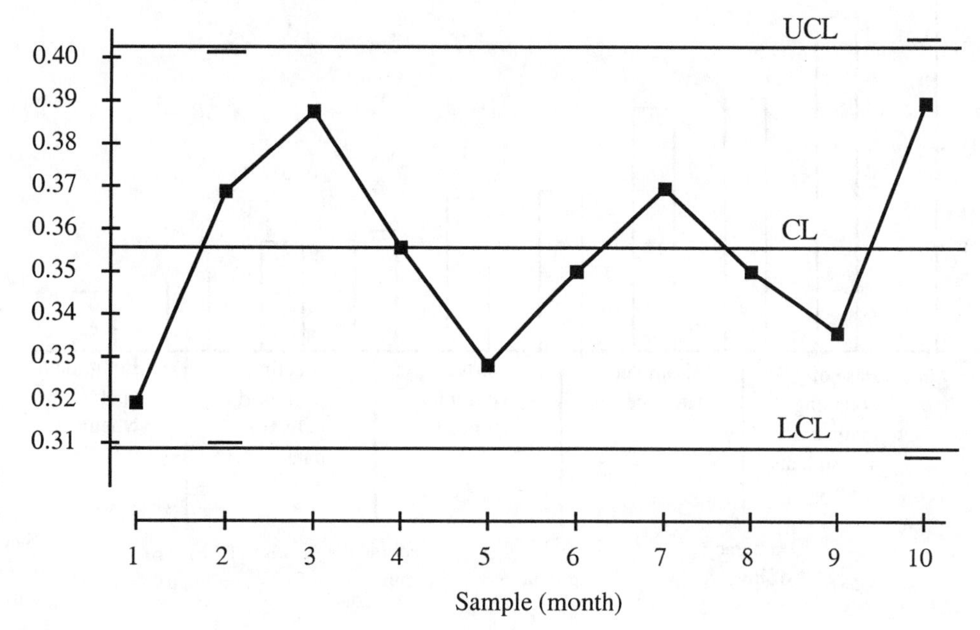

We can see that these exact limits do not affect our conclusions in this case.

12.65 a) The defect rate ("successes') is 8000 per million opportunities, so

$$\bar{p} = \frac{\text{total number of successes in past samples}}{\text{total number of opportunities in these samples}} = \frac{8{,}000}{1{,}000{,}000} = 0.008$$

If the manufacturer processes $n = 500$ orders per month, we would expect to see

$$n\,\bar{p} = 500(0.008) = 4$$

defective orders per month.

b) The center line and control limits for a p chart for plotting monthly proportions of defective orders is (because these proportions will be based on n = 500 orders)

$$\text{UCL} = \bar{p} + 3\sqrt{\frac{\bar{p}(1-\bar{p})}{n}} = 0.008 + 3\sqrt{\frac{0.008(1-0.008)}{500}} = 0.008 + 0.012 = 0.020$$

$$\text{CL} = \bar{p} = 0.008$$

$$\text{LCL} = \bar{p} - 3\sqrt{\frac{\bar{p}(1-\bar{p})}{n}} = 0.008 - 3\sqrt{\frac{0.008(1-0.008)}{500}} = 0.008 - 0.012 = 0$$ (negative values are set to 0 because a proportion can never be less than 0)

To be above the UCL, the proportion must be > 0.020. There are 500 orders per month, so the number N of defective orders needed to yield a proportion > 0.020 is N > 500(0.020) = 10.

CHAPTER 12 REVIEW EXERCISES

12.67 a) The percents add up to a value much larger than 100%, This is because customers can have more than one complaint.

b) A Pareto chart of the percent of complaints is given below.

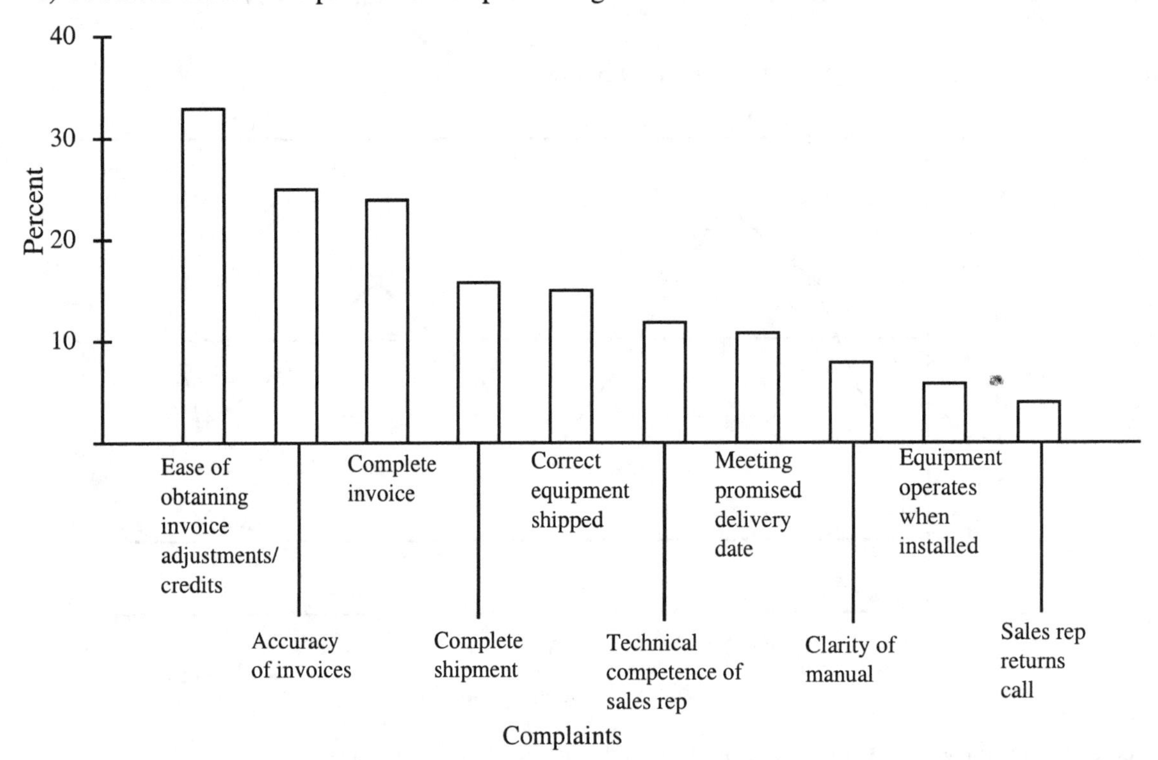

The category with the largest number of complaints in the ease of obtaining invoice adjustments/credits, so we might target this area for improvement.

12.69 a) Presumably, we are interested in monitoring and controlling the amount of time the system is unavailable. We might measure the time the system is not available in sample time periods (perhaps the average time the system is unavailable in a sample of days in a week or in a month) and use an $\bar{x}$ and s chart to monitor how this length of time the system is unavailable varies.

One could also simply count the number of times the system is unavailable on a sample of days in a week or in a month and monitor these counts using an $\bar{x}$ and s chart. This would be reasonable if we were concerned about the frequency of the unavailability of the system rather than the amount of time the system is unavailable.

b) To monitor the time to respond to requests for help we might measure the response times in a sample of time periods (perhaps several short time periods in a shift or in a day, or perhaps consecutive short time periods in a shift or day) and use an $\bar{x}$ and s chart to monitor how the response times vary.

c) Here we might examine samples of programming changes (perhaps in a sample of days in a week or a month) and record the proportion in the sample that are not properly documented. Because we are monitoring a proportion we would use a p chart.

12.71 We do not know σ so we use the data to estimate σ. Using software, we find the mean of the standard deviations of the 22 samples is

$$\bar{s} = 7.65$$

and so we estimate σ to be (using the fact that the samples are each of size $n = 3$ and according to Table 12.3, $c_4 = 0.8862$)

$$\hat{\sigma} = \frac{\bar{s}}{c_4} = \frac{7.65}{0.8862} = 8.63$$

Thus, the center line and control limits for an s chart for this process are (using Table 12.3 for the values of c_4, B_5, and B_6)

$$\text{CL} = c_4\,\hat{\sigma} = 0.8862(8.63) = 7.65$$
$$\text{UCL} = B_6\,\hat{\sigma} = 2.276(8.63) = 19.642$$
$$\text{LCL} = B_5\,\hat{\sigma} = 0(8.63) = 0$$

The resulting s chart is given below.

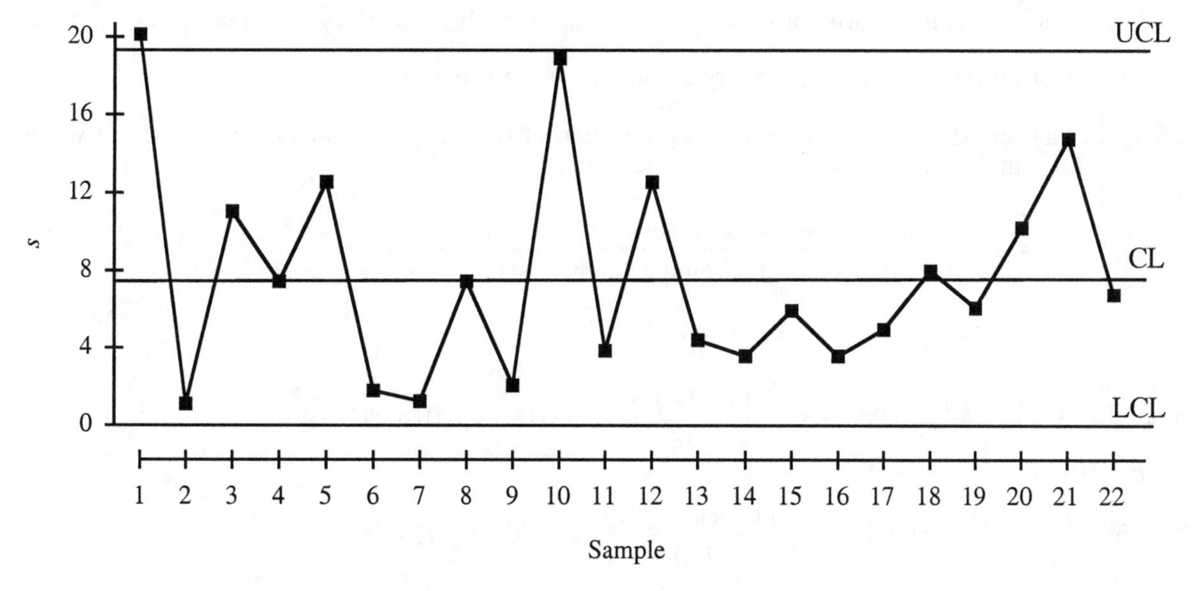

The first sample is out of control, perhaps reflecting an initial lack of skill or initial problems in using the new system. After the first sample, the process is in control with respect to short-term variation, although sample 10 is very close to the upper control limit.

12.73 a) If we remove samples 1 and 10, we find the mean of the standard deviations of the remaining 20 samples is

$$\bar{s} = 6.47$$

and so we estimate σ to be (using the fact that the samples are each of size $n = 3$ and according to Table 12.3, $c_4 = 0.8862$)

$$\hat{\sigma} = \frac{\bar{s}}{c_4} = \frac{6.47}{0.8862} = 7.30$$

From the specifications given, we know that USL = 805 and LSL = 855 in units of mm $\times 10^{-4}$. Thus, we estimate the capability ratio C_p to be

$$\hat{C}_p = \frac{\text{USL} - \text{LSL}}{6\hat{\sigma}} = \frac{855 - 805}{6(7.30)} = 1.14$$

This value tells us that the specification limits will lie just within 3 standard deviations of the process mean if the process mean is in the center of the specification limits (i.e., if the process mean is 830).

b) C_{pk} depends on the value of the process mean. If the process mean can easily be adjusted, it is easy to change the value of C_{pk}. Computing C_{pk} from the current process mean, therefore, is probably not completely informative about the capability of the process. A better measure of the process capability is to center the process mean within the specification limits (something that can be done easily) and then compute a capability index. But if the process is properly centered, C_{pk} and C_p will be the same, so using C_p at the outset is ultimately more informative about the process capability.

c) Because we had only the 22 sample standard deviations, we used $\frac{\bar{s}}{c_4}$ to estimate σ, the process variation. A better estimate would be to compute the sample standard deviation s of all $22 \times 3 = 66$ observations in the samples. This gives an estimate of all the variation in the output of the process (including sample to sample variation). Using $\frac{\bar{s}}{c_4}$ is likely to give a slightly too small estimate of the process variation and hence a slightly too large (optimistic) estimate of C_p.

12.75 a) In this setting, we are measuring the proportion of films that are satisfactory. Thus, we would use a p chart. From the information given we would estimate

$$\bar{p} = \frac{\text{total number of successes in past samples}}{\text{total number of opportunities in these samples}} = \frac{15}{5000} = 0.003$$

The control limits for a sample of n = 100 films would be

$$\text{UCL} = \bar{p} + 3\sqrt{\frac{\bar{p}(1-\bar{p})}{n}} = 0.003 + 3\sqrt{\frac{0.003(1-0.003)}{100}} = 0.003 + 0.016 = 0.019$$

$$\text{CL} = \bar{p} = 0.003$$

$$\text{LCL} = \bar{p} - 3\sqrt{\frac{\bar{p}(1-\bar{p})}{n}} = 0.003 - 3\sqrt{\frac{0.003(1-0.003)}{100}} = 0.003 - 0.016 = 0$$

Note that for LCL, negative values are set to 0 because a proportion can never be less than 0.

b) If the proportion of unsatisfactory films is 0.003, then in a sample of 100 films the expected number of unsatisfactory films is 0.3. Thus, unless we use very large samples (on the order of 1000), most of our samples will have no unsatisfactory films and plotting the sample values (most of which will be 0) will not be very informative.

CHAPTER 13

TIME SERIES FORECASTING

SECTION 13.1

OVERVIEW

Many data sets in business inovlve measuring a variable such as sales at regular intervals of time. The resulting data is called a **time series.** Many time series display a long-run trend, and may also exhibit a repeating seasonal pattern.

Regression methods can be used to model both the **trend** and **seasonal** aspects of a time series. A linear effect in a regression model gives a trend-only model, while adding either indicator variables or seasonality factors can be used to adjust the trend-only model for seasonal effects. When examining a time series, read the description carefully as many government agencies provide **seasonally adjusted** time series.

Although regression models can be used to describe a time series, inference using standard regression methods may be invalid as successive observations in a time series are often correlated. A **lagged residual plot** or the computation of the **autocorrelation** of the residuals will often show that the standard regression assumption of independent observations is not satisfied.

APPLY YOUR KNOWLEDGE

13.1 a) Each year, sales are lowest for the first two quarters, and then increase in the third and fourth quarters. Sales decrease from the fourth quarter of one year to the first quarter of the next. This pattern is repeated year after year.

b) A time plot of these data is given below.

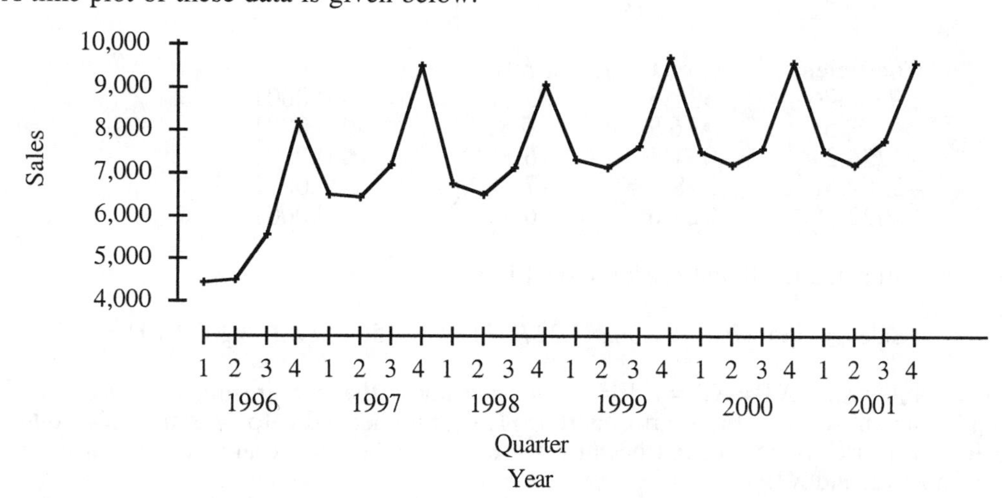

c) There appears to be a positive trend, although the trend levels off after 1998.

d) There is an obvious repeating pattern. Each year, sales in the first two quarters are low, increase in the third quarter, and then increase by an even larger amount in the fourth quarter. Sales then drop substantially in the first quarter of the following year.

13.3 a) Using statistical software, we get the following.

R squared = 35.0%
s = 1170 with 24 – 2 = 22 degrees of freedom

Source	Sum of Squares	df	Mean Square	F-ratio	P
Regression	16217509	1	16217509	11.9	0.002
Residual	30100626	22	1368210		

Variable	Coefficient	s.e. of Coeff	t-ratio	prob
Constant	5903.22	492.9	11.98	≤ 0.0001
x	118.75	34.5	3.44	0.0023

We see that the equation of the least-squares line is

$$\text{Sales} = 5903.22 + 118.75x$$

where sales are in millions of dollars and x takes on values 1, 2, 3,..., 24 as described in the statement of the problem.

b) The intercept corresponds to $x = 0$. $x = 1$ represents the first quarter of 1996, so $x = 0$ is the quarter preceding the first quarter of 1996. This means that $x = 0$ represents the fourth quarter of 1995.

c) The slope represents the increase in the response corresponding to a unit increase in the predictor variable x. In this case the response is sales, in millions of dollars, and a unit change in the predictor x corresponds to a change in time of one quarter. Thus, the slope is the increase in sales (in millions of dollars) that occurs from one quarter to the next. In particular, the least-squares regression line predicts a increase in sales of $118.75 million each quarter.

13.5 a) Using statistical software to estimate the trend-and-season model, we get the following results.

R squared = 86.8%
s = 566.7 with 24 – 5 = 19 degrees of freedom

Source	Sum of Squares	df	Mean Square	F-ratio
Regression	40215885	4	10053971	31.3
Residual	6102250	19	321171	

Variable	Coefficient	s.e. of Coeff	t-ratio	prob
Constant	7858.76	331.3	23.70	≤ 0.0001
x	99.54	16.9	5.88	≤ 0.0001
$X1$	–2274.21	331.1	–6.87	≤ 0.0001
$X2$	–2564.58	328.9	–7.80	≤ 0.0001
$X3$	–2022.79	327.6	–6.17	≤ 0.0001

We see that the estimated trend-and-season model is

$$\text{Sales} = 7858.76 + 99.54x - 2274.21X1 - 2564.58X2 - 2022.79X3$$

b) If we know that $X1 = X2 = X3 = 0$, then we know that we are not in any of the first three quarters, and hence must be in the fourth quarter. Thus, another indicator variable, $X4$, to tell us if we are in the fourth quarter is not needed because we can tell whether we are in the fourth quarter from the values of $X1$, $X2$, and $X3$.

c) The intercept corresponds to $x = X1 = X2 = X3 = 0$. $x = 0$ is the quarter before the first quarter of 1996, namely the fourth quarter of 1995. $X1 = X2 = X3 = 0$ means we make no seasonal adjustments for being in the first, second, or third quarter. Thus, the intercept again represents the fourth quarter of 1995.

The values of fourth-quarter sales in 1996, 1997, 1998, 1999, 2000, and 2002 are all above 8000, and given the clear pattern of seasonal variation the estimate of the intercept in part (a) appears to be a better estimate than that of Exercise 13 (b) for fourth-quarter sales in 1995.

13.7 a) In Exercise 13.3, we computed the trend to be

$$\text{Sales} = 5903.22 + 118.75x$$

For each quarter, we compute the value of this trend and then divide the actual sales by the trend. The results are summarized below. (Note that we have rounded off the estimates of the slope and intercept in our trend. If you did not round off, your results may differ slightly from ours.)

x	Sales	Trend	Sales/Trend
1	4452	6021.9700	0.739
2	4507	6140.7200	0.734
3	5537	6259.4700	0.885
4	8157	6378.2200	1.279
5	6481	6496.9700	0.998
6	6420	6615.7200	0.970
7	7208	6734.4700	1.070
8	9509	6853.2200	1.388
9	6755	6971.9700	0.969
10	6483	7090.7200	0.914
11	7129	7209.4700	0.989
12	9072	7328.2200	1.238
13	7339	7446.9700	0.986
14	7104	7565.7200	0.939
15	7639	7684.4700	0.994
16	9661	7803.2200	1.238
17	7528	7921.9700	0.950
18	7207	8040.7200	0.896
19	7538	8159.4700	0.924
20	9573	8278.2200	1.156
21	7522	8396.9700	0.896
22	7211	8515.7200	0.847
23	7729	8634.4700	0.895
24	9542	8753.2200	1.090

To get the seasonality factors, we average together the values of the last column corresponding to each of the four quarters. For example, the seasonality factor for the first quarter is the average of the entries in the last column corresponding to $x = 1, 5, 9, 13, 17$, and 21, i.e., the average of 0.739, 0.998, 0.969, 0.986, 0.950, and 0.896. We summarize the results below.

Quarter	Seasonality Factor
1	0.923
2	0.885
3	0.960
4	1.231

b) The average of the four seasonality factors is

$$(0.923 + 0.885 + 0.960 + 1.231)/4 = 3.999/4 = 0.999$$

and this is close to 1. The fourth-quarter seasonality factor of 1.231 tells us that fourth-quarter sales are typically 23.1% above the average for all four quarters.

c) The plot is given on the next page. Notice that it mimics the pattern of the seasonal variation seen in Exercise 13.1.

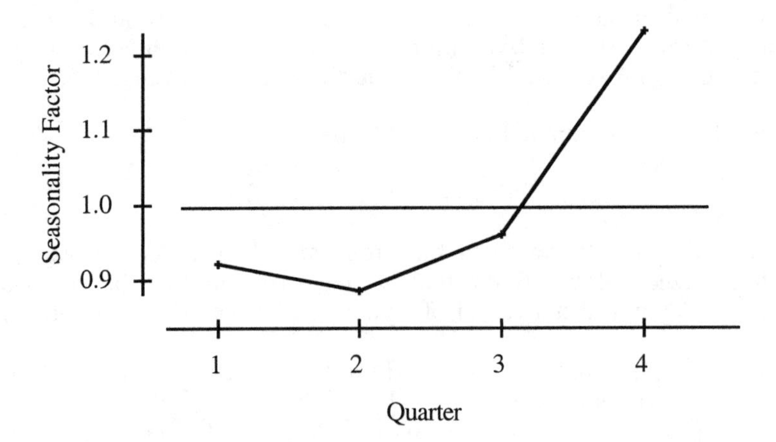

13.9 a) We calculated the seasonally adjusted values by dividing the sales data for each month by the corresponding seasonality factor for the month. Time plots of the unadjusted and seasonally adjusted data are given in the plot below. The time plot of the points plotted with the symbol x correspond to the unadjusted sales data. The time plot of the points plotted with the symbol o correspond to the seasonally adjusted sales data.

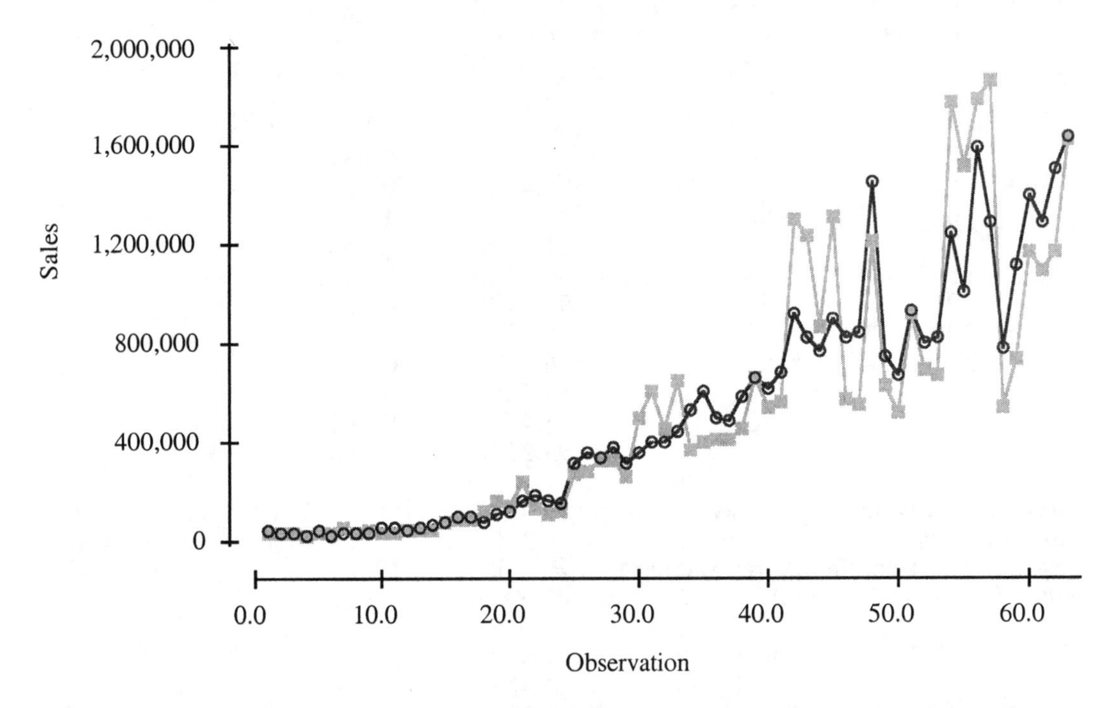

b) Seasonally adjusting the DVD player sales data smoothed the time series a little, but not to the degree that seasonally adjusting the sales data in Figure 13.7 did. This suggests that the seasonal pattern in the DVD player sales data is not as strong as in the monthly retail sales data.

SECTION 13.1 EXERCISES

13.11 a) The dashed line in the time plot on the next page corresponds to the least squares line. Using the trend-only model, the sales for the first three quarters tend to be overpredicted and the sales for the fourth quarter are underpredicted. Generally, the overpredictions in the first three quarters are of smaller magnitude than the underpredictions in the fourth quarter.

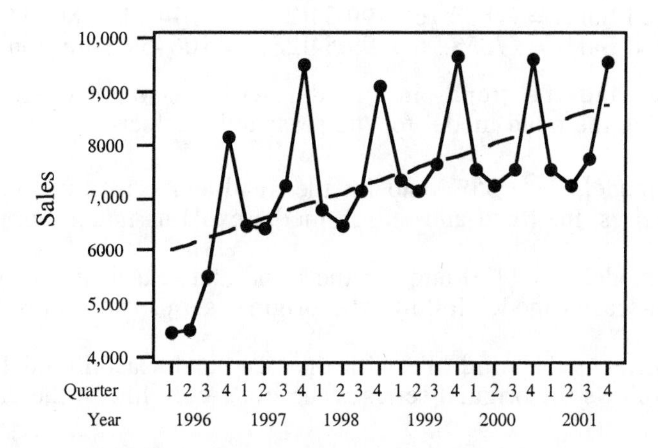

b) The equation of the least-squares line is

$$\text{Sales} = 5903.22 + 118.75x$$

where sales are in millions of dollars and x takes on values 1, 2, 3,...,24. The first quarter of 2002 corresponds to $x = 25$, and the fourth quarter of 2002 corresponds to $x = 28$. The predictions are

$x = 25$ Predicted Sales = 5903.22 + 118.75(25) = 8871.97 million dollars
$x = 28$ Predicted Sales = 5903.22 + 118.75(28) = 9228.22 million dollars

c) Based on previous history, we expect the first-quarter sales to be overpredicted and the fourth quarter to be underpredicted. Which forecast will be more accurate is difficult to guess, although in the past, predictions in the first three quarters tended to be slightly more accurate than in the fourth quarter.

13.13 a) For the linear-only model and the seasonality factors, the predictions are

$$\text{Predicted Sales} = (5903.22 + 118.75x) \times \text{SF}$$

where the seasonality factors were calculated in Exercise 13.7 as

Quarter	Seasonality Factor
1	0.923
2	0.885
3	0.960
4	1.231

The first quarter of 2002 corresponds to $x = 25$ and the fourth quarter of 2002 corresponds to $x = 28$. The predictions are

$x = 25$ Predicted Sales = (5903.22 + 118.75[25]) × 0.923 = 8188.83 million dollars
$x = 28$ Predicted Sales = (5903.22 + 118.75[28]) × 1.231 = 11359.94 million dollars

b) The first-quarter forecast has been multiplied by 0.923, as the trend-only model typically overpredicts the first quarter while the fourth-quarter forecast has been multiplied by 1.231 to account for the fact that the trend-only model typically underpredicts the fourth quarter. The seasonality factors are simple adjustments to the trend-only model predictions that are designed to account for the seasonality in the time series.

c) The estimated trend-and-season model from Exercise 13.5 is

$$\text{Sales} = 7858.76 + 99.54x - 2274.21X1 - 2564.58X2 - 2022.79X3$$

For the first quarter, we have $X1 = 1$ and $X2 = X3 = 0$, while for the fourth quarter we have $X1 = X2 = X3 = 0$. Thus, the first quarter of 2002 corresponds to $x = 25$, $X1 = 1$, and $X2 = X3 = 0$, and the fourth quarter of 2002 corresponds to $x = 28$ and $X1 = X2 = X3 = 0$. The predictions are

First quarter, 2002 Predicted Sales = (7858.76 + 99.54[25]) – 2274.21 = 8073.05 million dollars
Fourth quarter, 2002 Predicted Sales = (7858.76 + 99.54[28]) = 10645.88 million dollars

The trend-and-season model and the trend-only model with seasonality factors give similar predictions as both are adjusting the trend model for the seasonality effects.

13.15 a) For the trend-only model, R^2 = 35%, and for the trend-and-season model R^2 = 86.8%. As expected, from previous exercises, the trend-and-season model explains much more of the variability in the JCPenney sales.

b) For the trend-only model s = 1170 and for the trend-and-season model s = 566.7. Again this shows that the trend-and-season model follows the original series more closely than the trend-only model.

c) The original time series is the solid line with the trend-and-season model corresponding to the dashed line that closely follows the original series. The trend-only line is the dashed straight line in the plot.

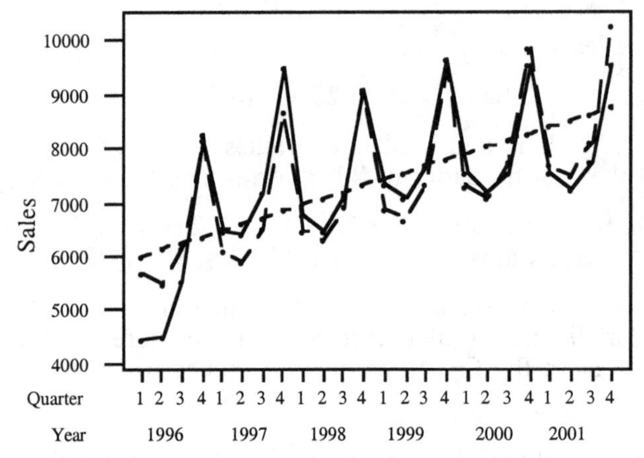

d) It is clear from the plot that the trend-and-season model is a substantial improvement over the trend-only model.

13.17 a) After fitting the linear trend-only model, the residual time plot still shows the seasonal pattern in the data. The residual for the fourth quarters are positive, and most of the residuals for the first three quarters are negative.

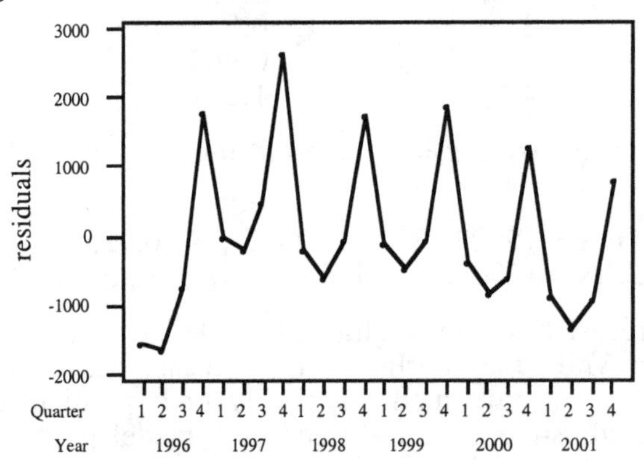

b) Autocorrelation is not apparent in the lagged residual plot on the next page. The correlation between successive residuals e_t and e_{t-1} is only 0.095. The biggest problem with the linear trend-only model is the omission of the seasonal effects and this shows up clearly in the time plot in (a) but not in the lagged residual plot.

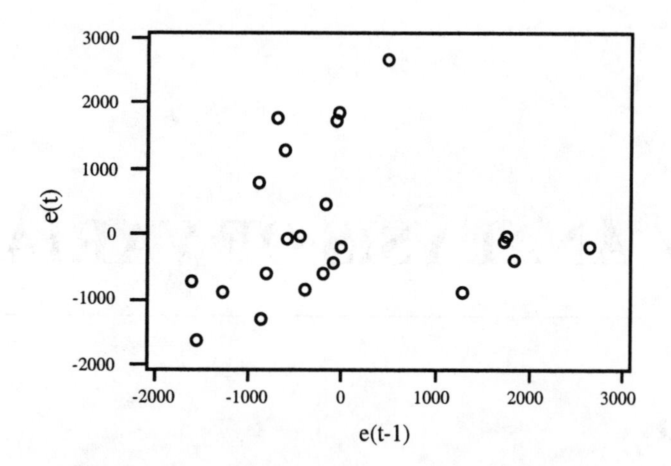

13.19 a) The other group of 10 points has December as the y-coordinate.

b) The correlation of 0.9206 suggests that there is a strong autocorrelation in much of the time series.

c) If we looked at the seasonally adjusted time series, the correlation would be closer to 0.9206. The outlying groups of points have the December sales as either the x- or y-coordinate, and this is what is reducing the correlation from 0.9206 to 0.4573. If there were a seasonal adjustment, these two sets of 10 points should no longer stand out from the remaining points.

CHAPTER 14

ONE-WAY ANALYSIS OF VARIANCE

SECTION 14.1

OVERVIEW

The **one-way analysis of variance** is a generalization of the two-sample t procedures. It allows comparison of more than two populations based on independent SRSs from each. As in the pooled two-sample t procedures, the populations are assumed to be normal with possibly different means, but with a common standard deviation. In the one-way analysis of variance we are interested in making formal inferences about the population means.

The simplest graphical procedure to compare the populations is to give side-by-side boxplots or stemplots (see Chapter 1). Normal quantile plots can be used to check for extreme deviations from normality or outliers. The summary statistics required for the analysis of variance calculations are the means and standard deviations of each of the samples. An informal procedure to check the assumption of equal variances is to make sure that the ratio of the largest to the smallest standard deviation is less than 2. If the standard deviations satisfy this criterion and the normal quantile plots seem satisfactory, then the one-way ANOVA is an appropriate analysis.

The data for one-way ANOVA are SRSs for each of I populations. The ith population has n_i observations $x_{i1}, x_{i2},\ldots, x_{in_i}$ one-way ANOVA model is

$$x_{ij} = \mu_i + \varepsilon_{ij}$$

for $i = 1,\ldots, I$ and $j = 1,\ldots, n_i$. The ε_{ij} are assumed to be from an $N(0, \sigma)$ distribution. The parameters of the model are the I population means $\mu_1, \mu_2,\ldots, \mu_I$ and the common standard deviation σ. The total number of observations is $N = n_1 + n_2 +\ldots+ n_I$. We estimate μ_i by the sample mean for the ith group

$$\bar{x}_i = \frac{1}{n_i}\sum_{j=1}^{n_i} x_{ij}$$

and the residuals $e_{ij} = x_{ij} - \bar{x}_i$ reflect the variation about the sample means that we see in the data. We estimate the variance σ^2 with the **pooled sample variance**

$$s_p^2 = \frac{(n_1-1)s_1^2+(n_2-1)s_2^2+\cdots+(n_I-1)s_I^2}{(n_1-1)+(n_2-1)+\cdots+(n_I-1)}$$

where s_i^2 is the sample variance of the n_i observations from population i. The pooled standard error is

$$s_p = \sqrt{s_p^2}$$

is the estimate of σ.

The ***F* statistic** computed in the ANOVA table can be used to test the **null hypothesis** that the population means are all equal. The **alternative hypothesis** is that at least two of the population means are not equal. Rejection of the null hypothesis does not provide any information as to which of the population means are different.

ANOVA is based on separating the total variation observed in the data into two parts: variation **among groups means** and variation **within groups.** If the variation among groups is large relative to the variation within groups, we have evidence against the null hypothesis.

An **analysis of variance table** organizes the ANOVA calculations. In one-way ANOVA there are three sources of variation: groups (represented by the **sum of squares for groups, SSG**), error (represented by the **sum of squares for error, SSE**), and total (represented by the **total sum of squares, SST**). These sums of squares are related by the formula SST = SSG + SSE. Associated with each sum of squares is a **degrees of freedom,** degrees of freedom for groups (denoted **DFG** and equal to $I - 1$) with SSG, degrees of freedom for error (denoted **DFE** and equal to $N - I$), and total degrees of freedom (denoted **DFT** and equal to $N - 1$). These are related in the same way as the sums of squares, namely DFT = DFG + DFE. To calculate a mean square for groups (denoted **MSG**), error (denoted **MSE**), or total (denoted **MST**) divide the corresponding sum of squares by its degrees of freedom. Degrees of freedom, sums of squares, and mean squares appear in the ANOVA table. The ***F* statistic** is

$$F = \frac{\text{MSG}}{\text{MSE}}$$

and it has the $F(I-1, N - I)$ distribution under the null hypothesis. Its ***P*-value** is used to test the null hypothesis.

APPLY YOUR KNOWLEDGE

14.1 There are $I = 3$ styles of covers that are to be compared. $N = 60$ stores are used. These are assigned at random to the three styles of covers so there are $n_1 = 20$ stores assigned to style 1, $n_2 = 20$ stores to style 2, and $n_3 = 20$ stores to style 3. We let x_{ij} denote the number of magazines sold in the jth ($j = 1, 2, \ldots, 20$) store assigned to style i. If μ_i denotes the mean number of magazines with cover style i that are sold in a one-week period by a store, then the ANOVA model is

$$x_{ij} = \mu_i + \varepsilon_{ij}$$

The ε_{ij} are assumed to be from an $N(0, \sigma)$ distribution where σ is the common standard deviation. The parameters of the model are the I population means μ_1, μ_2, and μ_3, and the common standard deviation σ.

14.3 a) The largest standard deviation is 120 (for cover C) and the smallest is 80 (for cover A). The ratio of these is 120/80 = 1.5. This ratio is less than 2, so by our rule of thumb, it is reasonable to pool the standard deviations for these data.

b) The parameters of the model are the I population means μ_1, μ_2, and μ_3, and the common standard deviation σ. There are three styles of cover, so $I = 3$. We estimate the population means as

Estimate of μ_1: $\bar{x}_1$ = mean sales for design A = 150
Estimate of μ_2: $\bar{x}_2$ = mean sales for design B = 175
Estimate of μ_3: $\bar{x}_3$ = mean sales for design C = 200

To estimate σ we first compute the pooled sample variance s_p^2. We the sample sizes are $n_1 = n_2 = n_3 = 20$ and the sample variances s_i^2 for each group are just the squares of the sample standard deviations given in the table. Thus,

$$s_p^2 = \frac{(n_1-1)s_1^2 + (n_2-1)s_2^2 + (n_3-1)s_3^2}{(n_1-1)+(n_2-1)+(n_3-1)} = \frac{(20-1)80^2 + (20-1)100^2 + (20-1)120^2}{(20-1)+(20-1)+(20-1)} = \frac{58,5200}{57} = 10,266.67$$

We estimate σ by $s_p = \sqrt{s_p^2} = \sqrt{10,266.67} = 101.32$.

14.5 Stem and leaf plots for the three groups are given below.

```
Basal          DRTA            Strat
0|4            0|              0|445
0|67           0|6777          0|666777
0|888999       0|888889999     0|889
1|01           1|000           1|111
1|22222223     1|2233          1|22333
1|45           1|5             1|44
1|6            1|6             1|
```

The distribution for the Basal group appears to be centered at a slightly larger score than that for the DRTA group and perhaps for the Strat group. The distribution of the DRTA scores shows some right skewness. There are no clear outliers in any of the groups.

14.7 In Figure 14.8, we see that SSG = 20.58, SSE = 572.45, and SST = 593.03. We note that

$$\text{SSG} + \text{SSE} = 20.58 + 572.45 = 593.03 = \text{SST}$$

14.9 In Figure 14.8, we see that DFG = 2, DFE = 63, and DFT = 65. We note that

$$\text{DFG} + \text{DFE} = 2 + 63 = 65 = \text{DFT}$$

14.11 From Figure 14.8, we find

$$\text{MST} = \text{SST/DFE} = 593.03/65 = 9.124$$

Using statistical software (we used Minitab), we find that

the mean of all 66 observations in Table 14.1 = 9.788
the variance of all 66 observations in Table 14.1 = 9.125

and we notice that (upto roundoff) MST = variance of all 66 observations in Table 14.1.

14.13 a) There are $I = 4$ groups. With six observations in each group there are a total of $N = 6 \times 4 = 24$ observations. Thus, the ANOVA F statistic has $I - 1 = 4 - 1 = 3$ numerator degrees of freedom and $N - I = 24 - 4 = 20$ denominator degrees of freedom.

b) If we look in Table E under the column labeled 3 and the row corresponding to 20 denominator degrees of freedom and p of .050, we find the entry 3.10. The F statistic would need to be larger than this value of 3.10 to have a P-value less than 0.05.

14.15 a) As the textbook points out (see the first paragraph of the subsection on the F test in Section 14.1) it is always true that s_p^2 = MSE. Hence, $s_p = \sqrt{\text{MSE}}$. Because of this, SAS calls s_p "Root MSE." You can verify this by looking at the output from SAS in Figure 14.9. In the row labeled "Error" and under the column labeled "Mean Square" the entry is 0.616003. This is the mean square error, denoted MSE, in ANOVA. The square root of MSE is called "Root MSE' by SAS. The square root of 0.616003 is 0.7849 (you can check this with a calculator). Thus "Root MSE" = 0.7849 = s_p.

b) As we saw in part (a), $s_p = \sqrt{\text{MSE}}$. In Excel, MSE is found in the ANOVA table in the entry of the row labeled "Within Groups" under the column labeled MS. Take the square root of this entry (0.616003 in this case) to get s_p.

14.17 Try the applet. You should find that the F statistic increases and the P-value decreases if the pooled standard error is kept fixed and the variation among the group means increases.

SECTION 14.2

OVERVIEW

The F statistic computed in the ANOVA table can be used to test the null hypothesis that the population means are all equal. The alternative hypothesis is that at least two of the population means are not equal. Rejection of the null hypothesis does not provide any information as to which of the population means are different.

If the researcher has specific questions about the population means before examining the data, these questions can often be expressed in terms of **contrasts.** Tests and confidence intervals about contrasts provide answers to these questions and allow the researcher to say more about which population means are different and what the sizes of these differences are. To be specific, a contrast is a combination of population means of the form

$$\psi = \sum a_i \mu_i$$

where the coefficients a_i sum to 0. The corresponding sample contrast is

$$c = \sum a_i \bar{x}_i$$

The standard error of c is

$$\text{SE}_c = s_p \sqrt{\sum \frac{a_i^2}{n_i}}$$

To test the null hypothesis H_0: $\psi = 0$, use the ***t* statistic**

$$t = \frac{c}{\text{SE}_c}$$

with degrees of freedom DFE that are associated with s_p.

A **level *C* confidence interval** for ψ is

$$c \pm t^* \text{SE}_c$$

where t^* is the value for the t(DFE) density curve with area C between $-t^*$ and t^*.

When there are no specific questions before examining the data, **multiple comparisons** are often used to follow up rejection of the null hypothesis in a one-way analysis of variance. These multiple comparisons are designed to determine which pairs of population means are different and to give confidence intervals for the differences. These methods are less powerful than contrasts, so use contrasts whenever a study is designed to answer specific questions.

To perform a **multiple comparisons procedure,** compute ***t* statistics** for all pairs of means using the formula

$$t_{ij} = \frac{\bar{x}_i - \bar{x}_j}{s_p \sqrt{\frac{1}{n_i} + \frac{1}{n_j}}}$$

If

$$|t_{ij}| \geq t^{**}$$

we declare the population means μ_i and μ_j different. The value of t^{**} depends on which multiple comparisons procedure we choose. Methods include the LSD method and the Bonferroni method.

Simultaneous confidence intervals for all differences $\mu_i - \mu_j$ have the form

$$(\bar{x}_i - \bar{x}_j) \pm t^{**} s_p \sqrt{\frac{1}{n_i} + \frac{1}{n_j}}$$

The critical values t^{**} depend on which multiple comparisons procedure we choose.

APPLY YOUR KNOWLEDGE

14.19 The contrast of interest in Exercise 14.18 is $0.5\mu_1 + 0.5\mu_2 - 0.5\mu_4 - 0.5\mu_5$. Thus, $a_1 = a_2 = 0.5$, $a_3 = 0$, and $a_4 = a_5 = -0.5$. We know that all $n_i = 25$ and that $s_p = 10$. The standard error for the contrast is

$$\mathrm{SE}_c = s_p \sqrt{\sum \frac{a_i^2}{n_i}} = 10\sqrt{\frac{0.5^2}{25} + \frac{0.5^2}{25} + \frac{0^2}{25} + \frac{(-0.5)^2}{25} + \frac{(-0.5)^2}{25}} = 10\sqrt{\frac{1}{25}} = 2$$

14.21 From Exercise 14.20, the difference between the average of the means of the first two groups and the average of the means of the last two groups is $c = 15$. From Exercise 14.19, $\mathrm{SE}_c = 2$. There are $N = 5 \times 25 = 125$ observations and $I = 5$ groups, so $\mathrm{DFE} = N - I = 120$. From Table D, $t^* = 1.984$ is the value for the $t(120)$ density curve with area 0.95 between $-t^*$ and t^*. The 95% confidence interval is thus,

$$c \pm t^*\mathrm{SE}_c = 15 \pm 1.984(2) = 15 \pm 3.968 = (11.032, 18.968)$$

14.23 From Figure 14.10, we see $n_2 = 22$, $\bar{x}_2 = 46.7273$, $n_3 = 22$, and $\bar{x}_3 = 44.2727$. From Example 14.21, we see that $s_p = 6.31$. Thus,

$$t_{23} = \frac{\bar{x}_2 - \bar{x}_3}{s_p\sqrt{\frac{1}{n_2} + \frac{1}{n_3}}} = \frac{46.7273 - 44.2727}{6.31\sqrt{\frac{1}{22} + \frac{1}{22}}} = \frac{2.4546}{1.9025} = 1.29.$$

14.25 From Example 14.22, we know that the value of t^{**} for the Bonferroni procedure when $\alpha = 0.05$ is $t^{**} = 2.46$ (this value can also be found with software or special tables). We know that $t_{23} = 1.29$. Because $|1.29| < t^{**} = 2.46$, using the Bonferroni procedure we would not reject the null hypothesis that the population means for groups 2 and 3 are different.

14.27 In the output in Figure 14.16, the difference in the means of groups 2 and 3 is $\bar{x}_2 - \bar{x}_3 = 2.4545$. The output also reports the standard error for this difference to be 1.90378. Thus,

$$t_{23} = \frac{\text{difference in means}}{\text{standard error}} = \frac{2.4545}{1.90378} = 1.29$$

14.29 The table with the means of groups that do not differ significantly are marked with the same letter in the table below.

Group	Mean	SD	n
Group 1	150.2[a]	19.1	30
Group 2	121.9[b]	18.3	30
Group 3	129.2[b, c]	18.4	30
Group 4	140.8[a, c]	22.1	30
Group 5	117.2[b]	20.6	30

We see that Groups 1 and 4 have the largest means. Group 1 differs from Groups 2, 3, and 5. Group 4 differs from Groups 2 and 5.

14.31 From Figure 14.16, the Bonferroni 95% confidence interval for the difference between the mean comprehension score for the DRTA (Group 2) method and the mean comprehension score for the Strat (Group 3) method is (after rounding to one decimal place) (–2.2, 7.1). This interval includes 0.

SECTION 14.3

OVERVIEW

The power of the F test depends upon the sample sizes, the variation among population means, and the within-group standard deviation. In order to carry out power calculations, software or special tables are needed. Some software, such as SAS and Minitab (the latest version), allows easy calculation of power.

APPLY YOUR KNOWLEDGE

14.33 a) In Exercise 14.32, we are told that $\sigma = 86$. The n_i are equal and we let n denote their common values. Because the n_i are equal, $\bar{\mu}$ is simply the average of the μ_i

$$\bar{\mu} = \frac{610+600+590+580}{4} = 595$$

The noncentrality parameter is therefore

$$\lambda = \frac{n\sum(\mu_i - \bar{\mu})^2}{\sigma^2} = \frac{n[(610-595)^2 + (600-595)^2 + (590-595)^2 + (580-595)^2]}{86^2} = 0.0676n$$

Also, because there are $I = 4$ groups and $N = nI = 4n$ observations,

$$\text{DFG} = I - 1 = 3,\ \text{DFE} = N - I = 4n - 4$$

For testing at level $\alpha = 0.05$, we use software to determine F^*, the upper 0.05 critical value of the F(DFG, DFE) = $F(3, 4n - 4)$ distribution. For various choices of n, using SAS (the latest version of Minitab can also be used to calculate the power using the same information), we get the following table.

n	DFG	DFE	F^*	λ	Power
10	3	36	2.8663	0.676	0.0883
20	3	76	2.7249	1.352	0.1366
30	3	116	2.6828	2.028	0.1891
40	3	156	2.6626	2.704	0.2444
50	3	196	2.6507	3.380	0.3010
100	3	396	2.6274	6.76	0.5684
150	3	596	2.6198	10.14	0.7645
200	3	796	2.6160	13.52	0.8829
250	3	996	2.6139	16.90	0.9458
500	3	1996	2.6095	33.80	0.9994

b) A plot of the power versus the sample size is given below.

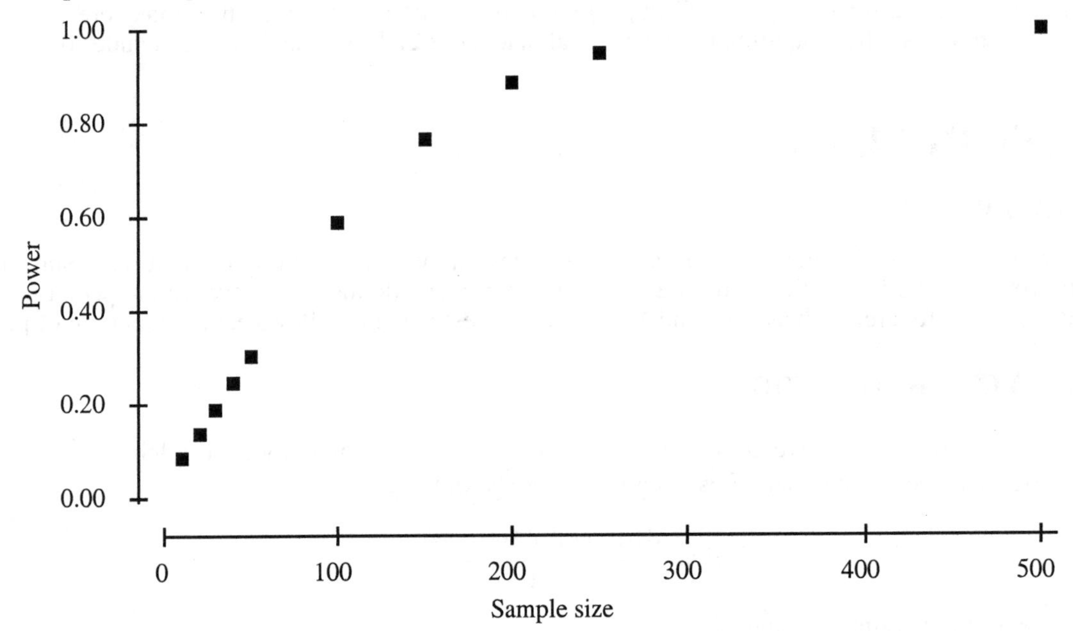

We that power increases as sample size increases, but the amount of increase gradually levels off as sample size increases.

c) Large sample sizes are needed to obtain high power. If the cost per student is high, one might consider a sample size of 150, with a power of 0.7645. If this is not adequate, one might consider a sample size of about 200, with a power of 0.8829. If it is important to have a fairly large power, one might consider a sample size of 250, with a power of 0.9458. Further improvement in the power comes at the cost of much larger sample sizes and so is probably not cost effective.

CHAPTER 14 EXERCISES

14.35 a) The response variable is the rating (on a 5-point scale) of the likelihood the household would buy the water treatment device. There are three populations, and these are the responses of all people to each of the three water treatment devices. There are three different water treatment devices that are to be compared, so $I = 3$. There are $N = 225$ households in the study. A third receive each of the devices, so $n_1 = n_2 = n_3 = (1/3)225 = 75$.

b) The response variable is the strength of the concrete. There are five populations, and these are the strengths of all possible batches of concrete for each of the five mixtures. There are five mixtures of concrete to be compared, so $I = 5$. Six batches of each mixture are to be prepared and measured so, $n_1 = n_2 = n_3 = n_4 = n_5 = 6$. The total number of observations is the sum of the n_i, which is $N = 5 \times 6 = 30$.

c) The score on a final exam is the response variable. There are three populations, and these are the final exam scores of all people after being taught by each of the three methods. There are three teaching methods, so $I = 3$. Twenty students are randomly assigned to each method, so $n_1 = n_2 = n_3 = 20$. The total number of observations is the sum of the n_i, which is $N = 3 \times 20 = 60$.

14.37 a) Using the results of part (a) of Exercise 14.35, we find for the first setting

$$\text{Degrees of freedom for groups} = \text{DFG} = I - 1 = 3 - 1 = 2$$
$$\text{Degrees of freedom for error} = \text{DFE} = N - I = 225 - 3 = 222$$
$$\text{Total degrees of freedom} = \text{DFT} = N - 1 = 225 - 1 = 224$$

Using the results of part (b) of Exercise 14.25, we find for the second setting

$$\text{Degrees of freedom for groups} = \text{DFG} = I - 1 = 5 - 1 = 4$$
$$\text{Degrees of freedom for error} = \text{DFE} = N - I = 30 - 5 = 25$$
$$\text{Total degrees of freedom} = \text{DFT} = N - 1 = 30 - 1 = 29$$

Using the results of part (c) of Exercise 14.25, we find for the third setting

Degrees of freedom for groups = DFG = $I - 1 = 3 - 1 = 2$
Degrees of freedom for error = DFE = $N - I = 60 - 3 = 57$
Total degrees of freedom = DFT = $N - 1 = 60 - 1 = 59$

b) Let μ_1, μ_2, and μ_3 be the population mean ratings for the three water treatment devices. For the first setting, the null and alternative hypotheses are

$$H_0: \mu_1 = \mu_2 = \mu_3$$
$$H_a: \text{not all of the } \mu_i \text{ are equal}$$

Let μ_1, μ_2, μ_3, μ_4, and μ_5 be the population mean strengths for the five concrete mixtures For the second setting, the null and alternative hypotheses are

$$H_0: \mu_1 = \mu_2 = \mu_3 = \mu_4 = \mu_5$$
$$H_a: \text{not all of the } \mu_i \text{ are equal}$$

Let μ_1, μ_2, and μ_3 be the population mean final exam scores for the three teaching methods. For the third setting, the null and alternative hypotheses are

$$H_0: \mu_1 = \mu_2 = \mu_3$$
$$H_a: \text{not all of the } \mu_i \text{ are equal}$$

c) For the first setting, using the results in part (a)

numerator degrees of freedom for the F statistic = DFG = 2
denominator degrees of freedom for the F statistic = DFE = 222

For the second setting, using the results in part (a)

numerator degrees of freedom for the F statistic = DFG = 4
denominator degrees of freedom for the F statistic = DFE = 25

For the third setting, using the results in part (a)

numerator degrees of freedom for the F statistic = DFG = 2
denominator degrees of freedom for the F statistic = DFE = 57

14.39 a) The largest standard deviation is 62 and the smallest is 40. The ratio of the largest to the smallest standard deviation is 62/40 = 1.55, which is less than 2. It is reasonable to assume equal standard deviations when we analyze these data.

b) We find

Group 1: $s_1 = 62$, so the variance is $s_1^2 = 62^2 = 3844$
Group 2: $s_2 = 40$, so the variance is $s_2^2 = 40^2 = 1600$
Group 3: $s_3 = 52$, so the variance is $s_3^2 = 52^2 = 2704$
Group 4: $s_4 = 48$, so the variance is $s_4^2 = 48^2 = 2304$

The sample sizes for the four groups are $n_1 = 20$, $n_2 = 220$, $n_3 = 18$, and $n_4 = 15$. The pooled variance is thus

$$s_p^2 = \frac{(n_1-1)s_1^2 + (n_2-1)s_2^2 + (n_3-1)s_3^2 + (n_4-1)s_4^2}{(n_1-1)+(n_2-1)+(n_3-1)+(n_4-1)}$$
$$= \frac{19(3844)+219(1600)+17(2704)+14(2304)}{19+219+17+14} = \frac{501660}{269} = 1864.91$$

c) The pooled standard error is $s_p = \sqrt{s_p^2} = \sqrt{1864.91} = 43.18$

d) The sample size in the second group is much larger than that for all other groups combined. Thus, the variance for this group is weighted more heavily in the formula for s_p^2 ,and so s_p^2 will be closer to the variance for the second group than to the variance for any other group. This means that the pooled standard error (the square root of the pooled variance) will be closer to the standard deviation of the second group than any other group.

14.41 a) There are $I = 5$ groups that are being compared. There are 13 observations per group, so the total number of observations is $N = 5\times13 = 65$. Thus, the F statistic has

$$\text{numerator degrees of freedom} = I - 1 = 5 - 1 = 4$$
$$\text{denominator degrees of freedom} = N - I = 65 - 5 = 60$$

The value of the F statistic is 1.61. We look in Table E at the entries corresponding to 4 numerator degrees of freedom and 60 denominator degrees of freedom. $F = 1.61$ is less than the smallest critical value of 2.04 in the Table, and 2.04 corresponds to a P-value of 0.100. Thus, the actual P-value is larger than 0.100. Using statistical software, we find the exact P-value is 0.1835.

b) There are $I = 10$ groups that are being compared. There are four observations per group, so the total number of observations is $N = 10\times4 = 40$. Thus, the F statistic has

$$\text{numerator degrees of freedom} = I - 1 = 10 - 1 = 9$$
$$\text{denominator degrees of freedom} = N - I = 40 - 10 = 30$$

The value of the F statistic is 4.68. We look in Table E at the entries corresponding to 9 numerator degrees of freedom and 30 denominator degrees of freedom. $F = 4.68$ is larger than the largest critical value of 4.39 in the Table, and 4.39 corresponds to a P-value of 0.001. Thus, the actual P-value is smaller than 0.001. Using statistical software we find the exact P-value is 0.0006.

14.43 a) As in Example 14.4, we let μ_1, μ_2, μ_3, and μ_4 be the average amounts spent on textbooks by all freshmen, sophomores, juniors, and seniors at this college for this semester. The hypotheses for ANOVA are

$$H_0\text{: } \mu_1 = \mu_2 = \mu_3 = \mu_4$$
$$H_a\text{: not all of the } \mu_i \text{ are equal}$$

b) We know that there are $I = 4$ groups, that the sample sizes are $n_i = 50$ students per group, and thus that there are a total of $N = 4\times50 = 200$ students in the study. Thus,

$$\text{Degrees of freedom for groups} = \text{DFG} = I - 1 = 4 - 1 = 3$$
$$\text{Degrees of freedom for error} = \text{DFE} = N - I = 200 - 4 = 196$$
$$\text{Total degrees of freedom} = \text{DFT} = N - 1 = 200 - 1 = 199$$

The outline of the ANOVA table is given below.

Source	Degrees of Freedom	Sum of Squares	Mean Square	F
Groups	DFG = 3	SSG	MSG = SSG/DFG	MSG/MSE
Error	DFE = 196	SSE	MSE = SSE/DFE	
Total	DFT = 199	SST	MST = SST/DFT	

c) Under the assumption that H_0 is true, the F statistic has the $F(I - 1, N - I) = F(3, 196)$ distribution.

d) Look in Table E for the entries corresponding to the column labeled 3 (the degrees of freedom in the numerator) and the rows labeled 196 (the degrees of freedom in the denominator). Because there are no rows labeled 196, we use the rows with the next closest value that is less than 196. In this case, these are the rows labeled 100 (but note that one could use the rows labeled 200 because these values will be close to the correct values for 196). In the row labeled by $p = .050$, we find the entry 2.70. This is the (approximate) critical value for an $\alpha = 0.05$ test (but note that the correct critical value will be between 2.70 and the entry of 2.65 in the rows labeled 200).

14.45 We use statistical software to run the Bonferroni multiple comparisons procedure. The results are summarized in the table below. Conditions are denoted by the number of days after baking.

Bonferroni Post Hoc Tests

Condition $i - j$	Difference	std. err.	P-value
1 – 0	–6.75000	1.321	0.036734
3 – 0	–26.9100	1.321	0.000053
3 – 1	–20.1600	1.321	0.000219
5 – 0	–36.2900	1.321	0.000012
5 – 1	–29.5400	1.321	0.000033
5 – 3	–9.38000	1.321	0.008542
7 – 0	–40.3850	1.321	0.000007
7 – 1	–33.6350	1.321	0.000017
7 – 3	–13.4750	1.321	0.001551
7 – 5	–4.09500	1.321	0.238149

The differences in the group means are all significantly different from 0 at the $\alpha = 0.05$ level except the difference after 5 and 7 days. At the $\alpha = 0.01$ level, all the differences are significantly different from 0, except the difference between 0 and 1 days after baking and the difference between 5 and 7 days after baking.

14.47 a) The ANOVA in Exercise 14.46 did not reject the hypothesis at the 0.05 level that any of the group means differed. Thus, no further analysis on which group means differed is appropriate.

b) We use statistical software to run the Bonferroni multiple comparisons procedure. The results are summarized in the table below. Conditions are denoted by the number of days after baking.

Bonferroni Post Hoc Tests

Condition $i - j$	Difference	std. err.	P-value
1 – 0	–0.110000	0.0608	0.752553
3 – 0	–0.140000	0.0608	0.514112
3 – 1	–0.030000	0.0608	0.999966
5 – 0	–0.045000	0.0608	0.998871
5 – 1	0.065000	0.0608	0.982858
5 – 3	0.095000	0.0608	0.861026
7 – 0	–0.385000	0.0608	0.014421
7 – 1	–0.275000	0.0608	0.061029
7 – 3	–0.245000	0.0608	0.096013
7 – 5	–0.340000	0.0608	0.025003

The only group means that are significantly different at the $\alpha = 0.05$ level are the difference between 0 and 7 days after baking and between 7 and 5 days after baking.

14.49 a) Normal probability plots for each of the four treatment groups are given below.

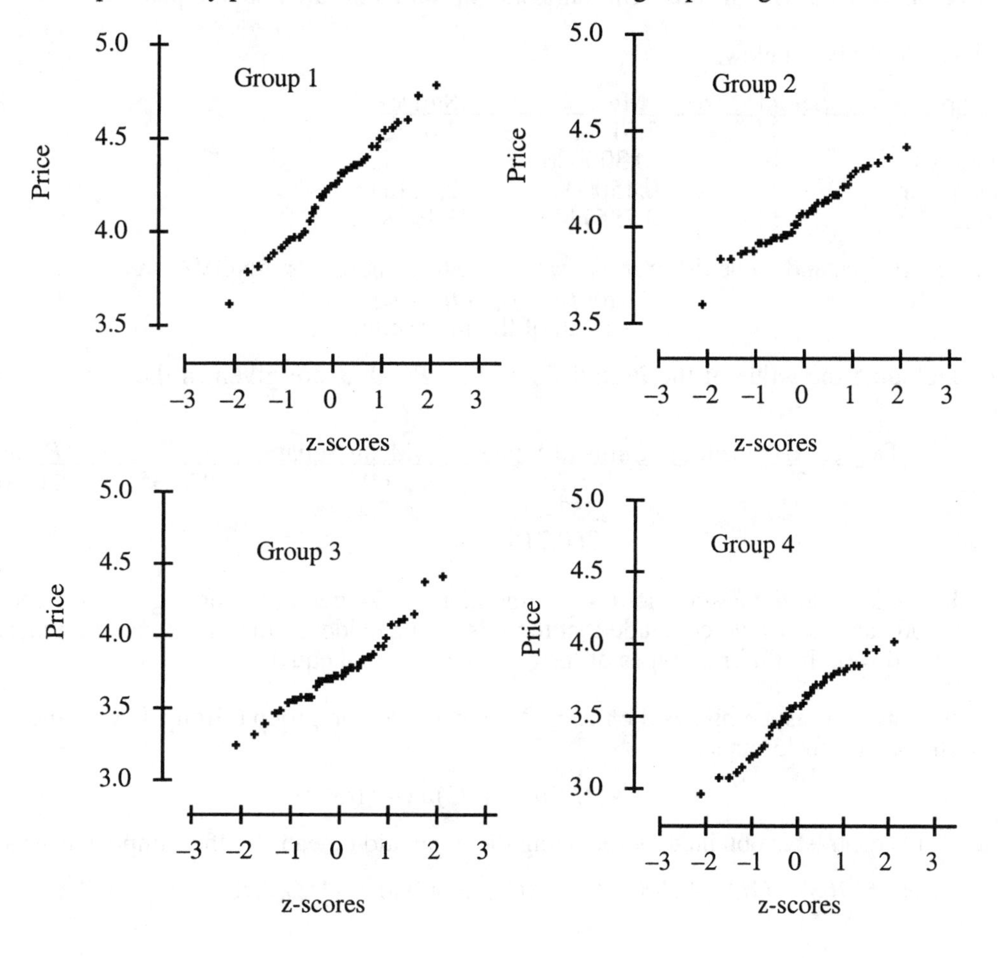

Group 2 shows a modest outlier, but otherwise there are no serious departures from normality. The assumption that the data are (approximately) normal is not unreasonable.

b) A table is given below.

Number of promotions	Sample Size	Mean	StdDev
1	40	4.22400	0.273410
3	40	4.06275	0.174238
5	40	3.75900	0.252645
7	40	3.54875	0.275031

c) The largest standard deviation is 0.275031 and the smallest is 0.174238. The ratio of the largest to the smallest is 0.275031/0.174238 = 1.58. This is less than 2, so it is not unreasonable to assume that the population standard deviations are equal.

d) Let μ_1, μ_2, μ_3, and μ_4 be the group means. The hypotheses for ANOVA are

$$H_0\text{: } \mu_1 = \mu_2 = \mu_3 = \mu_4$$
$$H_a\text{: not all of the } \mu_i \text{ are equal}$$

The results, including the value of the F statistic and its P-value, are given in the following ANOVA table.

Source	Degrees of Freedom	Sums of Squares	Mean Square	F	P-value
Group	3	10.9885	3.66285	59.903	≤ 0.0001
Error	156	9.53875	0.061146		
Total	159	20.5273			

From the table, we see that the F statistic has 3 numerator and 156 denominator degrees of freedom. The P-value is $\leq$ 0.0001, and we would conclude that there is strong evidence that the population mean expected prices associated with the different numbers of promotions are not all equal.

14.51 a) The table is given below.

Group	Sample Size	Mean	StdDev
Piano	34	3.61765	3.05520
Singing	10	-0.300000	1.49443
Computer	20	0.450000	2.21181
None	14	0.785714	3.19082

b) Let μ_1, μ_2, μ_3, and μ_4 be the group means. The hypotheses for ANOVA are

$$H_0\text{: } \mu_1 = \mu_2 = \mu_3 = \mu_4$$
$$H_a\text{: not all of the } \mu_i \text{ are equal}$$

The results, including the value of the F statistic and its P-value, are given in the following ANOVA table.

Source	Degrees of Freedom	Sums of Squares	Mean Square	F	P-value
Group	3	207.281	69.0938	9.2385	≤ 0.0001
Error	74	553.437	7.47887		
Total	77	760.718			

From the table, we see that the F statistic has 3 numerator and 74 denominator degrees of freedom. The P-value is $\leq$ 0.0001 and we would conclude that there is strong evidence that the population mean change is scores associated with the different types of instruction are not all equal.

14.53 The following contrast compares the mean of the piano lesson group (Group 1) with the average of the means of the other three groups.

$$\psi = \mu_1 - (1/3)\mu_2 - (1/3)\mu_3 - (1/3)\mu_4$$

An estimate of this contrast is obtained by replacing the population means by the sample means and is

$$c = (3.61765) - (1/3)(-0.300000) - (1/3)(0.450000) - (1/3)(0.785714) = 3.306$$

From Exercise 14.51, we get that the pooled standard deviation is

$$s_p = \sqrt{s_p^2} = \sqrt{\text{MSE}} = \sqrt{7.47887} = 2.735$$

The standard error of c is thus

$$\text{SE}_c = s_p\sqrt{\sum \frac{a_i^2}{n_i}} = 2.735\sqrt{\frac{1^2}{34} + \frac{(-\frac{1}{3})^2}{10} + \frac{(-\frac{1}{3})^2}{20} + \frac{(-\frac{1}{3})^2}{14}} = 2.735\sqrt{0.054} = 0.636$$

To test the null hypothesis H_0: $\psi = 0$, use the t *statistic*

$$t = \frac{c}{\text{SE}_c} = \frac{3.306}{0.636} = 5.20$$

with degrees of freedom DFE = 74. Reading row 60 in Table D (the row with the closest value for DFE that is smaller than 74), we see that 5.20 is larger than the largest entry 3.46 (with associated upper tail probability) in the table. Because we use a two-tailed test, the P-value $\leq 2\times0.0005 = 0.001$. We conclude that there is strong statistical evidence that the mean of the piano group differs from the average of the means of the other three groups.

14.55 a) A residual and normal probability plot of the residuals are given below.

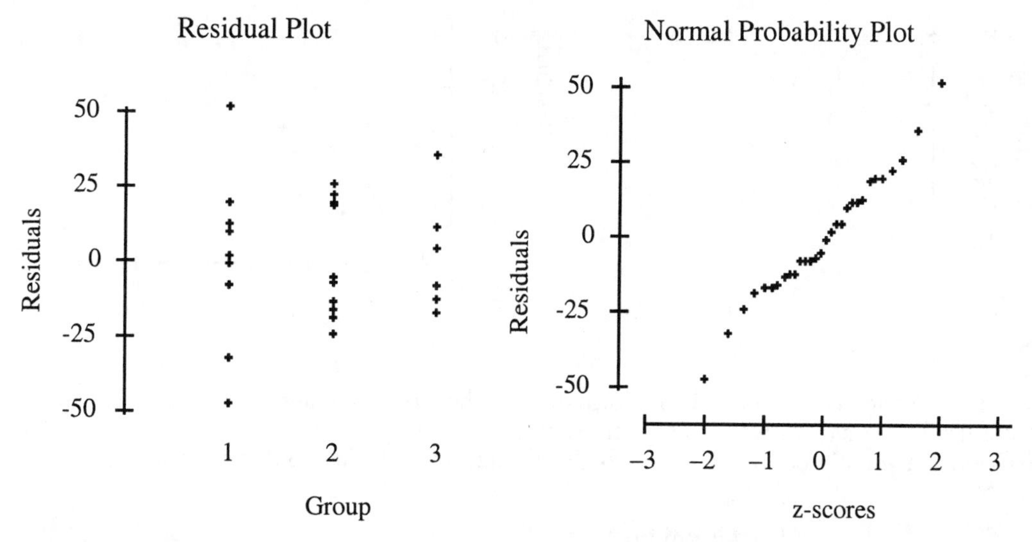

The plots show no serious departures from Normality, so the Normality assumption is reasonable.

b) We use statistical software to run the Bonferroni multiple comparisons procedure. The results are summarized in the table below. Group 1 is the control group, Group 2 is the low-jump group, and Group 3 is the high-jump group.

Bonferroni Post Hoc Tests

Group $i-j$	Difference	std. err.	P-value
2 – 1	11.4000	9.653	0.574592
3 – 1	37.6000	9.653	0.001750
3 – 2	26.2000	9.653	0.033909

At the $\alpha = 0.05$ level, we see that Group 3 (the high-jump group) differs from the other two. The other two groups (the control group and the low jump group) are not significantly different. It appears that the mean density after 8 weeks is different (higher) for the high jump group than for the other two. A boxplot of the data for the three groups helps one see the differences.

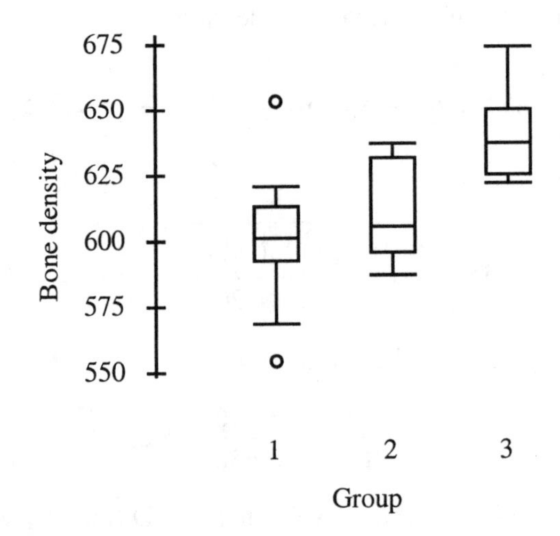

14.57 a) A residual and normal probability plot of the residuals are given below.

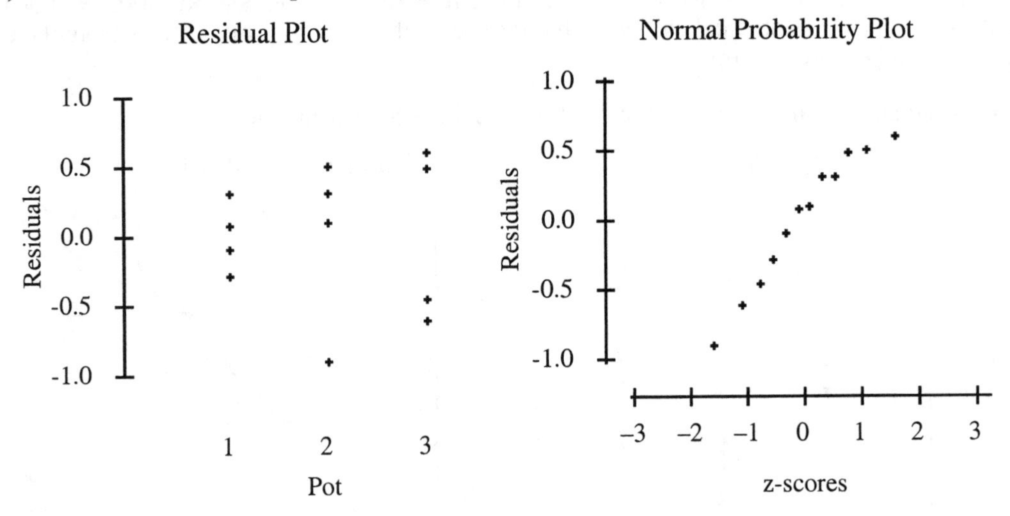

The plots show no serious departures from Normality, so the Normality assumption is reasonable.

b) We use statistical software to run the Bonferroni multiple comparisons procedure. The results are summarized in the table below. Group 1 is the aluminum pots, Group 2 is the clay pots, and Group 3 is the iron pots.

Bonferroni Post Hoc Tests

Group $i - j$	Difference	std. err.	P-value
2 – 1	0.120000	0.3751	0.985534
3 – 1	2.62250	0.3751	0.000192
3 – 2	2.50250	0.3751	0.000274

At the $\alpha = 0.05$ level, we see that Group 3 (the iron pots) differs from the other two. The other two groups (the aluminum and clay pots) are not significantly different. It appears that the mean iron content of yesiga wet' when cooked in iron pots is different (higher) than when cooked in the other two. A boxplot of the data for the three groups helps one see the differences.

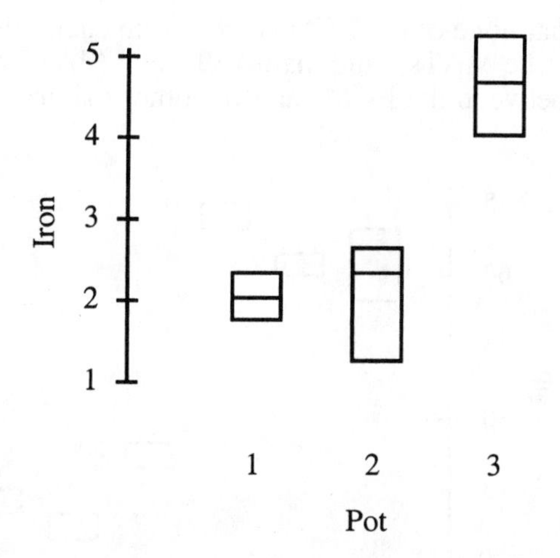

14.59 a) A residual and normal probability plot of the residuals are given below.

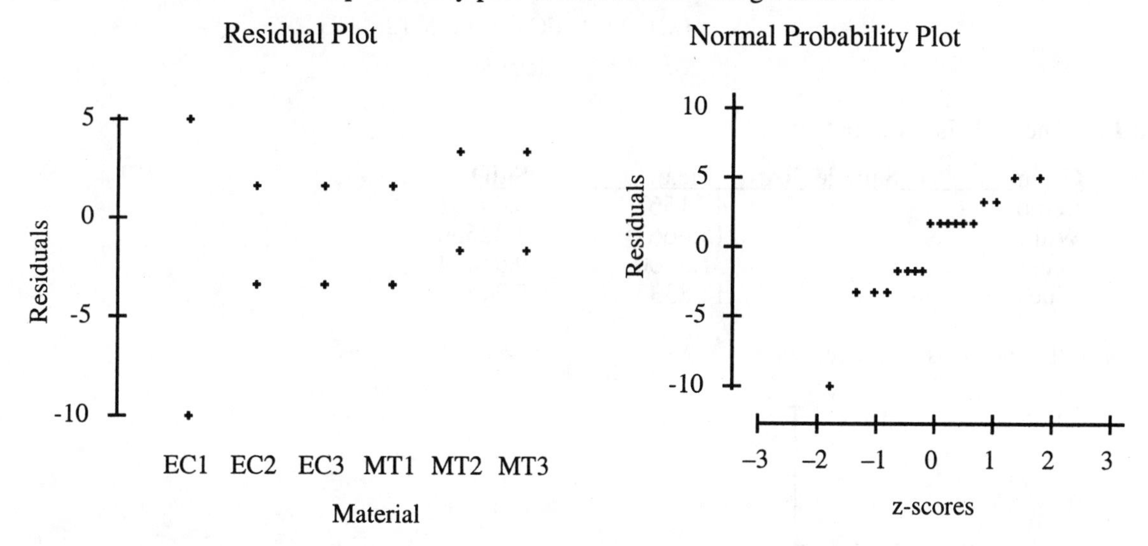

The plots show no serious departures from Normality (but note the granularity of the Normal probability plot. It appears that values are rounded to the nearest 5%), so the Normality assumption is not unreasonable.

b) We use statistical software to run the Bonferroni multiple comparisons procedure. The results are summarized in the table below.

Bonferroni Post Hoc Tests

Group $i - j$	Difference	std. err.	*P*-value
ECM2 – ECM1	−1.66667	3.600	1.00000
ECM3 – ECM1	8.33333	3.600	0.450687
ECM3 – ECM2	10.0000	3.600	0.223577
MAT1 – ECM1	−41.6667	3.600	0.000001
MAT1 – ECM2	−40.0000	3.600	0.000002
MAT1 – ECM3	−50.0000	3.600	0.000000
MAT2 – ECM1	−58.3333	3.600	0.000000
MAT2 – ECM2	−56.6667	3.600	0.000000
MAT2 – ECM3	−66.6667	3.600	0.000000
MAT2 – MAT1	−16.6667	3.600	0.008680
MAT3 – ECM1	−53.3333	3.600	0.000000
MAT3 - ECM2	−51.6667	3.600	0.000000
MAT3 – ECM3	−61.6667	3.600	0.000000
MAT3 – MAT1	−11.6667	3.600	0.101119
MAT3 – MAT2	5.00000	3.600	0.957728

At the $\alpha = 0.05$ level, we see that none of the ECMs differ from each other, that all the ECMs differ from all the other types of materials (the MATs), and that MAT1 and MAT2 differ from each other. The most striking differences are those between the ECMs and the other materials. This is evident in the boxplot below.

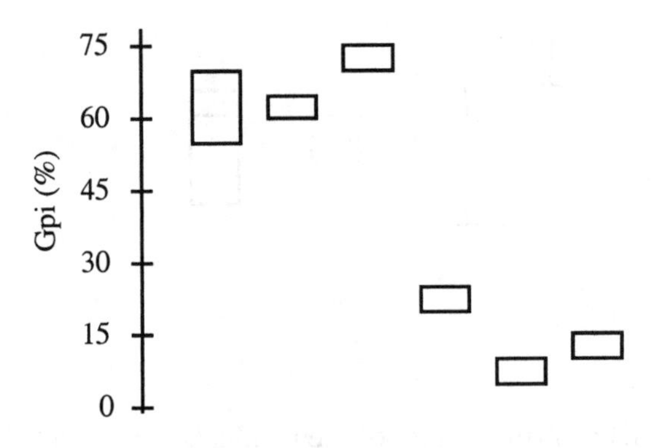

14.61 a) The table is given below.

Group	Sample Size	Mean	StdDev
Lemon	6	47.1667	6.79461
White	6	15.6667	3.32666
Green	6	31.5000	9.91464
Blue	6	14.8333	5.34478

A plot of the means is given below.

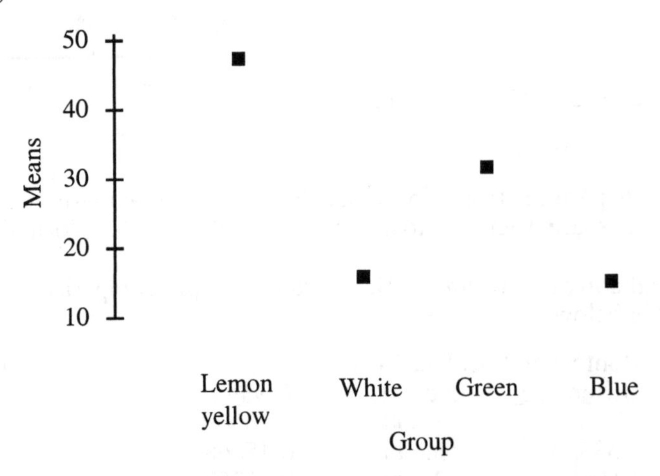

b) Let μ_1, μ_2, μ_3, and μ_4 be the mean number of all insects that would be trapped by the lemon yellow-, white-, green-, and blue -olored traps. The hypotheses for ANOVA are

$$H_0: \mu_1 = \mu_2 = \mu_3 = \mu_4$$
$$H_a: \text{not all of the } \mu_i \text{ are equal}$$

ANOVA tests whether the mean number of insects trapped by the different colors are the same or if they differ. If they differ, ANOVA does not tell us which ones differ.

c) The results, including the value of the F statistic and its P-value, are given in the following ANOVA table.

Source	Degrees of Freedom	Sums of Squares	Mean Square	F	P-value
Groups	3	4218.46	1406.15	30.552	≤ 0.0001
Error	20	920.500	46.0250		
Total	23	5138.96			

The value of s_p is

$$s_p = \sqrt{s_p^2} = \sqrt{\text{MSE}} = \sqrt{46.0250} = 6.78$$

We conclude that there is strong evidence of a difference in the mean number of insects trapped by the different colors.

14.63 a) The missing entries are given in the completed ANOVA table below.

Source	Degrees of Freedom	Sums of Squares	Mean Square	F
Groups	3	104,855.87	$\frac{\text{SSG}}{\text{DFG}} = \frac{104,855.87}{3} = 34,951.96$	$\frac{\text{MSG}}{\text{MSE}} = \frac{34,951.96}{2203.14} = 15.86$
Error	32	70,500.59	$\frac{\text{SSE}}{\text{DFE}} = \frac{70,500.59}{32} = 2203.14$	
Total	DFT = DFG + DFE = 3 + 32 = 35	SST = SSG + SSE = 175,356.46	$\frac{\text{SST}}{\text{DFT}} = \frac{175,356.46}{35} = 5010.18$	

b) Let Group 1 be those males on the treatment, Group 2 be those males in the control group, Group 3 be those males who jogged, and Group 4 be those males who were sedentary. Let μ_1, μ_2, μ_3, and μ_4 be the mean physical fitness score that would be obtained if all males were in each of the groups. The hypotheses for ANOVA are

$$H_0\text{: } \mu_1 = \mu_2 = \mu_3 = \mu_4$$
$$H_a\text{: not all of the } \mu_i \text{ are equal}$$

c) Under the assumption that H_0 is true, the F statistic has the $F(I-1, N-I) = F(\text{DFG, DFE}) = F(3, 32)$ distribution. We look in Table E at the entries corresponding to 3 numerator degrees of freedom and 32 denominator degrees of freedom. Because the table does not contain rows labeled 32, we use the next closest set of rows that is smaller than 32. These are the rows corresponding to 30 denominator degrees of freedom. $F = 15.86$ is larger than the largest critical value of 4.39 in these rows of the Table, and 4.39 corresponds to a P-value of 0.001. Thus, the actual P-value is smaller than 0.001.

d) The value of the within-group variance s_p^2 is

$$s_p^2 = \text{MSE} = 2203.14$$

The value of the pooled standard deviation s_p is

$$s_p = \sqrt{s_p^2} = \sqrt{\text{MSE}} = \sqrt{2203.14} = 46.94$$

14.65 a) If we let Egypt be Group 1, Kenya Group 2, and Mexico Group 3, then from the table we see that $n_1 = 46$, $n_2 = 111$, $n_3 = 52$ and $s_1 = 2.5$, $s_2 = 1.8$, $s_3 = 1.8$. The pooled estimate of the within-country variance s_p^2 is

$$s_p^2 = \frac{(n_1-1)s_1^2 + (n_2-1)s_2^2 + (n_3-1)s_3^2}{(n_1-1)+(n_2-1)+(n_3-1)} = \frac{(45)2.5^2 + (110)1.8^2 + (51)1.8^2}{(45)+(110)+(51)} = \frac{802.89}{206} = 3.90$$

This quantity corresponds to MSE (mean square error) in the ANOVA table.

b) A completed ANOVA table is the following.

Source	Degrees of Freedom	Sums of Squares	Mean Square	F
Groups	$I-1=2$	17.22	$\frac{SSG}{DFG}=\frac{17.22}{2}=8.61$	$\frac{MSG}{MSE}=\frac{8.61}{3.90}=2.21$
Error	$N-I=209-3=206$	$SSE = DFE \times MSE$ $= 206 \times 3.90$ $= 803.4$	3.90	
Total	$DFT = DFG + DFE$ $= 2 + 206 = 208$	$SST = SSG + SSE$ $= 175{,}356.46$	$\frac{SST}{DFT}=\frac{175{,}356.46}{35}=5{,}010.18$	

c) Let μ_1, μ_2, and μ_3 be the mean weight gain for the population of all pregnant women in each of the groups. The hypotheses for ANOVA are

$$H_0\colon\ \mu_1 = \mu_2 = \mu_3$$
$$H_a\colon \text{not all of the } \mu_i \text{ are equal}$$

d) Under the assumption that H_0 is true, the F statistic has the $F(I-1, N-I) = F(2, 206)$ distribution. We look in Table E at the entries corresponding to 2 numerator degrees of freedom and 206 denominator degrees of freedom. Because the table does not contain rows labeled 206, we use the next closest set of rows that is smaller than 206. These are the rows corresponding to 200 denominator degrees of freedom. $F = 2.21$ is smaller than the smallest critical value of 2.33 in these rows of the Table, and 2.33 corresponds to a P-value of 0.10. Thus, the actual P-value is greater than 0.100. We conclude that these data do not provide evidence that the mean weight gains of pregnant women in these 3 countries differ.

14.67 a) A contrast that represents a comparison of the average of the means of the first two groups with the mean of the third group is

$$\psi_1 = (0.5)\mu_1 + (0.5)\mu_2 - \mu_3$$

b) A contrast that compares the means of the first two groups is

$$\psi_2 = \mu_1 - \mu_2$$

14.69 a) Because computer science and engineering and other sciences require good mathematics skills, we might expect the SAT mathematics scores for students in these majors to be higher than those of students in most other majors. Thus, for the contrast $\psi_1 = (0.5)\mu_1 + (0.5)\mu_2 - \mu_3$ it would be reasonable to test the hypotheses

$$H_0\colon\ \psi_1 = 0$$
$$H_a\colon\ \psi_1 > 0$$

It is not clear whether one would expect computer science majors to, on average, have higher or lower SAT mathematics scores than engineering and other science majors. Thus, for the contrast $\psi_2 = \mu_1 - \mu_2$ it would be reasonable to test the hypotheses

$$H_0\colon \psi_2 = 0$$
$$H_a\colon\ \psi_2 \neq 0$$

b) Estimates of these contrasts are obtained by replacing the population means by the sample means. The sample means for each group are given in Example 3. Our estimates are

$$\text{Estimate of } \psi_1\colon c_1 = (0.5)\mu_1 + (0.5)\mu_2 - \mu_3 = (0.5)(619) + (0.5)(629) - 575 = 49$$
$$\text{Estimate of } \psi_2\colon c_2 = \mu_1 - \mu_2 = 619 - 629 = -10$$

c) Using the sample sizes given in Example 3, we estimate the standard errors of these contrasts to be

$$SE_{c_1} = s_p\sqrt{\sum \frac{a_i^2}{n_i}} = 82.5\sqrt{\frac{(0.5)^2}{103} + \frac{(0.5)^2}{31} + \frac{(-1)^2}{122}} = 11.278$$

$$SE_{c_2} = s_p\sqrt{\sum \frac{a_i^2}{n_i}} = 82.5\sqrt{\frac{(1)^2}{103} + \frac{(-1)^2}{31} + \frac{(0)^2}{122}} = 16.901$$

d) For testing the hypotheses

$$H_0: \psi_1 = 0$$
$$H_a: \psi_1 > 0$$

the test statistic is

$$t = \frac{c_1}{\mathrm{SE}_{c_1}} = \frac{49}{11.278} = 4.34$$

There are a total of $N = 256$ observations, and we are interested in $I = 3$ groups. Thus, the degrees of freedom associated with this test are DFE $= N - I = 256 - 3 = 253$. To give an approximate P-value, we look in Table D. There is no row corresponding to 253 degrees of freedom, so we read the next closest and smaller row, which is 100. $t = 4.34$ is larger than the largest entry in this row (corresponding to an upper tail probability of 0.0005), so we conclude the P-value < 0.0005. We conclude that there is strong evidence that the average of the mean SAT mathematics scores for computer science and engineering and other sciences majors is larger than the mean SAT mathematics scores for all other majors.

For testing the hypotheses

$$H_0: \psi_2 = 0$$
$$H_a: \psi_2 \neq 0$$

the test statistic is

$$t = \left|\frac{c_2}{\mathrm{SE}_{c_2}}\right| = \left|\frac{-10}{16.901}\right| = 0.59$$

The degrees of freedom associated with this test are DFE = 253. To give an approximate P-value, we look in Table D. There is no row corresponding to 253 degrees of freedom, so we read the next closest and smaller row, which is 100. $t = 0.59$ is smaller than the smallest entry in this row (corresponding to an upper tail probability of 0.25). The upper tail probability of 0.59 is therefore greater than 0.25. Because we are doing a two-tailed test, we must double the upper tail probability to get the P-value. Thus, P-value > 0.50. We conclude that there is not strong evidence that the average of the mean SAT mathematics scores for computer science majors differs from that of engineering and other sciences majors.

e) A **level C confidence interval** for ψ_1 is

$$c_1 \pm t^* \mathrm{SE}_{c_1}$$

where t^* is the value for the $t(253)$ density curve with area C between $-t^*$ and t^*. We are interested in a 95% confidence interval. There is no row in Table D corresponding to 253 degrees of freedom, so we read the next closest and smaller row which is 100. For 100 degrees of freedom, $t^* = 1.984$ for 95% confidence. Thus, our 95% confidence interval for ψ_1 is

$$c_1 \pm t^* \mathrm{SE}_{c_1} = 49 \pm 1.984(11.278) = 49 \pm 22.38 = (26.62, 71.38)$$

A level C confidence interval for ψ_{12} is

$$c_2 \pm t^* \mathrm{SE}_{c_2}$$

where t^* is the value for the $t(253)$ density curve with area C between $-t^*$ and t^*. As for the previous interval, we use $t^* = 1.984$ for 95% confidence. Thus, our 95% confidence interval for ψ_2 is

$$c_2 \pm t^* \mathrm{SE}_{c_2} = -10 \pm 1.984(16.901) = -10 \pm 33.53 = (-43.53, 23.53)$$

14.71 a) Let Group 1 be those males on the treatment, Group 2 be those males in the control group, Group 3 be those males who jogged, and Group 4 be those males who were sedentary. Let μ_1, μ_2, μ_3, and μ_4 be the mean physical fitness score that would be obtained if all males were in both groups. The appropriate contrasts and hypotheses are the following.

Question 1. Contrast: $\psi_1 = \mu_1 - \mu_2$ Hypotheses: H_0: $\psi_1 = 0$, H_a: $\psi_1 > 0$

Question 2. Contrast: $\psi_2 = \mu_1 - (0.5)\mu_2 - (0.5)\mu_4$ Hypotheses: H_0: $\psi_2 = 0$, H_a: $\psi_2 > 0$

Question 3. Contrast: $\psi_3 = \mu_3 - (1/3)\mu_1 - (1/3)\mu_2 - (1/3)\mu_4$ Hypotheses: H_0: $\psi_3 = 0$, H_a: $\psi_3 > 0$

b) In Exercise 14.63, we computed the value of the pooled standard deviation and found it to be s_p = 46.94. (This can also be computed using the standard deviations and sample sizes for each group given in this problems and the formula $s_p^2 = \frac{(n_1-1)s_1^2+(n_2-1)s_2^2+(n_3-1)s_3^2+(n_4-1)s_4^2}{(n_1-1)+(n_2-1)+(n_3-1)+(n_4-1)}$.) Thus, for each question, we estimate the contrast (by replacing the means by their sample values), its standard error, the test statistic, and the approximate *P*-value to be the following.

Question 1

Estimate of ψ_1: $c_1 = 291.91 - 308.97 = -17.06$

Estimate of standard error: $SE_{c_1} = s_p\sqrt{\sum \frac{a_i^2}{n_i}} = 46.94\sqrt{\frac{(1)^2}{10}+\frac{(-1)^2}{5}+\frac{(0)^2}{11}+\frac{(0)^2}{10}} = 25.71$

Test statistic: $t = \frac{c_1}{SE_{c_1}} = \frac{-17.06}{25.71} = -0.6635$

Approximate *P*-value: This is the upper tail probability because this is a one-sided test. There are a total of N = 36 observations and we are interested in I = 4 groups. Thus, the degrees of freedom associated with this test are DFE = $N - I$ = 36 – 4 = 32. There is no row corresponding to 32 degrees of freedom in Table D, so we use the next closest and smaller row of 30. The upper tail probability for 0.6635 is greater than 0.25 (but less than 0.5 because the t distribution is symmetric about 0). By the symmetry of the t distribution, we know that the upper tail probability of –0.6635 must be between 1 – 0.25 and 1 – 0.50 or between 0.50 and 0.75. The approximate *P*-value is therefore 0.50 < *P*-value < 0.75. There is not strong evidence that T is better than C.

Question 2

Estimate of ψ_2: $c_2 = 291.91 - (0.5)308.97 - (0.5)226.07 = 24.39$

Estimate of standard error: $SE_{c_2} = s_p\sqrt{\sum \frac{a_i^2}{n_i}} = 46.94\sqrt{\frac{(1)^2}{10}+\frac{(-0.5)^2}{5}+\frac{(0)^2}{11}+\frac{(-0.5)^2}{10}} = 19.64$

Test statistic: $t = \frac{c_2}{SE_{c_2}} = \frac{24.39}{19.64} = 1.24$

Approximate *P*-value: This is the upper tail probability because this is a one-sided test. The degrees of freedom associated with this test are 32. There is no row corresponding to 32 degrees of freedom in Table D, so we use the next closest and smaller row of 30. The upper tail probability for 1.24 is between 0.10 and 0.15. The approximate *P*-value is therefore 0.10 < *P*-value < 0.15. There is not strong evidence that T is better than the average of C and S.

Question 3

Estimate of ψ_3: $c_3 = 366.87 - (1/3)291.91 - (1/3)308.97 - (1/3)226.07 = 91.22$

Estimate of standard error: $SE_{c_3} = s_p\sqrt{\sum \frac{a_i^2}{n_i}} = 46.94\sqrt{\frac{(1)^2}{10}+\frac{(-1/3)^2}{5}+\frac{(-1/3)^2}{11}+\frac{(-1/3)^2}{10}} = 17.78$

Test statistic: $t = \frac{c_3}{SE_{c_3}} = \frac{91.22}{17.78} = 5.13$

Approximate *P*-value: This is the upper tail probability because this is a one-sided test. The degrees of freedom associated with this test are 32. There is no row corresponding to 32 degrees of freedom in Table D, so we use the next closest and smaller row of 30. The upper tail probability for 5.13 is larger than the largest entry of 3.646 in the row, corresponding to an upper tail probability of 0.0005. The approximate *P*-value is therefore *P*-value < 0.0005. There is strong evidence that J is better than the average of the other three groups.

c) This is an observational study. Males were not assigned at random to treatments. Thus, although the researchers tried to match those in groups with respect to age and other characteristics, there are reasons why people choose to jog or choose to be sedentary that may affect other aspects of their health. It is always risky to draw conclusions of causality from a single (small) observational study, no matter how well-designed it is in other respects.

14.73 To perform a multiple comparisons procedure, we compute t statistics for all pairs of means using the formula

$$t_{ij} = \frac{\bar{x}_i - \bar{x}_j}{s_p\sqrt{\frac{1}{n_i} + \frac{1}{n_j}}}$$

If

$$|t_{ij}| \geq t^{**} = 2.81$$

we declare the population means μ_i and μ_j different at level $\alpha = 0.05$. From Exercises 14.63 and 14.71, we know that $s_p = 46.94$. For each pair of means we get the following.

Group 1 (T) vs. Group 2 (C): $t_{12} = \frac{\bar{x}_1 - \bar{x}_1}{s_p\sqrt{\frac{1}{n_1} + \frac{1}{n_2}}} = \frac{291.91 - 308.97}{46.94\sqrt{\frac{1}{10} + \frac{1}{5}}} = \frac{-17.06}{25.71} = -0.66$

$|-0.66|$ is not larger than $t^{**} = 2.81$, so we do not have strong evidence that T and C differ.

Group 1 (T) vs. Group 3 (J): $t_{13} = \frac{\bar{x}_1 - \bar{x}_3}{s_p\sqrt{\frac{1}{n_1} + \frac{1}{n_3}}} = \frac{291.91 - 366.87}{46.94\sqrt{\frac{1}{10} + \frac{1}{11}}} = \frac{-74.96}{20.51} = -3.65$

$|-3.65|$ is larger than $t^{**} = 2.81$, so we have strong evidence that T and J differ.

Group 1 (T) vs. Group 4 (S): $t_{14} = \frac{\bar{x}_1 - \bar{x}_4}{s_p\sqrt{\frac{1}{n_1} + \frac{1}{n_4}}} = \frac{291.91 - 226.07}{46.94\sqrt{\frac{1}{10} + \frac{1}{10}}} = \frac{65.84}{20.99} = 3.14$

$|3.14|$ is larger than $t^{**} = 2.81$, so we have strong evidence that T and S differ.

Group 2 (C) vs. Group 3 (J): $t_{23} = \frac{\bar{x}_2 - \bar{x}_3}{s_p\sqrt{\frac{1}{n_2} + \frac{1}{n_3}}} = \frac{308.97 - 366.87}{46.94\sqrt{\frac{1}{5} + \frac{1}{11}}} = \frac{-57.9}{25.32} = -2.29$

$|-2.29|$ is not larger than $t^{**} = 2.81$, so we do not have strong evidence that C and J differ.

Group 2 (C) vs. Group 4 (S): $t_{24} = \frac{\bar{x}_2 - \bar{x}_4}{s_p\sqrt{\frac{1}{n_2} + \frac{1}{n_4}}} = \frac{308.97 - 226.07}{46.94\sqrt{\frac{1}{5} + \frac{1}{10}}} = \frac{82.9}{25.71} = 3.22$

$|3.22|$ is larger than $t^{**} = 2.81$, so we have strong evidence that C and J differ.

Group 3 (J) vs. Group 4 (S): $t_{34} = \frac{\bar{x}_3 - \bar{x}_4}{s_p\sqrt{\frac{1}{n_3} + \frac{1}{n_4}}} = \frac{366.87 - 226.07}{46.94\sqrt{\frac{1}{11} + \frac{1}{10}}} = \frac{140.8}{20.51} = 6.86$

$|6.86|$ is larger than $t^{**} = 2.81$, so we have strong evidence that J and S differ.

Our overall conclusions are that J and C do not differ, T and C do not differ, but all other pairs differ.

14.75 We are told that $\sigma = 2.4$. The n_i are equal and we let n denote their common value. Because the n_i are equal, $\bar{\mu}$ is simply the average of the μ_i

$$\bar{\mu} = \frac{2.6+3.0+3.4}{3} = 3.0$$

The noncentrality parameter is therefore

$$\lambda = \frac{n\sum(\mu_i - \bar{\mu})^2}{\sigma^2} = \frac{n[(2.6-3.0)^2 + (3.0-3.0)^2 + (3.4-3.0)^2]}{2.4^2} = 0.0556n$$

Also, because there are $I = 3$ groups and $N = nI = 3n$ observations,

$$\text{DFG} = I - 1 = 2,\ \text{DFE} = N - I = 3n - 3$$

For testing at level $\alpha = 0.05$, we use software to determine F^*, the upper 0.05 critical value of the F(DFG, DFE) = $F(2, 3n - 3)$ distribution. For the choices of n given, using SAS (the latest version of Minitab can also be used to calculate the power using the same information), we get the following table.

n	DFG	DFE	F^*	λ	Power
50	2	147	3.0576	2.78	0.2950
100	2	297	3.0261	5.56	0.5453
150	2	447	3.0158	8.34	0.7336
175	2	522	3.0130	9.73	0.8017
200	2	597	3.0108	11.12	0.8548

A sample size of 175 gives reasonable power. The gain in power by using 200 women per group may not be worthwhile unless it is easy to get women for the study. If it is difficult or expensive to include more women in the study, one might consider a sample size of 150 per group.

14.77 a) The table is given below.

Group	Sample size	Mean	StdDev	Std Error
0	2	76.1016	2.39753	1.695
1	2	65.5547	3.32561	2.352
3	2	34.0547	1.20429	0.852
5	2	19.3984	1.69043	1.195
7	2	13	0.419845	0.297

b) We use facts about transformations that we learned in Chapter 4. The sample sizes would be the same in both tables. The means, standard deviations, and standard errors above could have been obtained from those in Exercise 14.44 by dividing each by 64 and multiplying the result by 100. Means, standard deviations, and standard errors change the same way as individual values.

c) The ANOVA table for the transformed data with the F statistic, degrees of freedom, and P-value is given below.

Source	Degrees of Freedom	Sums of Squares	Mean Square	F	P-value
Group	4	6263.97	1565.99	367.74	≤ 0.0001
Error	5	21.2920	4.25840		
Total	9	6285.26			

We conclude that there is strong evidence that the group means differ, i.e., that the mean % vitamin C content is not the same for all conditions.

The degrees of freedom, the F statistic, and P-value are all the same as in Exercise 14.44.

14.79 a) The ANOVA table with the incorrect observation is given below.

Source	Degrees of Freedom	Sums of Squares	Mean Square	F	P-value
Group	3	40820.3	13606.8	2.0032	0.1460
Error	20	135853	6792.65		
Total	23	176673			

The P-value is larger than .10, so we would conclude that there is not strong evidence that the mean number of insects that will be trapped differs between the different color traps.

b) The results are very different. In Exercise 14.61 the P-value was less than 0.0001 and we concluded that there was strong evidence that the mean number of insects that will be trapped differs between the different color traps. The outlier increased the sum of squares of error considerably and this results in a much smaller value of F.

c) The table is given below.

Group	Count	Mean	StdDev
Lemon yellow	6	114.667	164.416
White	6	15.6667	3.32666
Green	6	31.5000	9.91464
Blue	6	14.8333	5.34478

The unusually large value of the mean and standard deviation might indicate that there was an error in the data recorded for the lemon yellow trap.

14.81 a) A scatterplot of expected price versus number of promotions is given below.

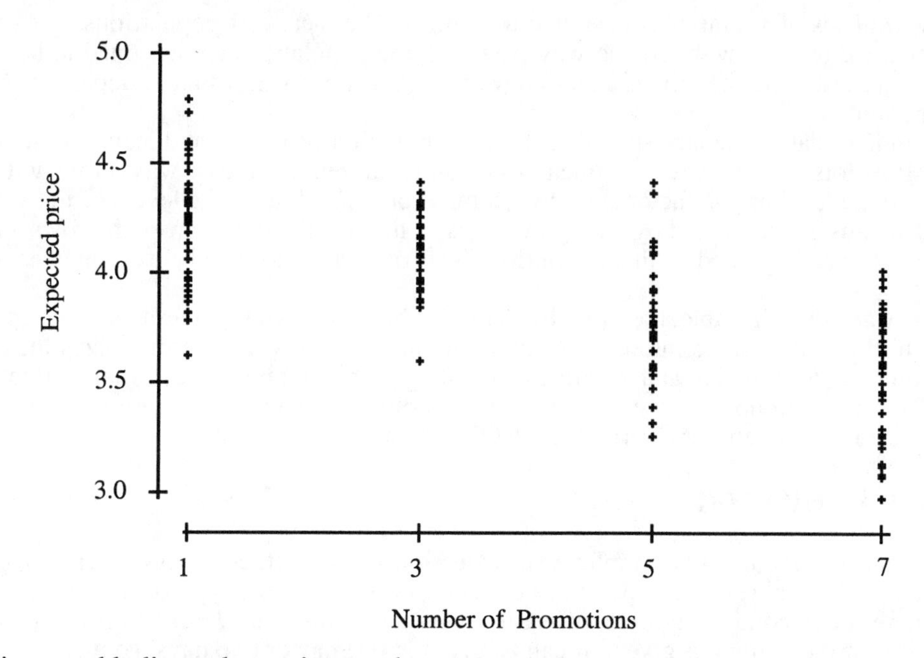

The pattern is a roughly linear decreasing trend.

b) The test in regression that tests the null hypothesis that the explanatory variable has no linear relationship with the response is the t test of whether or not the slope is 0.

c) Using software, we obtain the following.

Variable	Coefficient	s.e. of Coeff	t-ratio	P-value
Constant	4.36453	0.0401	109	≤ 0.0001
Number of promotions	–0.116475	0.0087	-13.3	≤ 0.0001

The t statistic for testing whether the slope is 0 is –133 and we see that the P-value is ≤ 0.0001. Thus, there is strong evidence that the slope is not 0. The ANOVA in Exercise 14.49 showed that there is strong evidence that the mean expected price is different for the different numbers of promotions. This is consistent with our regression results here because if the slope is different from 0, the mean expected price is changing as the number of promotions changes.

In this example the regression is more informative. It not only tells us that the means differ, but it also gives us information about how they differ.

CHAPTER 15

TWO-WAY ANALYSIS OF VARIANCE

SECTION 15.1

OVERVIEW

The two-way analysis of variance is designed to compare the means of populations that are classified according to two factors. As with the one-way ANOVA, the populations are assumed to be normal with possibly different means and the same standard deviation. The observations are independent SRSs drawn from each population.

The preliminary data summary should include examination of means and standard deviations, and normal quantile plots. Typically, the means are summarized in a two-way table with the rows corresponding to the level of one factor and the columns corresponding to the levels of the second factor. The **marginal means** are computed by taking averages of these cell means across the rows and columns. These means are typically plotted so that the **main effects** of each factor can be examined as well as their **interaction.**

In the two-way ANOVA table, the **model** variation is broken down into parts due to each of the main effects and a third part due to the interaction. In addition, the ANOVA table organizes the calculations required to compute F statistics and P-values to test hypotheses about these main effects and their interaction. The within-group variance is estimated by pooling the standard deviations from the cells and corresponds to the mean square for error in the ANOVA table.

APPLY YOUR KNOWLEDGE

15.1 The response variable is the employees evaluation of the effectiveness of the program. The response variable takes values on a 7-point scale. The first factor is the type of training program. The company is interested in comparing three different programs, so there are $I = 3$ levels of this factor. The second factor is how the training is given. It can be given in one day or two days, so there are $J = 2$ levels for this factor. There are a total of $N = 120$ observations to be taken.

15.3 The response variable is the comparison of the new lotion with the regular product. The response variable takes values on a 7-point scale indicating the level of preference for the new or regular product. The first factor is the formulation of the lotion. The research and marketing group is interested in comparing five different formulations, so there are $I = 5$ levels of this factor. The second factor is fragrance. The lotion can be made with three fragrances, so there are $J = 3$ levels for this factor. There are a total of $N = 1800$ observations to be taken.

15.5 The plot of the difference in means between men and women for each age group is given on the next page. When there are only two levels of a factor, plotting the differences is another way to look for interaction. If the factors did not interact, the population mean difference between salaries for men and women should be the same for each age group. A plot of the differences in sample means should then look reasonably close to a horizontal line. The plot on the next page shows that the difference in salaries between men and women does not appear to be the same for each age group. The difference is smallest for the youngest age group, and the gap widens for the next two age groups and narrows slightly for the age group 65+. This plot suggests that the two factors interact.

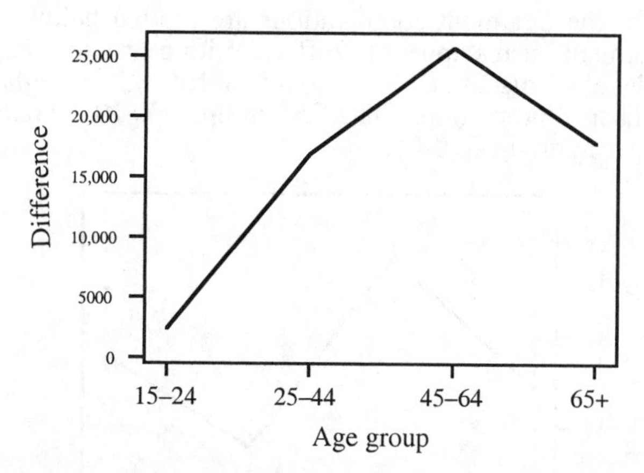

15.7 The plots of the four tables are given below. There is interaction in all plots with the exception of (c), for which the two lines are parallel. In (a), the response is increasing more rapidly as the level of A "increases" when B = 2 than when B = 1. In (b), the response is increasing as the level of A "increases" when B = 1 and decreasing when B = 2. This is the strongest interaction in the four plots. In (d), there is no difference in mean response when B = 1 and B = 2 for the first two levels of A, but there is a large difference at the third level of A.

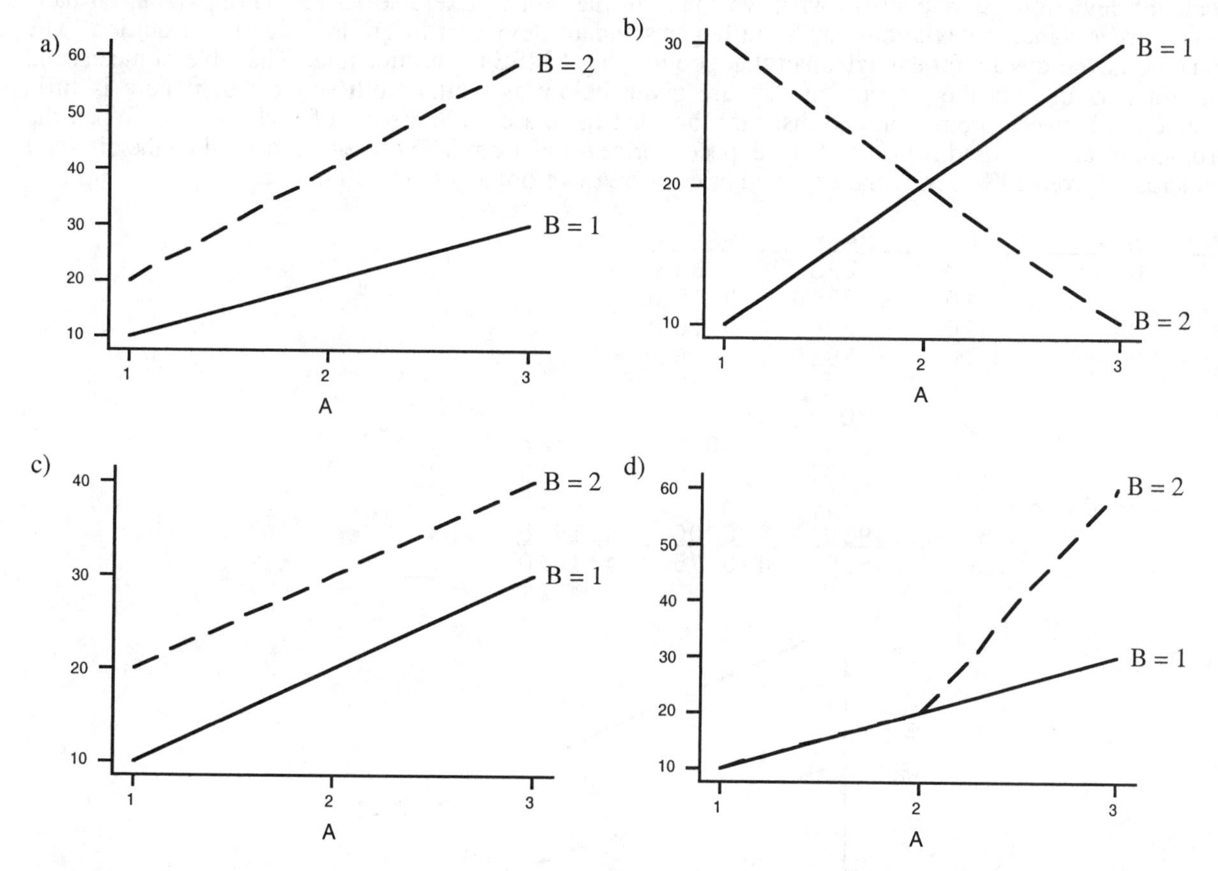

15.9 The two factors are type of packaging and color. Packaging comes in four different types so there are $I = 4$ levels of this factor. Three colors can be used for each type of packaging, so there are $J = 3$ levels for this factor. Forty subjects will be used for each combination of package type and color, so there are a total of $N = 12 \times 40 = 480$ observations to be taken. The F statistic used to test the null hypothesis of no main effect of packaging has an $F(I - 1, N - IJ) = F(4 - 1{,}480 - (3)(4)) = F(3, 468)$ distribution. The F statistic used to test the null hypothesis of no main effect of color has an $F(J - 1, N - IJ) = F(2, 468)$ distribution. The F statistic used to test the null hypothesis of no interaction has an $F([I - 1][J - 1], N - IJ) = F(6, 468)$ distribution.

15.11 The expected prices for the treatment combinations are plotted below and the group means are connected by lines. The treatment means appear to differ. With 5 promotions, the means are less than with 1 promotion for each level of discount. A discount of 10% gives higher expected price than a discount of 30% for each number of promotions. The first group, D30-P1 seems more spread out than the others, and there is a low outlier in group D10-P5.

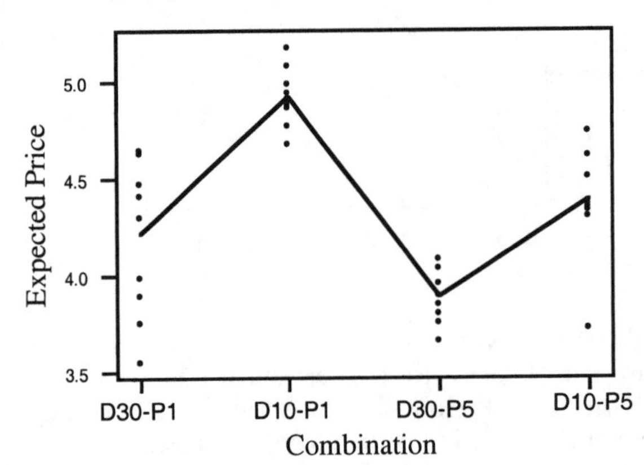

15.13 The means and standard deviations for the four groups are given in the table below. In terms of the standard deviations, they confirm what we found in the plot in Exercise 15.11. Groups 1 and 4 have much larger standard deviations, with the larger standard deviation in group 4 due to an outlier. The standard deviations are far enough apart that pooling to get MSE is questionable. The table of means and marginal means and the interaction plot are given below as well. Both suggest that there is little interaction between promotion and discount, but that there are main effects of both factors. When the promotions are increased from 1 to 5 the expected price drops from $4.57 to $4.14, and when the discount is increased from 10% to 30% the expected price decreases from $4.66 to $4.08.

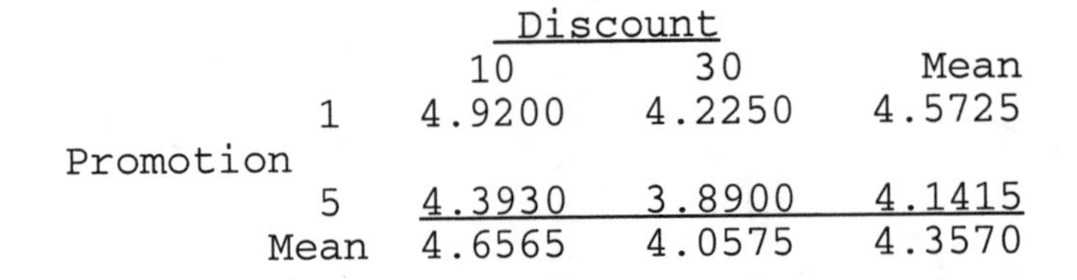

Variable	N	Mean	StDev
D1-P30	10	4.2250	0.3860
D1-P10	10	4.9200	0.1520
D5-P30	10	3.8900	0.1629
D5-P10	10	4.3930	0.2685

		Discount		
		10	30	Mean
Promotion	1	4.9200	4.2250	4.5725
	5	4.3930	3.8900	4.1415
	Mean	4.6565	4.0575	4.3570

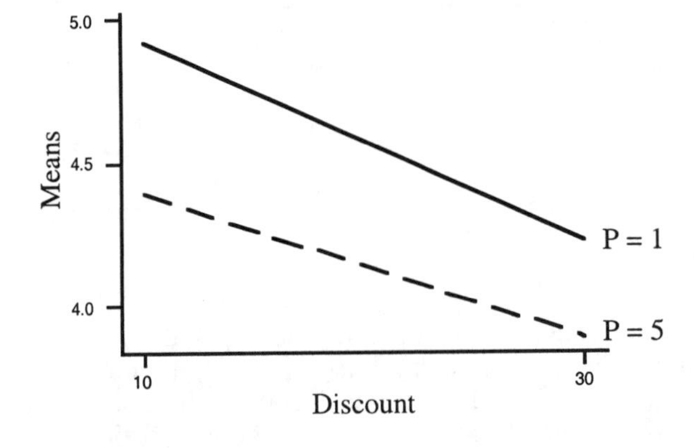

15.15 The information in the basic ANOVA table is contained in the output of all four packages, although the Excel and Minitab output presented are the most basic and simplest to read. To find the sums of squares and tests for main effects and interactions in SAS requires going to the second table in the output, while the SPSS output includes several lines at the beginning of the ANOVA table, which should not be familiar to you and are of limited use. Excel and SPSS include information on the means and standard deviations for the treatments and the marginal means as part of the basic output, but these can be gotten easily in both SAS and Minitab as well. For just the basic ANOVA table, Minitab produces output most similar to what you have seen in the text.

CHAPTER 15 REVIEW EXERCISES

15.17 a) The response variable is the number of hours of sleep the subject gets on a particular night. The first factor is smoking level—nonsmoker, moderate or heavy smoker—so there are $I = 3$ levels of this factor. The second factor is gender—male or female—so there are $J = 2$ levels for this factor. There are a total of $N = 80 \times 6 = 480$ observations to be taken.

b) The response variable is the measure of concrete strength. The first factor is the formula for the mixture, and there are $I = 6$ levels of this factor. The second factor is the number of cycles of freezing and thawing, 0, 100 or 500, so there are $J = 3$ levels for this factor. There are a total of $N = 9 \times 6 = 54$ observations to be taken.

c) The response variable is the score on the final exam. The first factor is the teaching method and there are $I = 4$ levels of this factor. The second factor is the major—special education or other—so there are $J = 2$ levels for this factor. There are a total of $N = 32$ students in the study.

15.19 a) There are $I = 3$ levels of the first factor and $J = 4$ levels of the second factor. Six observations will be used for each treatment combination, so there are a total of $N = 12 \times 6 = 72$ observations to be taken. The test for the first main effect has an $F(I - 1, N - IJ) = F(2, 60)$ distribution. The test for the second main effect has an $F(J - 1, N - IJ) = F(3, 60)$ distribution and the test for interaction has an $F([I - 1][J - 1], N - IJ) = F(6, 60)$ distribution.

b) At the 5% level, using software the three F critical values are 3.1504, 2.5781, and 2.2541 for the two main effects and interaction, respectively.

c) At the 1% level, using software the three F critical values are 4.9774, 4.1259, and 3.1187 for the two main effects and interaction, respectively.

15.21 a) The group means are plotted on the axes below.

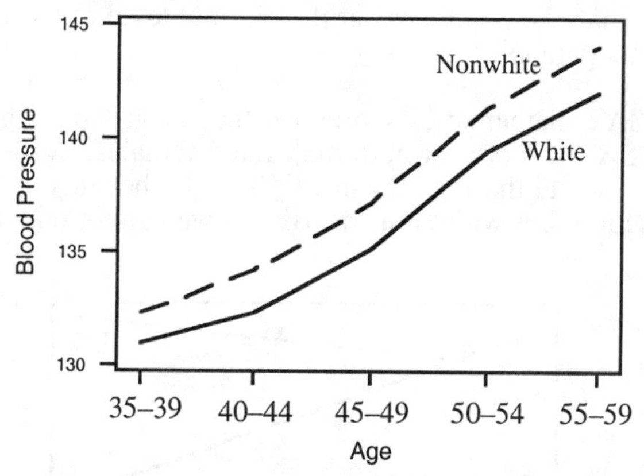

b) The means for the nonwhites are all slightly higher than for whites. The mean systolic blood pressure is increasing with age group and there does not seem to be an interaction as the increases in mean systolic blood pressure with age seem to be approximately the same for both groups.

c) The marginal means are 135.98 for whites and 137.82 for nonwhites. For the age groups the marginal means are 131.65 (35–39), 133.25 (40–44), 136.20 (45–49), 140.35 (50–54), and 143.05 (55–59). The mean systolic blood pressure is around 2 points higher for nonwhites than whites in each age group. The mean systolic blood pressure increases between 2 and 4 points for each increase in age group.

15.23 a) The interaction plot is given below

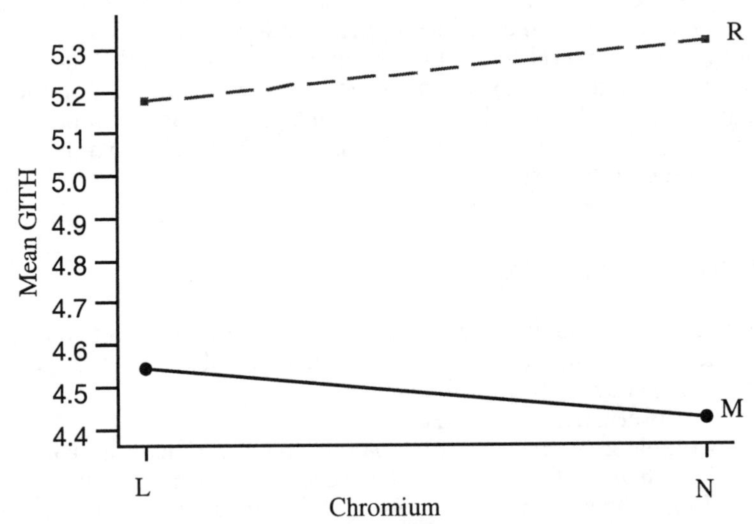

b) The plot suggests that there may be an interaction. The effect of Chromium when going from Low to Normal levels is to decrease mean GITH when the rats could eat as much as they wanted (M) and to increase the mean GITH when the total amount the rats could eat was restricted (R). Without a formal hypothesis test, we do not know if this apparent interaction is due to chance variation or whether the effect is real. In terms of the effect of Chromium, it appears to be small compared to the effect of Eat. The two lines are quite far apart, showing a larger effect of Eat. The change in going from the Low to Normal levels of Chromium is much smaller.

c)

	Eat		Mean
Chromium	M	R	
L	4.545	5.175	4.860
N	4.425	5.317	4.871
Mean	4.485	5.246	4.866

At the Low level of Chromium the difference between M and R is –0.63, and at the Normal level of Chromium the difference between M and R is –0.892. In the plot the two means at the Low level of Chromium are closer together than the two means at the Normal level of Chromium. This is reflected in the fact that the two lines are not parallel.

15.25 The plot of the mean SAT mathematics scores for the six groups is given below. The category "other" had the lowest mean SATM score for both males and females. Males had slightly higher mean SATM in CS than EO majors, while the females in EO have higher mean SATM than the CS majors. There appears to be an interaction, but without a formal test we cannot tell if it is real or due to chance variation.

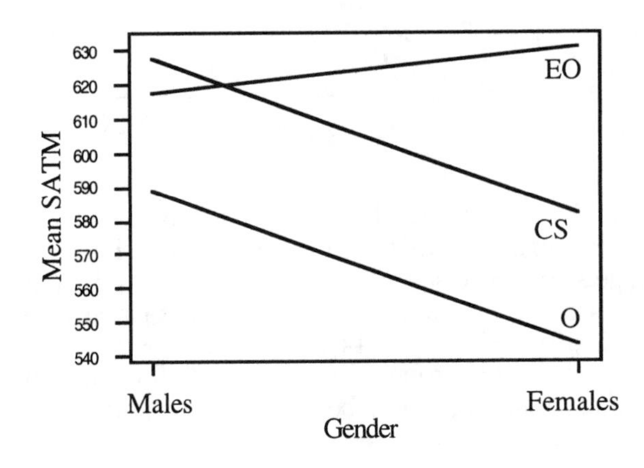

15.27 a) The table below gives the sample sizes, means, and standard deviations for the 12 material-time groups. The standard deviations vary considerably, but because each standard deviation is based on only three observations, we would expect a large amount of variability.

Variable	Group	N	Mean	StDev
% Gpi	ECM1,4	3	65.00	8.66
	ECM1,8	3	63.33	2.89
	ECM2,4	3	63.33	2.89
	ECM2,8	3	63.33	5.77
	ECM3,4	3	73.33	2.89
	ECM3,8	3	73.33	5.77
	MAT1,4	3	23.33	2.89
	MAT1,8	3	21.67	5.77
	MAT2,4	3	6.67	2.89
	MAT2,8	3	6.67	2.89
	MAT3,4	3	11.67	2.89
	MAT3,8	3	10.00	5.00

The table below gives the means for the different material-time groups and the marginal means for material and time.

ROWS: TIME COLUMNS: MATERIAL

	ECM1	ECM2	ECM3	MAT1	MAT2	MAT3	MEAN
4	65.000	63.333	73.333	23.333	6.667	11.667	40.556
8	63.333	63.333	73.333	21.667	6.667	10.000	39.722
MEAN	64.167	63.333	73.333	22.500	6.667	10.833	40.139

b)

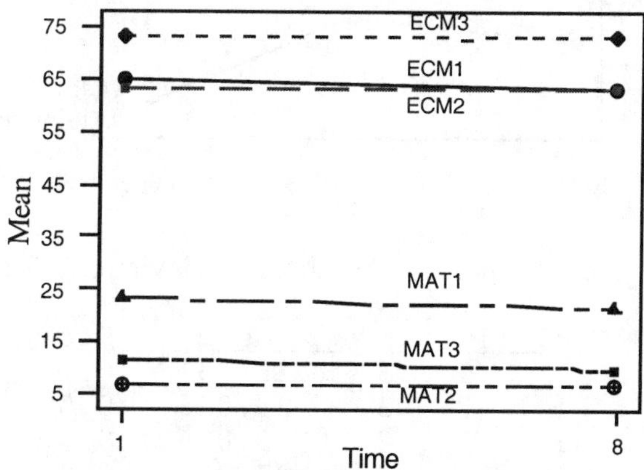

The most striking feature in the plot is the complete lack of a time effect. The mean % Gpi at 4 and 8 weeks are almost identical and as a result there is almost no interaction. The important effect is the material, with the ECM (extracellular) material having a much higher % Gpi than the MAT (inert) material.

c) The ANOVA table is reproduced below.

Analysis of Variance for % Gpi

Source	DF	SS	MS	F	P
Material	5	27,045.1	5409.0	251.26	0.000
Time	1	6.2	6.2	0.29	0.595
Material*Time	5	6.2	1.3	0.06	0.998
Error	24	516.7	21.5		
Total	35	27,574.3			

The ANOVA table supports the interaction plot in (b). There is no evidence of a time effect or an interaction between time and material. However, there is a highly significant effect of material. Note that the ANOVA table does not tell which materials are different or whether there is a difference between the ECM (extracellular) material and the MAT (inert) material. These types of conclusions require further multiple comparisons or examination of specific contrasts as in the one-way analysis of variance.

15.29 The F statistics for the one-way ANOVA's are 58.14 at two weeks, 137.94 at four weeks, and 115.54 at eight weeks. All are highly significant, with P-values less than 0.001. This suggests that there are differences between the materials at each time period. Multiple comparisons can be used to help determine which materials are better at each time period. The general finding from doing the Bonferroni multiple comparisons on all pairs of treatments is that the ECMs are superior to the inert materials. Instead of doing multiple comparisons, it would also be possible to look at contrasts that compare the average of the three ECMs with the average of the three inert materials at each time period, or to make specific comparisons among the ECMs or inert materials. The one-way ANOVA just tells us the materials are different. The choice of further analysis depends on the specific questions that the experimenter would like to answer.

15.31 The plot of the cell means is given below. Despite the fact that there is a highly significant interaction, for each type of food the mean iron content is higher when that food is cooked in an iron pot than in either clay or aluminum. The main effect of pot is highly significant, and it is clear from the plot that there is little difference between clay and aluminum, while iron pots are associated with higher iron content. The interaction only says the difference is not necessarily the same between type of pot for each food.

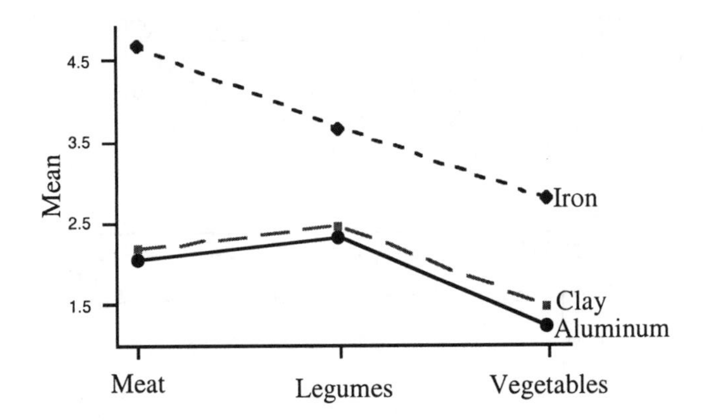

15.33 a) The table below gives the means and standard deviations for each of the 15 treatment combinations.

Tool	Time	N	Mean	Stdev
1	1	3	25.031	0.001
1	2	3	25.028	0.000
1	3	3	25.026	0.000
2	1	3	25.017	0.001
2	2	3	25.020	0.001
2	3	3	25.016	0.000
3	1	3	25.006	0.001
3	2	3	25.013	0.001
3	3	3	25.009	0.001
4	1	3	25.012	0.000
4	2	3	25.019	0.001
4	3	3	25.014	0.002
5	1	3	24.997	0.001
5	2	3	25.006	0.000
5	3	3	25.000	0.001

b) The plot below is for the means of each of the time and tool combinations. The plot suggests little interaction as the lines for each time appear fairly parallel. The biggest difference is between the tools, but this is of little importance as we expect the tools to have slightly different diameters and this can be adjusted. There is also a difference in times. With the exception of tool 1, time 2 has the highest diameters, time 3 has the middle diameter, and time 1 has the lowest diameter.

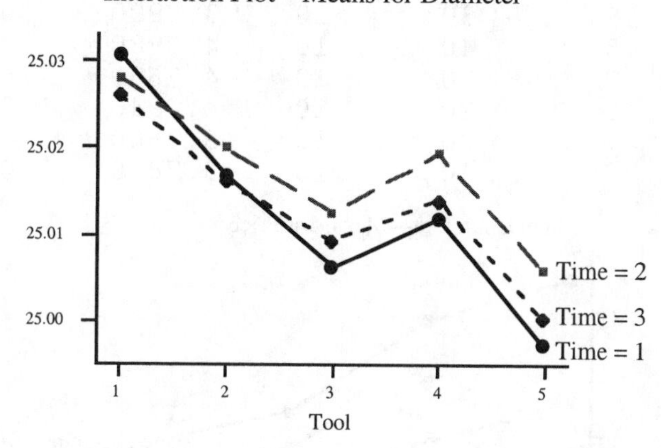

c) The analysis of variance table below was produced by Minitab. Both main effects and the interaction are statistically significant. The degrees of freedom for the main effect of tool are 4 and 30, for the main effect of time are 2 and 30 and for interaction are 8 and 30. The *F* statistics and *P*-values are given in the table.

Analysis of Variance for Diameter

Source	DF	SS	MS	F	P
Tool	4	0.00359714	0.00089928	412.98	0.000
Time	2	0.00018992	0.00009496	43.61	0.000
Tool*Time	8	0.00013324	0.00001665	7.65	0.000
Error	30	0.00006533	0.00000218		
Total	44	0.00398562			

d) Both main effects and the interaction are statistically significant, although the *F* for interaction is much smaller than the *F* for the main effect of time, which is much smaller than the *F* for the main effect of tool. The interaction occurs because tool 1's mean diameters changed differently over time compared to the other tools. The differences in the sizes of these effects was apparent in the plot in part (b). The reason that the interaction is statistically significant despite the appearance of the plot is that there is so little variability within each treatment. Many of the standard deviations were zero, as shown in part (a). When there is very little within group variability or the sample sizes are large it is possible to detect effects that while being statistically significant are of little practical importance. Because tool differences are of limited interest, the most important finding is the fact that the diameters vary with time.

15.35 a) The table below gives the means and standard deviations for each of the 16 treatment combinations. The plot on the next page is for the means of each of the discount and promotion combinations. As the number of promotions goes up the expected price tends to decrease, and as the percent discount increases, the expected price tends to decrease although the sample means of the expected prices for a discount of 40% were above those for a discount of 30% for each of the values of promotion. There appears to be little interaction in the plot.

Number of Promotions	Percent Discount	N	Mean Price	Standard Deviation
1	10	10	4.9200	0.0481
1	20	10	4.6890	0.0737
1	30	10	4.2250	0.1220
1	40	10	4.4230	0.0584
3	10	10	4.7560	0.0768
3	20	10	4.5240	0.0856
3	30	10	4.0970	0.0742
3	40	10	4.2840	0.0645

Number of Promotions	Percent Discount	N	Mean Price	Standard Deviation
5	10	10	4.3930	0.0849
5	20	10	4.2510	0.0838
5	30	10	3.8900	0.0515
5	40	10	4.0580	0.0557
7	10	10	4.2690	0.0854
7	20	10	4.0940	0.0761
7	30	10	3.7600	0.0828
7	40	10	3.7800	0.0678

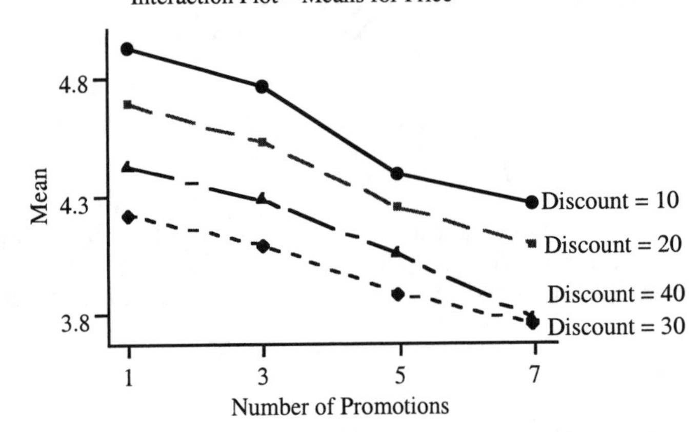

b) The ANOVA table produced by Minitab is given below. As expected, there is little interaction but the main effects of both discount and price are highly significant.

Analysis of Variance for Price

Source	DF	SS	MS	F	P
Promotions	3	8.3605	2.7868	47.73	0.000
Discount	3	8.3069	2.7690	47.42	0.000
Promote*Discount	9	0.2306	0.0256	0.44	0.912
Error	144	8.4087	0.0584		
Total	159	25.3067			

c) As the number of promotions goes up, the expected price tends to decrease. As the percent discount increases, the expected price tends to decrease although the sample means of the expected prices for a discount of 40% were above those for a discount of 30% for each of the values of promotion. There appears to be little interaction between the two factors.

15.37 a) The degrees of freedom for the main effects are the number of levels of the factor minus 1. The degrees of freedom for Chromium are $2 - 1 = 1$ and for Eat are $2 - 1 = 1$. The degrees of freedom for the interaction is the product of the degrees of freedom for the main effects and is $1 \times 1 = 1$. The total degrees of freedom is the number of subjects minus 1, which is $40 - 1 = 39$. The degrees of freedom for error can be gotten by subtraction. The mean squares are the sums of squares divided by their degrees of freedom, and the F for main effects and interaction are gotten by dividing the mean squares for each by the error mean square.

Source	Degrees of Freedom	Sum of Squares	Mean Square	F
A(Chromium)	1	0.00121	0.00121	0.04
B(Eat)	1	5.79121	5.79121	192.89
AB	1	0.17161	0.17161	5.72
Error	36	1.08084	0.03002	
Total	39	7.04487		

b) The F statistic used to test the null hypothesis of no interaction has an $F(1, 36)$ distribution. The numerical value of the F statistic is 5.72. Referring this to the critical values in Table E, we see the P-value is between 0.025 and 0.010. Since 36 degrees of freedom is not in the table, we need to look at the entries for 30 and 40. Computer software using the $F(1, 36)$ distribution gives the P-value as 0.0221.

c) The main effect of Chromium has an $F(1, 36)$ distribution and $F = 0.04$. From Table E, the P-value is greater than 0.10. Computer software using the $F(1, 36)$ distribution gives the P-value as 0.8426. The main effect of Eat has an $F(1, 36)$ distribution and $F = 192.89$. From Table E, the P-value is less than 0.001. Computer software using the $F(1, 36)$ distribution gives the P-value as 0.0000.

d) The within-group variance is the mean square for error, so $s_p^2 = 0.03002$, and $s_p = \sqrt{0.03002} = 0.173$.

e) The interaction between Eat and Chromium is statistically significant, but the effect is relatively small compared to the main effect of Eat. It would be up to the experimenter to determine whether the difference in the effect of going from Low to Normal levels of Chromium for the two levels of Eat is large enough to be of practical interest. By far, the biggest effect is the main effect of Eat. The mean GITH scores are much smaller when the rats could eat as much as they wanted (M) than when the total amount the rats could eat was restricted (R).

15.39 a) All three of the F statistics have degrees of freedom of 1 and 945. The P-value for interaction is 0.1477, and the P-values for the main effects of handedness and gender were both less than 0.0001. We concluded that there is no significant interaction and that both handedness and gender have an effect on mean lifetime.

b) We conclude from the marginal means that woman live on the average about six years longer than men, while right-handed people live on the average about nine years longer than left-handed people. No interaction means that the increase in life expectancy for right-handedness is the same for both men and women.

15.41 The group and marginal means, the plot of the mean heart rate for the four groups and the ANOVA table are given below and on the next page. The plot shows that males have lower mean heart rates than females in both the control and runner groups (by about 15 beats), while the runners have lower heart rates than the controls for both genders (by about 30 beats). Both main effects are highly significant and the differences reported in parentheses are the differences in the marginal means. The plot would suggest little interaction, but the ANOVA table shows the interaction is statistically significant. Comparing the F value for interaction with the F values for main effects we see that it is much smaller. The reason for the statistical significance of the interaction is that the sample sizes are very large, and with large samples statistical procedures have the ability to detect very small differences. In this case, the mean decrease in heart rate associated with running for females is estimated as 148 – 115.99 = 32.01, while the estimated mean decrease for men is 130 – 103.97 = 26.03. The analysis tells us the difference in these decreases is statistically significant, as this is the meaning of interaction. Whether this difference is of practical importance is another issue. If the interaction is judged unimportant, then just summarize this data in terms of the marginal means.

	Male	Female	Mean
Controls	130.00	148.00	139.00
Runners	103.97	115.99	109.98
Mean	116.99	131.99	124.49

Analysis of Variance for Heartrate

Source	DF	SS	MS	F	P
Group	1	168432	168432	695.65	0.000
Gender	1	45030	45030	185.98	0.000
Group*Gender	1	1794	1794	7.41	0.007
Error	796	192730	242		
Total	799	407986			

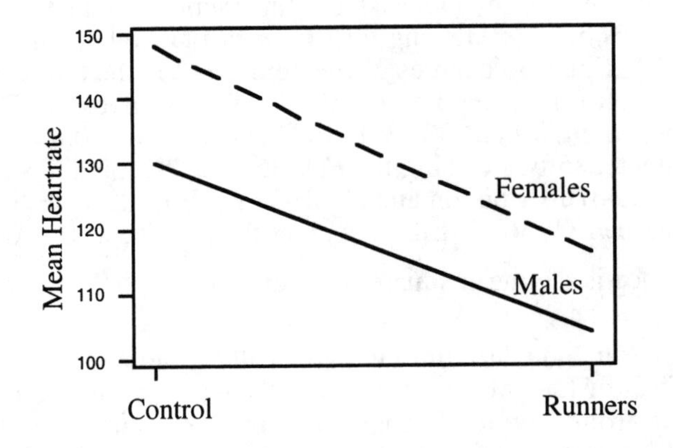
150
140
130
120
110
100
Mean Heartrate
Females
Males
Control
Runners

CHAPTER 16

NONPARAMETRIC TESTS

SECTION 16.1

OVERVIEW

Many of the statistical procedures described in previous chapters assumed that the samples were drawn from normal populations. **Nonparametric tests** do not require any specific form for the distributions of the populations from which the samples were drawn. Many nonparametric tests are **rank tests;** that is, they are based on the **ranks** of the observations rather than on the observations themselves. When ranking the observations from smallest to largest, tied observations receive the average of their ranks.

The **Wilcoxon rank sum test** compares two distributions. The objective is to determine if one distribution has systematically larger values than the other. The observations are ranked, and the **Wilcoxon rank sum statistic** W is the sum of the ranks of one of the samples. The Wilcoxon rank sum test can be used in place of the **two-sample** t **test** when samples are small or the populations are far from normal.

Exact P-values require special tables and are produced by some statistical software. However, many statistical software packages give only approximate P-values based on a normal approximation, typically with a continuity correction employed. Many packages also make an adjustment in the normal approximation when there are ties in the ranks.

APPLY YOUR KNOWLEDGE

16.1 We give the ordered observations, from lowest to highest, and their ranks. The observations in bold are the 12 Asian nations.

Observation:	–12.1	–5.2	–1.7	–1.5	–1.0	1.0	**1.3**	1.4	**2.3**	**2.9**	3.2	3.5	3.7	**3.9**	**4.1**	**4.2**	4.9	**5.1**
Rank:	1	2	3	4	5	6	7	8	9	10	11	12	13	14	15	16	17	18

Observation:	**5.6**	**5.9**	6.0	**6.2**	**6.4**	7.0	**8.8**
Rank:	19	20	21	22	23	24	25

Looking at the pattern of bold-faced numbers (Asian nations) it appears that the Asian nations generally have higher ranks.

16.3 We wish to test the hypotheses

H_0: no difference in the distribution in growth rates in Asia and Eastern Europe
H_a: growth rates in Asia are systematically higher

Because we wish to see if the growth rates in Asia are higher, we take Asia as the first population and Eastern Europe as the second. The sum of the ranks for the Asian nations (see the solution to Exercise 16.1 where the ranks are given) is

$$W = 7 + 9 + 10 + 14 + 15 + 16 + 18 + 19 + 20 + 22 + 23 + 25 = 198$$

There are n_1 = 12 Asian nations and n_2 = Eastern European nations for a total of N = 25 observations. Thus, the Wilcoxon rank sum statistic W has mean and standard deviation

$$\mu_W = \frac{n_1(N+1)}{2} = \frac{12(25+1)}{2} = 156 \text{ and } \sigma_W = \sqrt{\frac{n_1 n_2(N+1)}{12}} = \sqrt{\frac{12\times 13(25+1)}{12}} = 18.38$$

To find the P-value, we need to calculate $P(W \geq 198)$. We calculate an approximate P-value using the Normal approximation with continuity correction. With the continuity correction, we act as if the whole number 198 covers the entire interval 197.5 to 198.5, so we calculate the P-value $P(W \geq 198)$ as $P(W \geq 197.5)$. We find

$$P\text{-value} = P(W \geq 197.5) = P(\frac{W-\mu_W}{\sigma_W} \geq \frac{197.5-156}{18.38}) = P(Z \geq 2.26) = 0.0119$$

The P-value obtained from statistical software (Minitab) is 0.0120

16.5 We give the ordered observations, from lowest to highest, and their ranks.

Observation:	22	23	23	24	24	24	24	24	24	24	25	25	25	26	26	26	27	27
Rank:	1	2.5	2.5	7	7	7	7	7	7	7	12	12	12	15	15	15	17.5	17.5

Observation:	28	28	28	28	28	28	29	30	30	30	30	31	32	33
Rank:	21.5	21.5	21.5	21.5	21.5	21.5	25	27.5	27.5	27.5	27.5	31	32	33

SECTION 16.1 EXERCISES

16.7 a) We give the ordered observations, from lowest to highest, and their ranks. The observations in bold are those from September.

Observation:	2.9175	2.9200	2.9250	2.9325	2.9675	**3.5725**	**3.5825**	**3.5900**	**3.5950**	**3.6150**
Rank:	1	2	3	4	5	6	7	8	9	10

b) The sum of the ranks for September prices is $W = 6 + 7 + 8 + 9 + 10 = 40$. There are n_1 = 5 observations from September and n_2 = 5 observations from July for a total of N = 10 observations. Under the null hypothesis that prices in July and September do not differ systematically, W has mean and standard deviation

$$\mu_W = \frac{n_1(N+1)}{2} = \frac{5(10+1)}{2} = 27.5 \text{ and } \sigma_W = \sqrt{\frac{n_1 n_2(N+1)}{12}} = \sqrt{\frac{5\times 5(10+1)}{12}} = 4.79$$

c) First, we need to calculate $P(W \geq 40)$. We use the Normal approximation with continuity correction. With the continuity correction, we act as if the whole number 40 covers the entire interval 39.5 to 40.5, and we calculate $P(W \geq 40)$ as $P(W \geq 39.5)$. We find

$$P(W \geq 39.5) = P(\frac{W-\mu_W}{\sigma_W} \geq \frac{39.5-27.5}{4.79}) = P(Z \geq 2.51) = 0.0060$$

The two-sided P-value is actually $P(|Z| \geq 2.51) = 2P(Z \geq 2.51) = 2(0.0060) = 0.0120$ because of the symmetry of the normal distribution about 0. Thus, the P-value = 0.0120.

16.9 a) A back-to-back stemplot is given below.

16 weeks		2 weeks
8	09	
	10	
00	11	8
4	12	0669
	13	
0	14	

These two distributions do not appear to have the same shape. The distribution after two weeks has much less spread than the distribution after 16 weeks.

b) Even though the two distributions do not have the same shape, we can use the Wilcoxon test to test the hypotheses

H_0: no difference in the distribution of breaking strengths after 2 and 16 weeks
H_a: breaking strengths are systematically larger for strips buried two weeks

Notice that we have rephrased the alternative hypothesis given in the problem to its equivalent in terms of which distribution is systematically *larger.* To test these hypotheses, we first arrange the observations in order and assign ranks. The observations in bold are those after 2 weeks.

Observation:	98	110	110	**118**	**120**	124	**126**	**126**	**129**	140
Rank:	1	2	3	4	5	6	7	8	9	10

The sum of the ranks of breaking strengths after 2 weeks is

$$W = 4 + 5 + 7 + 8 + 9 = 33$$

There are $n_1 = 5$ observations after 2 weeks and $n_2 = 5$ observations after 16 weeks for a total of $N =$ 10 observations. Under the null hypothesis that the breaking strengths do not systematically differ, W has mean and standard deviation

$$\mu_W = \frac{n_1(N+1)}{2} = \frac{5(10+1)}{2} = 27.5 \text{ and } \sigma_W = \sqrt{\frac{n_1 n_2(N+1)}{12}} = \sqrt{\frac{5\times 5(10+1)}{12}} = 4.79$$

To find the P-value, we need to calculate $P(W \geq 33)$. We use the Normal approximation with continuity correction. With the continuity correction, we act as if the whole number 33 covers the entire interval 32.5 to 33.5, and we calculate the P-value $P(W \geq 33)$ as $P(W \geq 32.5)$. We find

$$P\text{-value} = P(W \geq 32.5) = P(\frac{W-\mu_W}{\sigma_W} \geq \frac{32.5-27.5}{4.79}) = P(Z \geq 1.04) = 0.1492$$

We conclude that there is not strong evidence that the breaking strengths are higher for strips buried 2 weeks (or equivalently, there is not strong evidence that the breaking strengths are lower for strips buried longer).

16.11 a) A back-to-back stemplot of these data using split stems is given below.

```
 Unlogged    Logged
          | 0 |4
          | 0 |
       333| 1 |024
     99855| 1 |55788
      2210| 2 |
          | 2 |
```

It appears that the species counts for unlogged forest plots are systematically a bit larger than for logged forest plots.

b) We use the Wilcoxon test to test the hypotheses

H_0: no difference in the distribution of species counts for unlogged and logged forest plots
H_a: species counts are systematically larger for unlogged forest plots

Notice that we have rephrased the question given in the problem to its equivalent in terms of which distribution is systematically *larger.* There are many ties, and we need to use statistical software to analyze data with tied values. Output from Minitab follows.

```
Mann-Whitney Confidence Interval and Test

unlogged    N =   12      Median =      18.500
logged      N =    9      Median =      15.000
```

```
Point estimate for ETA1-ETA2 is         3.500
95.7 pct c.i. for ETA1-ETA2 is (-0.002,7.000)
W = 159.0
Test of ETA1 = ETA2  vs.  ETA1 > ETA2 is significant at 0.0298
The test is significant at 0.0290 (adjusted for ties)
```

From this output, we see that the P-value, adjusted for ties, is P-value = 0.0290 We conclude that there is reasonably strong evidence (significant at the 0.05 level but not at the 0.01 level) that species counts are systematically larger for unlogged forest plots (or equivalently, species counts for logged plots are systematically reduced after 8 years).

16.13 We use the Wilcoxon test to test the hypotheses

H_0: no difference in the distribution of the responses of men and women to the question about the safety of food served at restaurants
H_a: responses of women to the question about the safety of food served at restaurants are systematically larger than those of men

Larger responses indicate greater concern. There are many ties, and we need to use statistical software to analyze data with tied values. Output from Minitab is given below.

```
Mann-Whitney Confidence Interval and Test

srest Females      N = 196      Median =     2.0000
srest Males        N = 107      Median =     2.0000
Point estimate for ETA1-ETA2 is           -0.0000
95.0 Percent C.I. for ETA1-ETA2 is (0.0001, 0.0001)
W = 32267.5
Test of ETA1 = ETA2  vs.  ETA1 > ETA2 is significant at 0.0003
The test is significant at 0.0001 (adjusted for ties)
```

From this output, we see that the P-value, adjusted for ties, is P-value = 0.0001. There is strong evidence that the responses of women to the question about the safety of food served at restaurants are systematically larger than those of men. Thus, there is evidence that women are more concerned than men about the safety of food at restaurants.

16.15 a) We used software (Minitab) to carry out the chi-square test. The output is given below.

```
Expected counts are printed below observed counts

         City 1   City 2    Total
    1        70       62      132
          69.31    62.69

    2        52       63      115
          60.38    54.62

    3        69       50      119
          62.48    56.52

    4        22       19       41
          21.53    19.47

    5        28       24       52
          27.30    24.70

Total       241      218      459

ChiSq =   0.007 +  0.008 +
          1.163 +  1.286 +
          0.680 +  0.752 +
          0.010 +  0.011 +
          0.018 +  0.020 = 3.955

df = 4
```

We see that the value of the X^2 statistic for testing whether there is a relationship between city and income has value $X^2 = 3.955$ with df = 4. Using Minitab, we find the exact P-value is 0.4121. Thus, there is not strong evidence that there is a relationship between city and income.

b) We use the Wilcoxon test to test the hypotheses

H_0: no difference in the distribution of income codes for cities 1 and 2
H_a: income codes for one city are systematically larger than those for the other

Note that the alternative is two-sided. There are many ties, and we need to use statistical software to analyze data with tied values. Output from Minitab is given below.

```
Mann-Whitney Confidence Interval and Test

Income1      N = 241      Median =     2.0000
Income2      N = 218      Median =     2.0000
Point estimate for ETA1-ETA2 is       -0.0000
95.0 Percent C.I. for ETA1-ETA2 is (0.0000,0.0000)
W = 56370.0
Test of ETA1 = ETA2  vs.  ETA1 ≠ ETA2 is significant at 0.5080
The test is significant at 0.4949 (adjusted for ties)

Cannot reject at alpha = 0.05
```

From this output, we see that the P-value, adjusted for ties, is P-value = 0.4949. Thus, there is not strong evidence that income codes for one city are systematically larger than those for the other.

SECTION 16.2

OVERVIEW

The **Wilcoxon signed rank test** is a nonparametric test for matched pairs. It tests the null hypothesis that there is no systematic difference between the observations within a pair against the alternative that one observation tends to be larger.

The test is based on the **Wilcoxon signed rank statistic W^+,** which provides another example of a nonparametric test using ranks. The absolute values of the observations are ranked, and the sum of the ranks of the positive (or negative) differences gives the value of W^+. The **matched pairs t test** and the **sign test** are two other alternative tests for this setting.

P-values can be found from special tables of the distribution or a normal approximation to the distribution of W^+. Some software computes the exact P-value and other software uses the normal approximation, typically with a ties correction. Many packages make an adjustment in the normal approximation when there are ties in the ranks.

SECTION 16.2 EXERCISES

16.17 We want to determine whether the low stepping rate raises heart rate significantly. Let m be the median of the increases in heart rate (final heart rate at the end of the exercise – resting heart rate) for the population of all adults. If the heart rate tends to be raised by the low stepping rate, then for most people the increase in heart rate will be positive (a negative increase means the heart rate was lower after the exercise) and the median m should be positive. Thus, in terms of the median, we test the hypotheses

H_0: m = 0
H_a: m > 0

If we look at the data, we see that for subjects 2 and 4 the increase in heart rate is 9. Because we have tied values for the differences, we must use software to carry out the test. The output from Minitab is given on the next page.

```
TEST OF MEDIAN = 0.000000 VERSUS MEDIAN > 0.000000

                 N FOR   WILCOXON              ESTIMATED
            N    TEST   STATISTIC    P-VALUE      MEDIAN
Diff        5       4        10.0      0.050       7.500
```

The P-value is 0.05, so we would conclude that there is (some) evidence that the median increase in heart rate is positive, i.e., the low rate raises heart rate.

16.19 Presumably, one wishes to show that the intensive French course improves one's ability to understand spoken French. Thus, we would test the hypotheses

H_0: scores on the pretest and posttest have the same distribution
H_a: scores on the posttest are systematically higher than on the pretest

We use software (Minitab) to test these hypotheses. The output is given below.

```
TEST OF MEDIAN = 0.000000 VERSUS MEDIAN > 0.000000

                 N FOR   WILCOXON              ESTIMATED
            N    TEST   STATISTIC    P-VALUE      MEDIAN
diffs      20      17       138.5      0.002       3.000
```

From the output, we see that the test statistic is $W^+ = 138.5$ and the P-value = 0.002. We would conclude that there is strong evidence that the scores on the posttest are systematically higher than on the pretest.; i.e., that the intensive French course improves scores between the pretest and posttest.

16.21 To see if there is evidence of a systematic loss in vitamin C in shipping and storage, we test the hypotheses

H_0: vitamin C has the same distribution at both times
H_a: vitamin C is systematically higher at the factory

We use software (Minitab) to test these hypotheses. We first compute the differences in the vitamin C content for each bag and then use the Wilcoxon signed rank test. The output is given below.

```
TEST OF MEDIAN = 0.000000 VERSUS MEDIAN > 0.000000

                 N FOR   WILCOXON              ESTIMATED
            N    TEST   STATISTIC    P-VALUE      MEDIAN
difference 27      27       341.0      0.000       5.500
```

Form the output, the P-value = 0.000. With the full data set, we see there is strong evidence that vitamin C is systematically higher at the factory.

16.23 To see if there is a systematic difference between the level of concern about food safety at outdoor fairs and at fast-food restaurants, we test the hypotheses

H_0: responses expressing the level of concern about food served at outdoor fairs and at fast-food restaurants have the same distribution
H_a: responses expressing the level of concern about food served at outdoor fairs and at fast-food restaurants are systematically higher for one than the other

We use software (Minitab) to test these hypotheses. We first compute the differences in the responses for each of the 303 subjects and then use the Wilcoxon signed rank test. The output is given below.

```
TEST OF MEDIAN = 0.000000 VERSUS MEDIAN not equal to 0.000000

                 N FOR   WILCOXON              ESTIMATED
            N    TEST   STATISTIC    P-VALUE      MEDIAN
sfair-sfast 303   129      4730.5      0.206   0.000E+00
```

The P-value is 0.206, and we conclude that there is not strong evidence that there is a systematic difference between the level of concern about food safety at outdoor fairs and at fast-food restaurants.

16.25 We wish to see if there is good evidence that the Gasaver improves median gas mileage by 22% or more. There are two ways we might proceed.

1. We could test the hypotheses

$$H_0\text{: median} = 22\%$$
$$H_a\text{: median} > 22\%$$

These do not exactly represent the claim, because these are really testing if there is good evidence that the Gasaver improves the median gas mileage by *more than* 22%. However, if there is good evidence that the Gasaver improves the median gas mileage by more than 22%, then there is certainly good evidence that the Gasaver improves the median gas mileage by 22% or more.

To test these hypotheses, we transform the data by subtracting 22% from each value. For this transformed data, our hypotheses become

$$H_0\text{: median} = 0$$
$$H_a\text{: median} > 0$$

We use software (Minitab) to transform the data and test these hypotheses using the Wilcoxon signed rank test. Output from Minitab is given below.

```
TEST OF MEDIAN = 0.000000 VERSUS MEDIAN > 0.000000

                      N FOR    WILCOXON                 ESTIMATED
                N     TEST    STATISTIC    P-VALUE         MEDIAN
change-22      15       15         88.0      0.059          7.450
```

The P-value is 0.059, and so there is not very strong (only moderate) evidence that the gas mileage is increased by *more than* 22%.

2. We could test the hypotheses

$$H_0\text{: median} = 22\%$$
$$H_a\text{: median} < 22\%$$

These better represent the claim, but the traditional interpretation of the P-value here would be whether it indicates there is good evidence that the Gasaver improves the median gas mileage by *less than* 22%. A large P-value would only mean that there is not good evidence that the Gasaver improves the median gas mileage by less than 22%. This is not exactly the same thing as saying that there is good evidence that the Gasaver improves the median gas mileage by 22% or more.

To test these hypotheses, we transform the data by subtracting 22% from each value. For this transformed data, our hypotheses become

$$H_0\text{: median} = 0$$
$$H_a\text{: median} < 0$$

We use software (Minitab) to transform the data and test these hypotheses using the Wilcoxon signed rank test. Output from Minitab is given below.

```
TEST OF MEDIAN = 0.000000 VERSUS MEDIAN < 0.000000

                      N FOR    WILCOXON                 ESTIMATED
                N     TEST    STATISTIC    P-VALUE         MEDIAN
change-22      15       15         88.0      0.947          7.450
```

The P-value is 0.947, indicating that there is essentially no evidence that the Gasaver improves the median gas mileage by less than 22%.

If we combine the results of both approaches, we would conclude that there is good evidence that the Gasaver improves the median gas mileage by 22% or more.

16.27 We use Minitab to calculate a 95% confidence interval for the median amount spent by all shoppers at this market. The output is given below.

```
Wilcoxon Signed Rank Confidence Interval

                         ESTIMATED      ACHIEVED
                  N        MEDIAN     CONFIDENCE    CONFIDENCE INTERVAL
Amt.Spent     50           32.1           95.0       (26.3,      38.6)
```

The 95% confidence interval for the median is (26.3, 38.6).

SECTION 16.3

OVERVIEW

The **Kruskal-Wallis test** is the nonparametric test for the **one-way analysis of variance** setting. In comparing several populations, it tests the null hypothesis that the distribution of the response variable is the same in all groups and the alternative hypothesis that some groups have distributions of the response variable that are systematically larger than others.

The **Kruskal-Wallis statistic** H compares the average ranks received for the different samples. If the alternative is true, some of these should be larger than others. Computationally, it essentially arises from performing the usual one-way ANOVA to the ranks of the observations rather than the observations themselves.

P-values can be found from special tables of the distribution or a chi-square approximation to the distribution of H. When the sample sizes are not too small, the distribution of H for comparing I populations has approximately a chi-square distribution with $I - 1$ degrees of freedom. Some software computes the exact P-value and other software uses the chi-square approximation, typically with an adjustment in the chi-square approximation when there are ties in the ranks.

SECTION 16.3 EXERCISES

16.29 a) We use the Kruskal-Wallis test to test the hypotheses

H_0: vitamin C has the same distribution for all conditions after baking the bread
H_a: vitamin C is systematically higher for some conditions after baking the bread than for other conditions

We use software (Minitab) to carry out the Kruskal-Wallis test. The results are given below.

```
LEVEL      NOBS      MEDIAN   AVE. RANK     Z VALUE
    0         2      48.705         9.5        2.09
    1         2      41.955         7.5        1.04
    3         2      21.795         5.5        0.00
    5         2      12.415         3.5       -1.04
    7         2       8.320         1.5       -2.09
OVERALL      10                     5.5

H = 8.73   d.f. = 4   p = 0.069
```

The test statistic is $H = 8.73$ with 4 df, and the P-value for the test is 0.069. Using Table F, one can verify that this P-value comes from the chi-square approximation to the distribution of H.

The data show a pronounced decrease in vitamin C as the number of days after baking increases. However, the P-value of 0.069 does not indicate there is strong evidence (only modest evidence) that vitamin C is systematically higher for some conditions after baking the bread than for other conditions.

b) The P-value in part (a) is larger that the exact P-value of 0.0011. The exact P-value suggests that there is strong evidence that vitamin C is systematically higher for some conditions after baking the bread than for other conditions. The difference between the approximate and exact P-values is large enough to affect our conclusions.

16.31 a) ANOVA tests the hypotheses

H_0: the population *means* for all four groups are the same
H_a: not all four group *means* are equal

The Kruskal-Wallis test tests the hypotheses

H_0: number of insects trapped has the same *distribution* for all colors used
H_a: number of insects trapped is systematically higher for some colors than for others

b) We use software (Minitab) to compute the median number of insects trapped by boards of each color and to do the Kruskal-Wallis test. The output is given below.

LEVEL	NOBS	MEDIAN	AVE. RANK	Z VALUE
Yellow	6	46.50	21.2	3.47
White	6	15.50	7.3	-2.07
Green	6	34.50	14.8	0.93
Blue	6	15.00	6.7	-2.33
OVERALL	24		12.5	

```
H = 16.95   d.f. = 3   p = 0.001
H = 16.98   d.f. = 3   p = 0.001 (adj. for ties)
```

The medians for yellow and green are larger than those for white and blue, so yellow and green appear to be the more effective colors. The *P*-value for the Kruskal-Wallis test is 0.001, so there is strong evidence that the number of insects trapped is systematically higher for some colors than for others. The test does not tell us which colors systematically trap more insects, but the data suggests these are yellow and green.

16.33 a) Software (Minitab) gives the following summary information about breaking strength for each time.

Time	N	MEAN	MEDIAN	STDEV	SEMEAN
2	5	123.80	126.00	4.60	2.06
4	5	123.60	126.00	6.54	2.93
8	5	134.40	136.00	9.53	4.26
16	5	116.40	110.00	16.09	7.19

The ratio of the largest to smallest standard deviation is 16.09/4.60 = 3.5, which is larger than 2. According to our rule of thumb for ANOVA (see Chapter 14), because this ratio is greater than 2 it would not be safe to apply ANOVA to these data.

b) The medians for the four samples are given in the output in part (a). The hypotheses for the Kruskal-Wallis test in terms of medians are

H_0: the population medians for all four groups are the same
H_a: not all four group medians are equal

c) We use software (Minitab) to carry out the Kruskal-Wallis test. The output is given below.

LEVEL	NOBS	MEDIAN	AVE. RANK	Z VALUE
2	5	126.0	9.7	-0.35
4	5	126.0	10.2	-0.13
8	5	136.0	15.4	2.14
16	5	110.0	6.7	-1.66
OVERALL	20		10.5	

```
H = 5.60   d.f. = 3   p = 0.134
H = 5.63   d.f. = 3   p = 0.132 (adj. for ties)
```

The *P*-value is 0.132 (after adjusting for ties) and we would conclude that there is not strong evidence that the four group medians are not all equal, i.e., there is not strong evidence that the median breaking strength differs for the different lengths of times the strips were buried. To the extent that breaking strength indicates the presence of decay, there is not strong evidence that the fabric decays over time periods of 2, 4, 8, and 16 weeks.

16.35 a) Side-by-side stem and leaf plots are given below.

```
Unlogged     Logged 1 yr. ago    Logged 8 yrs. ago
  0|               0|2               0|
  0|               0|                0|4
  0|               0|7               0|
  0|               0|8               0|
  1|               1|11              1|0
  1|333            1|23              1|2
  1|55             1|4555            1|455
  1|               1|                1|7
  1|899            1|8               1|88
  2|01             2|                2|
  2|22             2|                2|
```

The three distributions appear to be skewed, and there are outliers in the plots logged 1 and 8 years ago. These features might suggest that ANOVA may not be safe to use.

The median number of trees per plot for the three groups is given in the output in part (b)

b) The hypotheses we test with the Kruskal-Wallis test are

H_0: species count in plots have the same distribution for all plot types
H_a: species count in plots are systematically higher for some plot types than for others

Using software (Minitab) to carry out the Kruskal-Wallis test, we get the following output.

```
LEVEL                  NOBS     MEDIAN   AVE. RANK    Z VALUE
Unlogged                 12      18.50        23.4       2.88
Logged 1 yr. ago         12      12.50        11.5      -2.47
Logged 8 yrs. ago         9      15.00        15.8      -0.44
OVERALL                  33                   17.0

H = 9.31   d.f. = 2   p = 0.010
H = 9.44   d.f. = 2   p = 0.009  (adj. for ties)
```

We see that the test statistics (adjusted for ties) is $H = 9.44$ and the P-value = 0.009. We conclude that there is strong evidence that species count in plots are systematically higher for some plot types than for others.

Chapter 16 Review Exercises

16.37 Below are back-to-back stemplots of right-hand thread (clockwise) and left-hand thread (counterclockwise) times.

```
Clockwise     Counterclockwise
      85| 07 |688
   99875| 08 |47
       6| 09 |3
 7543310| 10 |13578
 8866311| 11 |256
       2| 12 |33
      80| 13 |357
        | 14 |5678
        | 15 |
        | 16 |6
        | 17 |0
```

These plots suggest that it takes systematically longer to turn the knob counterclockwise than clockwise. The outliers in the counterclockwise data suggest that a test that does not require normality may be appropriate. We use the Wilcoxon test to test the hypotheses

H_0: times to turn the knob clockwise and counterclockwise have the same distribution
H_a: times to turn the knob counterclockwise are systematically higher to turn it clockwise

We use software (Minitab) to carry out the test. The output is given below.

```
Mann-Whitney Confidence Interval and Test

left        N =   25      Median =          115.00
right       N =   25      Median =          104.00
Point estimate for ETA1-ETA2 is              12.00
95.2 Percent C.I. for ETA1-ETA2 is (-2.01,28.00)
W = 727.0
Test of ETA1 = ETA2  vs.  ETA1 > ETA2 is significant at 0.0421
The test is significant at 0.0420 (adjusted for ties)
```

The P-value of the test (adjusted for ties) is 0.0420. This indicates there is evidence that times to turn the knob counterclockwise are systematically higher to turn it clockwise.

16.39 a) A stemplot of the data is given below. The distribution is right skewed with a large outlier and clearly Nonnormal.

```
0| 89
1| 00234
1| 55558899
2| 00000000000001111222222223
2| 59
3| 0
3| 5
4| 00
4|
5| 0
```

b) We wish to test the hypotheses

H_0: median cost $= 20$
H_a: median cost $\neq 20$

To test these hypotheses, we transform the data by subtracting 20 from each value. For this transformed data, our hypotheses become

H_0: median cost $= 0$
H_a: median cost $\neq 0$

We use software (Minitab) to transform the data and test these hypotheses using the Wilcoxon signed rank test. Output from Minitab is given below.

```
Wilcoxon Signed Rank Test
TEST OF MEDIAN = 0.000000 VERSUS MEDIAN not equal to 0.000000

                  N FOR   WILCOXON              ESTIMATED
            N     TEST   STATISTIC   P-VALUE       MEDIAN
Cost-20    50      37       378.0      0.695       0.5000
```

The P-value is 0.695, thus there is not strong evidence that the median cost differs from $20.

c) We test the hypotheses

H_0: mean cost $= 20$
H_a: mean cost $\neq 20$

using the t test. We use Minitab to carry out the test. The output is given below.

```
T-Test of the Mean
Test of mu = 20.00 vs. mu not = 20.00

Variable     N       Mean     StDev    SE Mean        T     P-Value
Cost        50      20.90      7.65       1.08     0.83        0.41
```

The P-value is 0.41, and we see there is not strong evidence that the mean cost differs from $20. This is consistent with the results in part (b) for the median.

16.41 a) Below are side-by-side stemplots of the calorie contents for the three types of hot dogs.

```
     Beef           Meat           Poultry
   8|             8|             8| 67
   9|             9|             9| 49
  10|            10| 7          10| 226
  11| 1          11|            11| 3
  12|            12|            12| 9
  13| 1259       13| 5689       13| 25
  14| 1899       14| 067        14| 2346
  15| 2378       15| 3          15| 2
  16|            16|            16|
  17| 56         17| 2359       17| 0
  18| 146        18| 2          18|
  19| 00         19| 015        19|
```

Five number summaries for each type are given below

Type	N	MIN	Q1	MEDIAN	Q3	MAX
Beef	20	111.00	139.50	152.50	179.75	190.00
Meat	17	107.00	138.50	153.00	180.50	195.00
Poultry	17	86.00	100.50	129.00	143.50	170.00

The data suggest that the distributions of the calorie content for beef and meat hot dogs are similar, while the distribution for poultry hot dogs is systematically smaller than those for beef and meat hot dogs.

b) Beef and meat hot dogs display the most striking Nonnormal features. Each has a low outlier. Both appear to be a mixture of two distributions, with several values in the 130 to 160 range, a gap of almost 20 calories, and then several values in the 170 to 195 range.

c) To test for differences in the calorie content for the three types of hot dogs, we use the Kruskal-Wallis test to test the hypotheses

H_0: calorie content has the same distribution for all three types of hot dots
H_a: calorie content is systematically higher for some types of hot dogs than for others

Using software (Minitab) to carry out the Kruskal-Wallis test, we get the following output.

LEVEL	NOBS	MEDIAN	AVE. RANK	Z VALUE
Beef	20	152.5	33.1	2.02
Meat	17	153.0	33.5	1.89
Poultry	17	129.0	14.9	-3.99
OVERALL	54		27.5	

```
H = 15.89  d.f. = 2  p = 0.000
H = 15.90  d.f. = 2  p = 0.000 (adj. for ties)
```

The test indicates (P-value = 0.000) there is strong evidence that the calorie content is systematically higher for some types of hot dogs than for others. The test does not tell us which types differ, but the data suggest that poultry hot dogs have systematically lower calories than beef and meat hot dogs.

16.43 a) Below are side-by-side stemplots of the sodium contents for the three types of hot dogs.

```
Beef           Meat           Poultry
1|             1|4            1|
1|             1|             1|
2|             2|             2|
2|59           2|             2|
3|011223       3|3            3|
3|778          3|67889        3|5557889
4|024          4|002          4|23
4|7789         4|579          4|
5|             5|0014         5|112244
5|8            5|             5|88
6|4            6|             6|
```

Five number summaries for each type are given below

	Type	N	MIN	Q1	MEDIAN	Q3	MAX
Sodium	Beef	20	253.0	319.8	380.5	478.5	645.0
	Meat	17	144.0	379.0	405.0	501.0	545.0
	Poultry	17	357.0	379.0	430.0	535.0	588.0

The data do not show dramatic differences in sodium content. Poultry hot dogs appear systematically to have slightly higher sodium content than the other two types, and meat hot dogs appear systematically to have slightly higher sodium content than beef.

b) All three types display some Nonnormal features. Beef hot dogs have two large outliers, meat hot dogs have one very low outlier, and poultry hot dogs appear to be a mixture of two distributions (a group consisting of sodium contents between 350 and 430 followed by a gap of about 180, and then a group of hot dogs with sodium content between 510 and 600

c) To test for differences in the sodium content for the three types of hot dogs, we use the Kruskal-Wallis test to test the hypotheses

H_0: sodium content has the same distribution for all three types of hot dots
H_a: sodium content is systematically higher for some types of hot dogs than for others

Using software (Minitab) to carry out the Kruskal-Wallis test, we get the following output.

```
LEVEL      NOBS     MEDIAN  AVE. RANK    Z VALUE
    1        20      380.5       22.0      -1.95
    2        17      405.0       28.1       0.20
    3        17      430.0       33.3       1.83
OVERALL      54                  27.5
H = 4.71   d.f. = 2   p = 0.095
H = 4.71   d.f. = 2   p = 0.095  (adj. for ties)
```

The test indicates (P-value = 0.095) there is not strong evidence that the sodium content is systematically higher for some types of hot dogs than for others.

16.45 We use software (Minitab) to carry out the necessary tests. First, we test the hypotheses

H_0: no difference in the distribution in iron content of meat cooked in aluminum and clay pots
H_a: the iron content of meat cooked in one of the types of pots is systematically higher than the other

We use the Wilcoxon rank sum test (Mann-Whitney test in Minitab). The output is given below.

```
Mann-Whitney Confidence Interval and Test

Iron-meat(aluminum)      N =    5      Median =     2.1400
Iron-meat(clay)          N =    5      Median =     2.4100
Point estimate for ETA1-ETA2 is    -0.2700
96.3 pct c.i. for ETA1-ETA2 is (-0.7100, 0.8602)
W = 22.0
```

```
Test of ETA1 = ETA2  vs.  ETA1 not equal to ETA2 is significant at
0.2963
```

The P-value for the test is 0.2963, so we do not have strong evidence that the iron content of meat cooked in one of the types of pots is systematically higher than the other.

Next, we test the hypotheses

H_0: no difference in the distribution in iron content of legumes cooked in aluminum and clay pots
H_a: the iron content of legumes cooked in one of the pots is systematically higher than the other

We use the Wilcoxon rank sum test (Mann-Whitney test in Minitab). The output is given below.

```
Mann-Whitney Confidence Interval and Test

Iron-legumes(aluminum)     N =    4      Median =          2.255
Iron-legumes(clay)         N =    4      Median =          2.455
Point estimate for ETA1-ETA2 is         -0.195
97.0 pct c.i. for ETA1-ETA2 is (-1.540,0.860)
W = 13.0
Test of ETA1 = ETA2  vs.  ETA1 not equal to. ETA2 is significant
at 0.1939
```

The P-value for the test is 0.1939, so we do not have strong evidence that the iron content of legumes cooked in one of the types of pots is systematically higher than the other.

16.47 NOTE: Answers will vary according to the particular sample selected.

a) From our sample of 200 individuals, we had the following number of people in each of the six education levels.

Education level	Count
1	9
2	12
3	52
4	67
5	37
6	23

There are 67 people of education level 4 and 37 people of education level 5 in our sample.

b) Summary statistics for the incomes of the two groups are given below.

Level	Count	Mean	Median	StdDev	Min	Max	Q1	Q3
4	67	35,304.7	29,160	34,228.3	327	229,843	15,533.2	42,219
5	37	46,069.4	35,831	36,200.5	1	170,373	18,987.5	71,812.8

We see that the median and the quartiles for those with a bachelor's degree are higher than the median and corresponding quartiles for those without a degree. From the values of the median, Q1, and Q3 it appears that among people who entered college, those with a bachelor's degree earn systematically more than those without a degree.

Histograms of the incomes for the two groups are given on the next page. Other than the large outlier in the histogram for those without a degree, the distribution for those with a bachelor's degree appears to be shifted somewhat to the right of the histogram for those without a degree. Thus, it again appears that among people who entered college, those with a bachelor's degree earn systematically more than those without a degree.

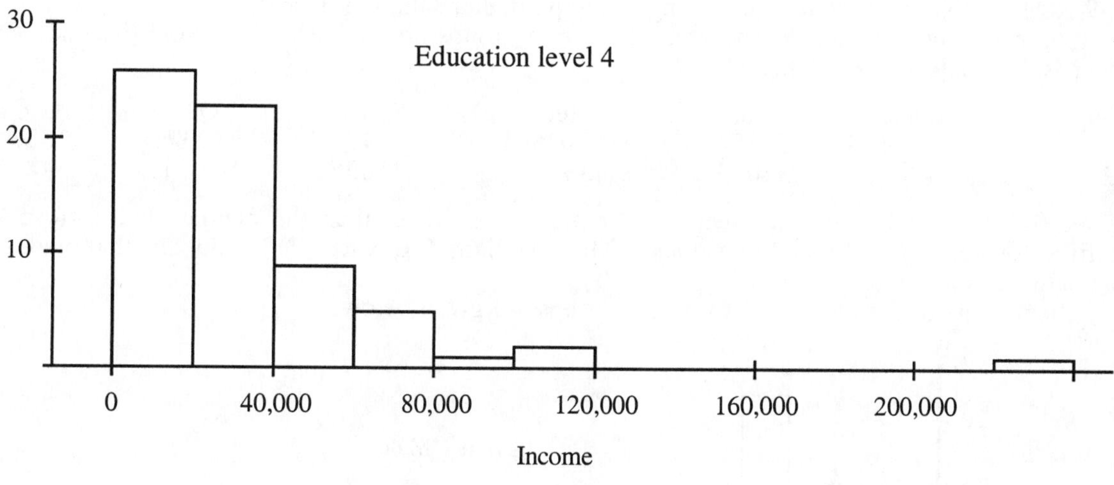

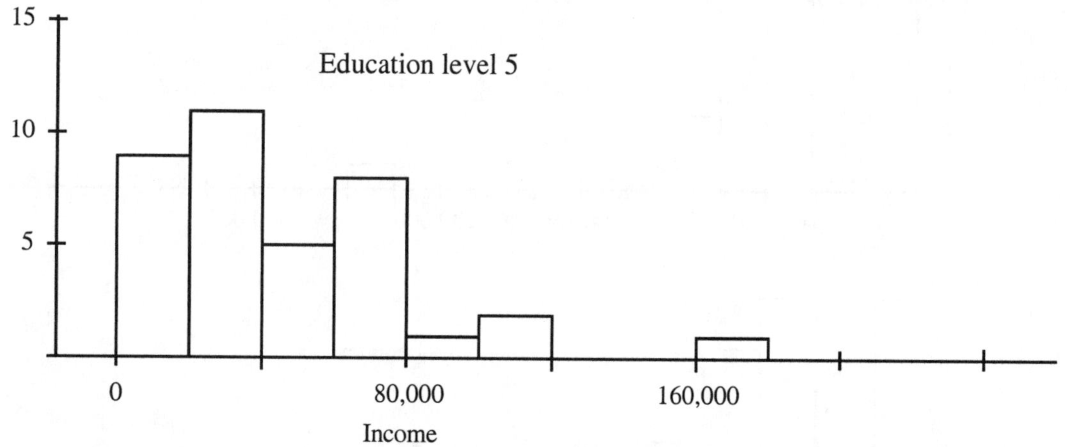

c) Both histograms in part (b) are skewed right, have large outliers, and are clearly Nonnormal.
d) Because the data appear nonNormal, we use the Wilcoxon test to test the hypotheses

H_0: no difference in the distribution of incomes for the bachelor's degree and nondegree groups
H_a: the incomes of the bachelor's degree group are systematically higher than the nondegree group

Using software we obtain the following result

```
Mann-Whitney Confidence Interval and Test

bachelor's          N =   37      Median =    35831
nondegree           N =   67      Median =    29160
Point estimate for ETA1-ETA2 is        8201
95.0 Percent C.I. for ETA1-ETA2 is (-1189, 19917)
W = 2188.0
Test of ETA1 = ETA2  vs.  ETA1 > ETA2 is significant at 0.0481
The test is significant at 0.0481 (adjusted for ties)
```

The *P*-value adjusted for ties is 0.0481 and so there is evidence at the $\alpha = 0.05$ level that the incomes of the bachelor's degree group are systematically higher than the nondegree group.

e) This is an observational study, not a designed experiment. It is usually unwise to draw conclusions about causality from observational studies. The factors that cause people who enter college to fail to complete the degree may also cause people to earn less (for example, less intelligent students or students with poor work habits may fail to complete the degree and these same traits may affect later job performance and hence income).

16.49 NOTE: Answers will vary according to the particular sample selected.

Summary statistics for the incomes of the two groups are given below. Note that the code for gender is 1 = male and 2 = female.

Group	Count	Mean	Median	StdDev	Q1	Q3
1	53	50254.1	31750	61481.1	21937.5	52343
2	47	36033.1	30122	25855.5	16100	48750

We see that the median and the quartiles for males are higher than the median and corresponding quartiles for females. From the values of the median, Q1, and Q3 it appears that males earn systematically more than females.

Histograms of the incomes for the two groups are given below.

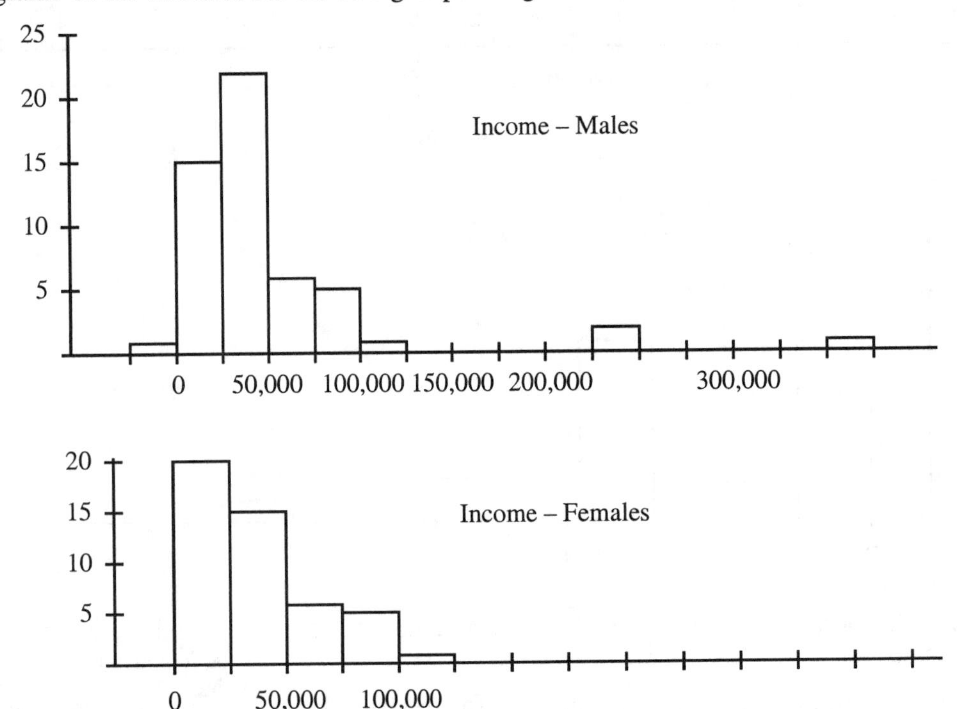

The histogram for males appears to be shifted slightly to the right of the histogram for those without a degree. The shift is slight, but it appears that males earn systematically more than females

We also perform a statistical test to determine if males earn significantly more than females. The data are right-skewed and appear to be Nonnormal. Thus we use the Wilcoxon test to test the hypotheses

H_0: no difference in the distribution of incomes for males and females
H_a: the incomes of males are systematically higher than the incomes of females

Using software, we obtain the following result

```
Mann-Whitney Confidence Interval and Test

Income Males        N =   53       Median =       31750
Income Females      N =   47       Median =       30122
Point estimate for ETA1-ETA2 is         4000
95.0 Percent C.I. for ETA1-ETA2 is (-5000, 12558)
W = 2807.5
Test of ETA1 = ETA2  vs.  ETA1 > ETA2 is significant at 0.1837
The test is significant at 0.1837 (adjusted for ties)

Cannot reject at alpha = 0.05
```

The P-value adjusted for ties is 0.1837, and so there is not strong evidence at the $\alpha = 0.05$ level that the incomes of males are systematically higher than the incomes of females. The shift observed in the summary statistics and plots is not statistically significant.

16.51 a) We assume there are no ties. The mean of the Wilcoxon rank sum statistic W_1 for the first sample is $\mu_{W_1} = \frac{n_1(N+1)}{2}$ and the mean of the Wilcoxon rank sum statistic W_2 for the second sample is $\mu_{W_2} = \frac{n_2(N+1)}{2}$. All the ranks from 1 to N must be represented exactly once in the combined samples, so $W_1 + W_2$ must be the sum of the ranks 1 to N. From Chapter 4, we know that the mean (expected value) of a sum of random variables is the sum of their means, hence

$$\mu_{W_1} + \mu_{W_2} = \mu_{W_1+W_2} = \text{the mean of the sum of the ranks 1 to } N = \text{the sum of the ranks 1 to } N$$

where the last equality follows from the fact that the mean of a constant is just the constant. Substituting the formulas for μ_{W_1} and μ_{W_2} in the above, we have

$$\text{the sum of the ranks 1 to } N = \mu_{W_1} + \mu_{W_2} = \frac{n_1(N+1)}{2} + \frac{n_2(N+1)}{2} = \frac{(n_1+n_2)(N+1)}{2} = \frac{N(N+1)}{2}$$

b) $N = 27$, and using the result in part (a) we find

$$\text{the sum of the ranks 1 to 27} = \frac{N(N+1)}{2} = \frac{27(28)}{2} = 378$$

c) The sum of the ranks sums for the four groups in Figure 16.10 is 308 + 350 + 745 + 550 = 1953. Using part (a), we also see that

$$\text{the sum of the ranks 1 to } N = 62 = \frac{N(N+1)}{2} = \frac{62(63)}{2} = 1953.$$

CHAPTER 17

LOGISTIC REGRESSION

OVERVIEW

In Chapter 8, we studied random variables y, which can take on only two values (yes or no, success or failure, live or die, acceptable or not). It is usually convenient to code the two values as 0 (failure) and 1 (success) and let p denote the probability of a 1 (success). If y is observed on n independent trials, the total number of 1's can often be modeled as binomial with n trials and probability of success p. In this chapter, we consider methods that allow us to investigate how y depends on one or more explanatory variables. The methods of simple linear and multiple linear regression do not directly apply because the distribution of y is binomial rather than normal. However, it is possible to use a metho, called **logistic regression,** which is similar to simple and multiple linear regression.

We define the **odds** as $p/(1-p)$, the ratio of the probability that the event happens to the probability that the event does not happen. If $\hat{p}$ is the sample proportion, the sample odds are $\hat{p}(1-\hat{p})$. The **logistic regression model** relates the (natural) log of the odds to the explanatory variables. In the case of a single explanatory variable x, the logistic regression model is

$$\log\left(\frac{p_i}{1-p_i}\right) = \beta_0 + \beta_1 x_i$$

where the responses y_i, for i = 1,..., n, are n independent binomial random variables with parameters 1 and p_i, i.e., they are independent with distributions $B(1, p_i)$. The **parameters** of this logistic regression model are β_0 (the intercept of the logistic model) and β_1 (the slope of the logistic model). The quantity e^{β_1} is called the **odds ratio.**

The formulas for estimates of the parameters of the logistic regression model are, in general, very complicated and, in practice, the estimates are computed using statistical software. Such software typically gives estimates b_0 for β_0 and b_1 for β_1 along with estimates SE_{b_0} and SE_{b_1} for the standard errors of these estimates. A **level C confidence interval for the intercept** β_0 is then determined by the formula

$$b_0 \pm z^* \text{SE}_{b_0}$$

where z^* is the upper $(1 - C)/2$ quantile of the standard normal distribution. Similarly, a **level C confidence interval for the slope** β_1 is determined by the formula

$$b_1 \pm z^* \text{SE}_{b_1}$$

A **level C confidence interval for the odds ratio** e^{β_1} is obtained by transforming the confidence interval for the slope, yielding the formula

$$\left(e^{b_1 - z^* \text{SE}_{b_1}},\ e^{b_1 + z^* \text{SE}_{b_1}}\right)$$

To **test the hypotheses** H_0: $\beta_1 = 0$ and H_a: $\beta_1 \neq 0$, compute the **test statistic**

$$X^2 = \left(\frac{b_1}{SE_{b_1}} \right)^2$$

Because the random variable X^2 has an approximate χ^2 distribution with 1 degree of freedom, the *P*-value for this test is $P(\chi^2 \geq X^2)$. This is the same as testing the null hypothesis that the odds ratio is 1.

In **multiple logistic regression,** the response variable again has two possible values, but there can be more than one explanatory variable. Multiple logistic regression is analogous to multiple linear regression. Fitting multiple logistic regression models is done, in practice, with statistical software.

APPLY YOUR KNOWLEDGE

17.1 The proportion of exclusive terrritory firms that are successful is

$$\hat{p} = \frac{\text{\# in exclusive territory group that are successful}}{\text{total number in exclusive territory group}} = \frac{108}{142} = 0.761$$

The odds are

$$\text{Odds} = \frac{\hat{p}}{1-\hat{p}} = \frac{108/142}{1-(108/142)} = 3.176$$

The proportion of nonexclusive terrritory firms that are successful is

$$\hat{p} = \frac{\text{\# in nonexclusive territory group that are successful}}{\text{total number in nonexclusive territory group}} = \frac{15}{28} = 0.536$$

The odds are

$$\text{Odds} = \frac{\hat{p}}{1-\hat{p}} = \frac{15/28}{1-(15/28)} = 1.154$$

17.3 For the exclusive territories, the odds of being successful was found in Exercise 17.1 to be 3.176. The log odds are

$$\log\text{(ODDS)} = \log(3.176) = 1.156$$

For the nonexclusive territories, the odds of being successful was found in Exercise 17.1 to be 1.154. The log odds are

$$\log\text{(ODDS)} = \log(1.154) = 0.143$$

17.5 The estimate b_0 is the log (ODDS) for no exclusive territory because no exclusive territory corresponds to $x = 0$. The estimate b_1 is the difference in the log (ODDS) for exclusive and nonexlusive territories. Using the answers from Exercise 17.3, it follows that

$$b_0 = \log\left(\frac{\hat{p}_0}{1-\hat{p}_0}\right) = 0.143 \text{ and}$$

$$b_1 = \log\left(\frac{\hat{p}_1}{1-\hat{p}_1}\right) - \log\left(\frac{\hat{p}_0}{1-\hat{p}_0}\right) = 1.156 - 0.143 = 1.013$$

The estimated odds ratio for exlusive territory versus no exclusive territory is

$$\frac{\text{Odds}_{\text{exclusive}}}{\text{Odds}_{\text{no exclusive}}} = e^{1.013} = 2.754$$

so that the odds for firms with exclusive territories are 2.754 times the odds for firms with no exlusive territories.

17.7 Using the fitted model in Example 17.5, the odds ratio for increasing x by 1 satisfies

$$\frac{\text{Odds}_{x+1}}{\text{Odds}_x} = \frac{e^{-13.71} \times e^{2.25(x+1)}}{e^{-13.71} \times e^{2.25x}} = e^{2.25} = 9.49$$

which shows that if we increase x by one unit, we increase the odds that the cheese will be acceptable by about 9.5 times.

17.9 The output below was produced by Minitab, where Type is the predictor with no exclusive territory corresponds to $x = 0$ and exclusive territories corresponding to $x = 1$. The estimates for the constant and type agree with the values of b_0 and b_1 obtained in Exercise 17.5. The test of H_0: Odds ratio = 1 has a P-value of 0.018, so there is good evidence that the odds for the two groups are different. Alternatively, the 95% confidence interval for the odds ratio (1.19, 6.36) does not include 1, so we conclude the odds for the two groups are different at significance level 0.05.

Logistic Regression Table

Predictor	Coef	SE Coef	Z	P	Odds Ratio	95% CI Lower	Upper
Constant	0.1431	0.3789	0.38	0.706			
Type	1.0127	0.4269	2.37	0.018	2.75	1.19	6.36

17.11 The confidence interval for the slope requires the estimate $b_1 = 3.1088$ and its standard error $SE_{b_1} = 0.3879$, which are given in all three outputs. For a 95% confidence interval, use $z^* = 1.96$ to get

$$b_1 \pm z^* SE_{b_1} = 3.1088 \pm (1.96)(0.3879) = 3.1088 \pm 0.7603 = (2.3485, 3.8691)$$

17.13 a) The value of "Z" in the Minitab output is the estimated coefficient divided by its standard error

$$Z = \frac{b_1}{SE_{b_1}} = \frac{3.1088}{0.3879} = 8.014$$

b) $Z^2 = 8.014^2 = 64.224$, the value of the chi-square statistic reported by SPSS and SAS. The two-sided P-value for the z will always be the same as for the chi-square.

CHAPTER 17 REVIEW EXERCISES

17.15 a) The proportion of high-tech companies that offer stock options to their key employees is

$$\hat{p} = \frac{\text{\# high-tech companies offering stock options}}{\text{total number of high-tech companies}} = \frac{73}{91} = 0.802$$

The odds are

$$\text{Odds} = \frac{\hat{p}}{1-\hat{p}} = \frac{73/91}{1-(73/91)} = 4.056$$

b) The proportion of non-high-tech companies that offer stock options to their key employees is

$$\hat{p} = \frac{\text{\# non-high-tech companies offering stock options}}{\text{total number of non-high-tech companies}} = \frac{75}{109} = 0.688$$

The odds are

$$\text{Odds} = \frac{\hat{p}}{1-\hat{p}} = \frac{75/109}{1-(75/109)} = 2.206$$

c) The estimated odds ratio for high-tech companies versus non-high-tech companies is

$$\frac{\text{Odds}_{\text{high-tech}}}{\text{Odds}_{\text{non-high-tech}}} = \frac{4.056}{2.206} = 1.839$$

so that the odds of offering stock options for high-tech companies is 1.839 times the odds for non-high-tech companies.

17.17 a) Letting $x = 1$ for high-tech companies and $x = 0$ for non-high-tech companies, the estimate b_1 is

$$b_1 = \log\left(\frac{\hat{p}_1}{1-\hat{p}_1}\right) - \log\left(\frac{\hat{p}_0}{1-\hat{p}_0}\right) = \log(4.056) - \log(2.206) = 0.609$$

The estimate of the standard error is given as $\text{SE}_{b_1} = 0.3347$, $z^* = 1.96$ and the 95% confidence interval is

$$b_1 \pm z^*\text{SE}_{b_1} = 0.609 \pm (1.96)(0.3347) = 0.609 \pm 0.656 = (-0.047, 1.265)$$

b) A 95% confidence interval for the odds ratio e^{β_1} is obtained by transforming the confidence interval for the slope, yielding the formula

$$\left(e^{b_1 - z^*\text{SE}_{b_1}}, e^{b_1 + z^*\text{SE}_{b_1}}\right) = (e^{-0.047}, e^{1.265}) = (.954, 3.543)$$

c) Because the value 1 is included in the 95% confidence interval for the odds ratio, there is no evidence at the 5% level of significance that the odds of offering stock options is different for high-tech and non-high-tech companies.

17.19 a) The proportion of bicyclists that tested positive for alcohol is

$$\hat{p} = \frac{\text{\# testing positive for alcohol}}{\text{total number of fatally injured bicyclists}} = \frac{542}{1711} = 0.317$$

b) The odds of a fatally injured bicyclist testing positive for alcohol is

$$\text{Odds} = \frac{\hat{p}}{1-\hat{p}} = \frac{542/1711}{1-(542/1711)} = 0.4636$$

c) The proportion of bicyclists that did not test positive for alcohol is

$$\hat{p} = \frac{\text{\# not testing positive for alcohol}}{\text{total number of fatally injured bicyclists}} = \frac{1169}{1711} = 0.683$$

d) The odds of a fatally injured bicyclist not testing positive for alcohol is

$$\text{Odds} = \frac{\hat{p}}{1-\hat{p}} = \frac{1169/1711}{1-(1169/1711)} = 2.1568$$

e) The answer in (c) is one minus the answer in (a) and the answer in (d) is one divided by the answer in (b).

17.21 a) The proportion of men who died from cardiovascular disease in the high-blood-pressure group is

$$\hat{p} = \frac{\text{\# with cardiovascular diseases in high-pressure group}}{\text{total number in high-blood-pressure group}} = \frac{55}{3338} = 0.0165$$

The odds are

$$\text{Odds} = \frac{\hat{p}}{1-\hat{p}} = \frac{55/3338}{1-(55/3338)} = 0.0168$$

b) The proportion of men who died from cardiovascular disease in the low-blood-pressure group is

$$\hat{p} = \frac{\text{\# with cardiovascular diseases in low-pressure group}}{\text{total number in low-blood-pressure group}} = \frac{21}{2676} = 0.0078$$

The odds are

$$\text{Odds} = \frac{\hat{p}}{1-\hat{p}} = \frac{21/2676}{1-(21/2676)} = 0.0079$$

c) The estimated odds ratio for the high-blood-pressure group vs. the low-pressure-group is

$$\frac{\text{Odds}_{\text{high-blood-pressure}}}{\text{Odds}_{\text{low-blood-pressure}}} = \frac{0.0168}{0.0079} = 2.127$$

so that the odds of dying from cardiovascular disease for the high-blood-pressure group is 2.127 times the odds for the low-blood-pressure group.

17.23 a) We are told that the estimated slope is $b_1 = 0.7505$ and its standard error is $SE_{b_1} = 0.2578$. The estimate of the odds ratio is $e^{b_1} = e^{0.7505} = 2.118$, which agrees with the answer obtained in Exercise 17.23 except for roundoff error. A 95% confidence interval for the odds ratio e^{β_1} is obtained by transforming the confidence interval for the slope, yielding the formula

$$\left(e^{b_1 - z^* SE_{b_1}},\ e^{b_1 + z^* SE_{b_1}}\right) = (e^{0.7505-(1.96)(0.2578)}, e^{0.7505+(1.96)(0.2578)}) = (1.278, 3.511)$$

b) The odds of dying from cardiovascular disease for the high-blood-pressure group is different than the odds for the low-blood-pressure group at the 5% level of significance because the value 1 is not included in the confidence interval. The odds of dying from cardiovascular disease for the high-blood-pressure group is estimated to be 2.127 times the odds for the low-blood-pressure group.

17.25 a) The estimated slope is $b_1 = 1.8171$ and its standard error is $SE_{b_1} = 0.3686$. For a 95% confidence interval, use $z^* = 1.96$ to get

$$b_1 \pm z^* SE_{b_1} = 1.8171 \pm (1.96)(0.3686) = 1.8171 \pm 0.7225 = (1.0946, 2.5396)$$

b) The value of X^2 can be obtained from the estimate and its standard error through the formula

$$X^2 = \left(\frac{b_1}{SE_{b_1}}\right)^2 = \left(\frac{1.8171}{0.3686}\right)^2 = 24.30$$

and has a *P*-value of 0.000.

c) $X^2 = 24.30$ which has a *P*-value of 0.000. There is very strong evidence that the slope is not equal to zero which supports the alternative hypothesis that there is gender bias in syntax textbooks. The odds of the female reference being juvenile is estimated to be 6.15 times the odds for the male reference.

17.27 a) The logistic regression model is for the log odds of a termination is

$$\log\left(\frac{p_i}{1-p_i}\right) = \beta_0 + \beta_1 x_i$$

where $x = 1$ for over 40 years of age and $x = 0$ for under 40 years of age.

b) The assumptions are that there are two populations: those over 40 and those under 40. The probabilities of being terminated in these two populations are p_1 and p_2, respectively. The probability of being terminated should be constant within each population, and the termination of individuals should be independent. If we are working within a single company or group of similar companies, these assumptions should be approximately satisfied. If the workers are from a variety of industries or a hetergeneous population, then the assumption of constant probability of termination within each group may be violated.

c) We are told that the estimated slope is $b_1 = 1.3504$ and its standard error is $\mathrm{SE}_{b_1} = 0.4130$. The estimate of the odds ratio $e^{b_1} = e^{1.3504} = 3.859$. A 95% confidence interval for the odds ratio e^{β_1} is obtained by transforming the confidence interval for the slope, yielding the formula

$$\left(e^{b_1 - z^*\mathrm{SE}_{b_1}},\ e^{b_1 + z^*\mathrm{SE}_{b_1}}\right) = \left(e^{1.3504-(1.96)(0.4130)}, e^{1.3504+(1.96)(0.4130)}\right) = (1.718, 8.670)$$

Because the interval doesn't include 1, there is evidence at the 5% level that the odds of being terminated are different for the two groups. The odds of being terminated for those over 40 is estimated to be 3.859 times the odds for those under 40.

d) If additional explanatory variables were available they could be included in a multiple logistic regression model. Then we could determine if there was still evidence of a difference in the odds of being terminated for the two age groups, after adjusting for the effects of these other variables. This would make a conclusion of age discrimination more convincing as it could reduce the effects of some obvious confounding variables.

17.29 The output below was produced by Minitab, where "user" is the predictor with Non Internet user for travel arrangements corresponding to $x = 0$ and Internet user for travel arrangements corresponding to $x = 1$. The test of H_0: Odds ratio = 1 has a P-value of 0.000, so there is very strong evidence that the odds of having completed college are different for the two groups. The estimated odds ratio is 1.90, so the odds of having completed college for those using the internet to make travel arrangements is estimated to be 1.90 times the odds for those who do not use the Internet to make travel arrangements.

Logistic Regression Table

Predictor	Coef	SE Coef	Z	P	Odds Ratio	95% CI Lower	95% CI Upper
Constant	-0.36552	0.06967	-5.25	0.000			
user	0.63930	0.09194	6.95	0.000	1.90	1.58	2.27

17.31 The output below was produced by Minitab, where "gender" is the predictor, with female corresponding to $x = 0$ and male corresponding to $x = 1$. The test of H_0: Odds ratio = 1 has a P-value of 0.000, so among those having fatal bicycle accidents there is very strong evidence that the odds of having a positive test for alcohol are different for males and females. The estimated odds ratio is 3.11, so the odds of testing positive for alcohol for males is estimated to be 3.11 times the odds for females.

Logistic Regression Table

Predictor	Coef	SE Coef	Z	P	Odds Ratio	95% CI Lower	95% CI Upper
Constant	-1.8040	0.2077	-8.69	0.000			
gender	1.1355	0.2146	5.29	0.000	3.11	2.04	4.74

CHAPTER 18

BOOTSTRAP METHODS AND PERMUTATION TESTS

NOTE: In those questions in this chapter for which the answer is obtained by resampling, your answer may differ slightly from the one we give. This is because samples are chosen at random, so the particular set of samples you select will differ form ours.

SECTIONS 18.1 and 18.2

OVERVIEW

To **bootstrap** a statistic (such as the sample mean), we draw hundreds of **resamples** with replacement from the original data, calculate the statistic for each resample, and inspect the **bootstrap distribution** of the resampled statistics. This bootstrap distribution approximates the sampling distribution of the statistic. Notice that we use a quantity (the bootstrap distribution) based on the sample to approximate a similar quantity (the sampling distribution) from the population; this is called the **plug-in principle.**

In most cases, bootstrap distributions have approximately the same shape and spread as the sampling distribution. However, they are centered at the statistic computed from the original data. This is in contrast to the sampling distribution, which is centered at the parameter.

We use graphs and numerical summaries to determine whether the bootstrap distribution is approximately normal, centered at the original statistic, and to estimate the spread. The **bootstrap standard error** is the standard deviation of the bootstrap distribution.

The bootstrap is used to estimate the variation in a statistic based on the original data. It does not add to or replace the original data.

APPLY YOUR KNOWLEDGE

18.1 a) and b) The 20 resamples are given below, along with their means.

	0.22	1.57	0.22	1.57	2.20	19.67	0.22	3.12	3.12	19.67	1.57	0.00	2.20	2.20	0.00	1.57	19.67	0.00	1.57	19.67
	0.00	1.57	19.67	0.00	1.57	0.22	3.12	1.57	0.00	1.57	0.00	0.22	0.00	3.12	1.57	0.00	0.22	0.00	19.67	0.00
	19.67	0.22	0.00	2.20	19.67	2.20	0.22	2.20	19.67	2.20	0.22	0.00	3.12	2.20	1.57	0.22	0.22	0.00	2.20	3.12
	19.67	3.12	0.22	2.20	2.20	0.22	0.22	2.20	0.22	1.57	19.67	2.20	1.57	0.00	2.20	19.67	2.20	1.57	0.22	0.00
	1.57	3.12	19.67	0.00	3.12	0.00	0.22	1.57	19.67	2.20	2.20	0.22	0.22	3.12	0.22	19.67	19.67	3.12	2.20	3.12
	0.00	0.22	2.20	19.67	0.22	3.12	2.20	3.12	1.57	0.00	1.57	1.57	0.00	1.57	3.12	0.00	19.67	0.00	0.22	0.00
mean:	6.86	1.64	7.00	4.27	4.83	4.24	1.03	2.30	7.38	4.54	4.21	0.70	1.19	2.04	1.45	6.86	10.28	0.78	4.35	4.32

c) A stemplot of the means (after rounding to one decimal place) is given below.

```
 0|7 8
 1|0 2 4 6
 2|0 3
 3|
 4|2 2 3 3 3 5 8
 5|
 6|9 9
 7|0 4
 8|
 9|
10|3
```

d) The bootstrap standard error is the standard deviation of the 20 means. Using software, we find that this is

$$\text{bootstrap standard error} = 2.658$$

SECTION 18.2 EXERCISES

18.3 a) A histogram of the dollar amounts spent by the 50 shoppers is given below.

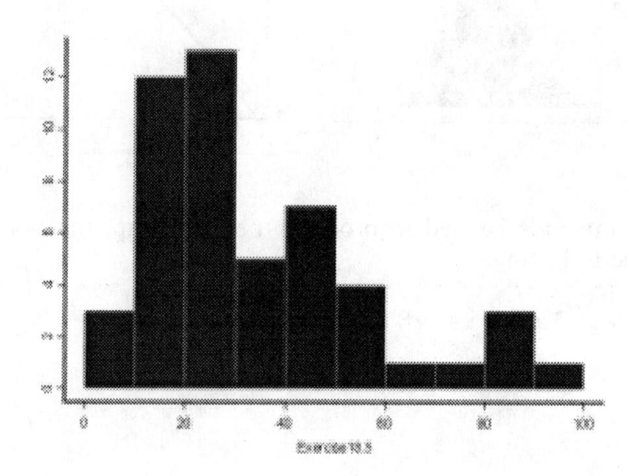

Notice that the histogram is right-skewed. The S-PLUS command needed to produce this histogram is given below.

```
> hist(Exercise18.3)
```

b) The S-PLUS commands needed to produce the bootstrap, the bootstrap distribution, and a normal quantile plot are the following.

```
> boot3 <- bootstrap(Exercise18.3, mean)
> title("Bootstrap Distribution of the Mean")
> plot(boot3)
> qqnorm(boot3)
```

The bootstrap distribution and a normal quantile plot are given on the next page. We see that the distribution is symmetric and bell-shaped, and that the quantile plot shows positive skewness. While positive skewness would not be a concern in raw data, here it occurs in a bootstrap distribution, after the central limit theorem has had a chance to work. Later in the chapter we learn ways to get more accurate confidence intervals in cases like this.

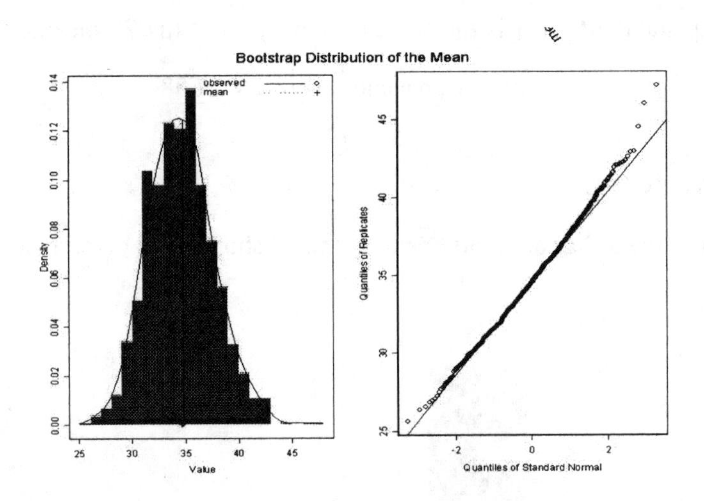

18.5 a) The S-PLUS commands needed to produce the bootstrap, the bootstrap distribution, and a normal quantile plot are the following.

```
> boot5 <- bootstrap(Exercise18.5, mean)
> plot(boot5)
> qqnorm(boot5)
> plot(boot3)
> qqnorm(boot3)
> boot5
```

The bootstrap distribution and a normal quantile plot are given below.

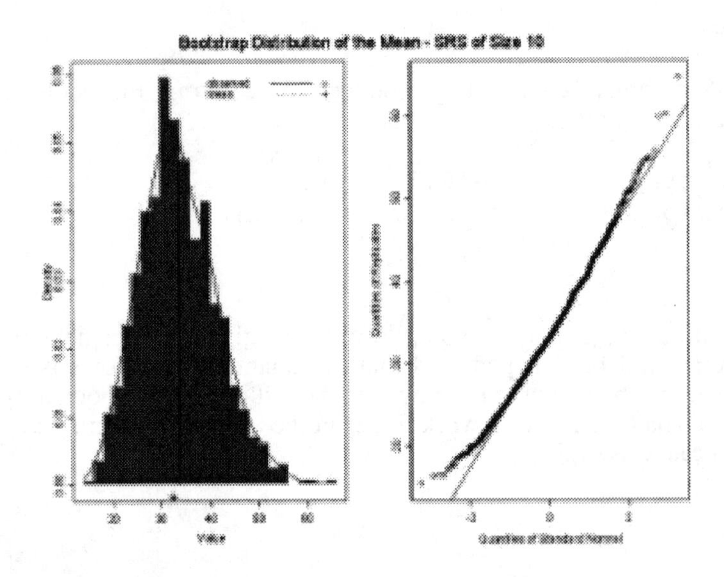

The distribution is roughly bell-shaped, but is less so than the bootstrap distribution in Exercise 18.3. The distribution here looks slightly right-skewed, and the quantile plot is less close to lying along a straight line than the quantile plot in Exercise 18.3.

b) The bootstrap summary statistics from S-PLUS for Exercise 18.3 and for the data from the 10 shoppers in this exercise are given below.

```
Call:
bootstrap(data = Exercise18.3, statistic = mean)

Number of Replications: 1000

Summary Statistics:
     Observed Mean     Bias    SE
mean     34.7 34.6 -0.09917 3.068

Call:
bootstrap(data = Exercise18.5, statistic = mean)

Number of Replications: 1000

Summary Statistics:
     Observed  Mean   Bias    SE
mean     33.4 33.51 0.1084 7.703
```

Recall that for a sample of size n, the standard error of the sample mean is $s/\sqrt{n}$, where s is the sample standard deviation. In Exercise 18.3, $n = 50$. Here, $n = 10$. Thus, the standard error of the sample mean in Exercise 18.3 is smaller than the standard error of the sample mean in this exercise. We would expect the bootstrap standard errors to have similar characteristics, and we see that this is the case.

18.7 For the data for the 50 shoppers in Exercise 18.3, we find (using software) that

$$s = 21.694$$

Thus, we find

$$s/\sqrt{50} = 21.694/\sqrt{50} = 3.068476$$

Recall that in Exercise 18.5 we saw that the bootstrap standard error based on these 50 shoppers is

$$\text{bootstrap standard error} = 3.068,$$

which agrees closely with $s/\sqrt{50} = 3.068476$.

SECTION 18.3

OVERVIEW

The bootstrap distribution of most statistics mimics the shape, spread, and bias of the actual sampling distribution. The **bootstrap standard error** is the standard deviation of the bootstrap distribution. It measures how much a statistic varies under random sampling. The bootstrap estimate of **bias** is the mean of the bootstrap distribution minus the value of the statistic for the original data. The bias is small when the bootstrap distribution is centered at the value of the statistic for the original data. Small bias suggests that the sampling distribution of the statistic is centered at the value of the population parameter.

The bootstrap can estimate the sampling distribution, bias, and standard error of many statistics. One example is the trimmed mean.

If the bootstrap distribution is approximately Normal (as seen in a histogram or quantile plot) and the bias is small, we can compute a **bootstrap *t* confidence interval** of the form

$$\text{statistic} \pm t^*\text{SE}$$

for the parameter. Do not use this *t* interval if the bootstrap distribution is not Normal or if it shows substantial bias.

APPLY YOUR KNOWLEDGE

18.9 The S-PLUS commands and the bootstrap summary statistics from S-PLUS for Exercise 18.4 are given below.

```
> boot4 <- bootstrap(Exercise18.4, mean)

> boot4
Call:
bootstrap(data = Exercise18.4, statistic = mean)

Number of Replications: 1000

Summary Statistics:
     Observed  Mean     Bias    SE
mean    141.8 141.3 -0.5164 12.83
```

The bias is –0.5164. This is small compared to the observed mean of 141.8. The bootstrap distribution of most statistics (such as the sample mean) mimics the shape, spread, and bias of the actual sampling distribution. Thus, we expect that the bias encountered is using $\bar{x}$ to estimate the mean survival time for all guinea pigs that receive the same experimental treatment is also small.

18.11 a) The 25% trimmed mean and the sample mean for the 50 shoppers can be computed using S-PLUS. The output from S-PLUS is given below.

```
> mean(Exercise18.3, trim = 0.25)
[1] 30.09577
> mean(Exercise18.3)
[1] 34.7022
```

The 25% trimmed mean is 30.09577, which is smaller than the sample mean of 34.7022. If we examine the histogram of the data for the 50 shoppers (refer to Exercise 18.3 in this Study Guide to see the histogram), we see that the data are strongly right-skewed. The trimmed mean eliminates much of the large right tail (i.e., the very large values that cause the sample mean to be large), and hence the trimmed mean is smaller than the sample mean.

b) We use S-PLUS to generate 1000 resamples, computing the trimmed mean of each sample. The S-PLUS commands to do this are given below.

```
> boot11 <- bootstrap(Exercise18.3, mean(Exercise18.3, trim = 0.25))
```

The standard deviation of the trimmed mean for these 1000 bootstrap samples is found to be SE = 3.16266. The upper 0.025 percentile of the *t* distribution with 49 df is found, using statistical software, to be $t^* = 2.0096$ (if you use Table D, use the value $t^* = 2.009$ that corresponds to 50 df). Thus, 95% confidence interval for the 25% trimmed mean spending in the population of all shoppers is

bootstrap trimmed mean ± t^*SE = 30.09577 ± (2.0096)(3.16266) = 30.09577 ± 6.35568
= (23.740, 36.451)

18.13 The formula for the standard error of $\bar{x}_1 - \bar{x}_2$ is

$$\sqrt{\frac{s_1^2}{n_1} + \frac{s_2^2}{n_2}}$$

where n_1 is the size of the sample from population 1, s_1 is the sample standard deviation for the sample from population 1, n_2 is the size of the sample from population 2, and s_2 is the sample standard deviation for the sample from population 2. In this case, $n_1 = 1664$, $s_1 = 14.7$, $n_2 = 23$, $s_2 = 19.5$, so the formula-based standard error is

$$\sqrt{\frac{s_1^2}{n_1} + \frac{s_2^2}{n_2}} = \sqrt{\frac{14.7^2}{1664} + \frac{19.5^2}{23}} = \sqrt{16.66} = 4.08$$

The bootstrap standard error in Example 18.7 is 4.052, which is close to the formula-based value.

18.15 The S-PLUS command to compute the difference in means (healthy minus failed) and the subsequent output are given below.

```
> mean(Exercise18.15[1:68, 2]) - mean(Exercise18.15[69:101, 2])
[1] 0.9019519
```

The difference in means is 0.901959

a) The S-PLUS commands to bootstrap the difference in means are the following.

```
> boot15 <- bootstrap(Exercise18.15, mean(ratio[status == "Healthy"])
- mean(ratio[status == "Failed"]), group = status)
```

The summary statistics are as follows.

```
Call:
bootstrap(data = Exercise18.15, statistic = mean(ratio[status ==
"Healthy"]) - mean(ratio[status == "Failed"]), group = status)

Number of Replications: 1000

Summary Statistics:
      Observed   Mean      Bias     SE
Param    0.902 0.9084 0.006432 0.1099
```

The S-PLUS commands to produce the bootstrap distributuion and a quantile plot are as follows.

```
> title(Bootstrap Difference in Means - Failed vs. Healthy Firms)
> plot(boot15)
> qqnorm(boot15)
```

The resulting histogram and quantile plot are given below.

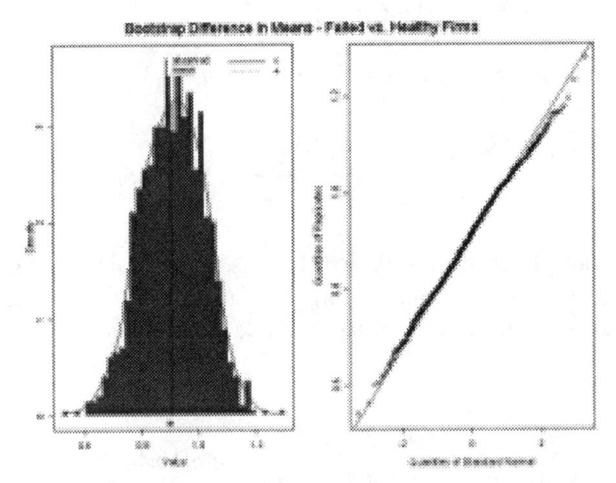

The distribution looks approximately normal and the bias is small. Thus, it meets the conditions for a bootstrap t confidence interval.

b) From part (a), we see that the bootstrap mean is 0.9084 and the standard error is SE = 0.1099. The upper 0.025 percentile of the t distribution with 99 df is found, using statistical software, to be t^* = 1.9842 (if you use Table D, use the value t^* = 1.984 that corresponds to 100 df). Thus,

$$\text{bootstrap mean} \pm t^*\text{SE} = 0.9084 \pm (1.9842)(0.1099) = 0.9084 \pm 0.2181 = (0.6903,\ 1.1265)$$

c) The two-sample t confidence interval reported on page 488 is (0.653, 1.151). This interval is wider than the bootstrap t confidence interval in part (b).

SECTION 18.3 EXERCISES

18.17 a) We repeat the resampling of the data in Table 18.1 to get another bootstrap distribution for the mean. The S-PLUS commands we used are the following.

```
> boot17 <- bootstrap(Exercise18.17, mean)
> plot(boot17)
> qqnorm(boot17)
```

The plots produced by these commands are given below.

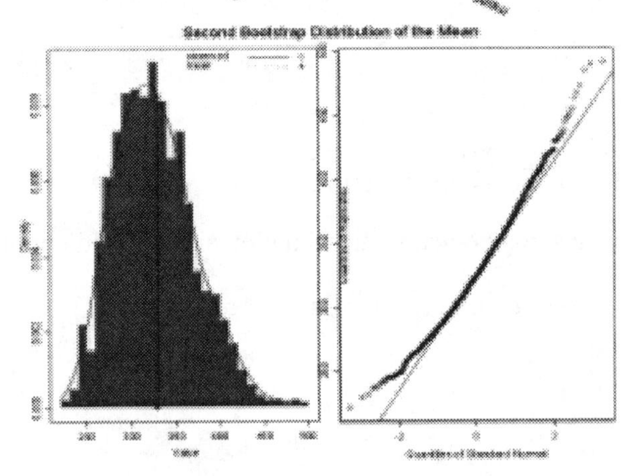

These plots look very similar to those in Figure 18.7.

b) The bootstrap summary statistics from the S-PLUS commands in part (a) are given below.

```
bootstrap summary statistics
> boot17
Call:
bootstrap(data = Exercise18.17, statistic = mean)

Number of Replications: 1000

Summary Statistics:
     Observed  Mean   Bias    SE
mean    329.3 326.7 -2.591 44.87
```

The bootstrap standard error of the mean is 44.87. In Example 18.5, the bootstrap standard error of the 25% trimmed mean is 16.83. We see that the bootstrap standard error of the mean is almost three times as large as the bootstrap standard error of the 25% trimmed mean. This is reflected in Figures 18.7 and 18.8. The bootstrap distribution for the mean (Figure 18.7) has greater spread (the histogram covers a range from about 225 to 475) than the bootstrap distribution for the 25% trimmed mean (the histogram covers a range from about 200 to 300).

c) Examining the bootstrap distribution of the mean in Figure 18.7 and in part (a) of this solution, we see that the bootstrap distribution is right-skewed and hence non-Normal. We should not use the bootstrap *t* interval if the bootstrap distribution is not Normal.

18.19 a) A histogram and Normal quantile plot of the data are given below.

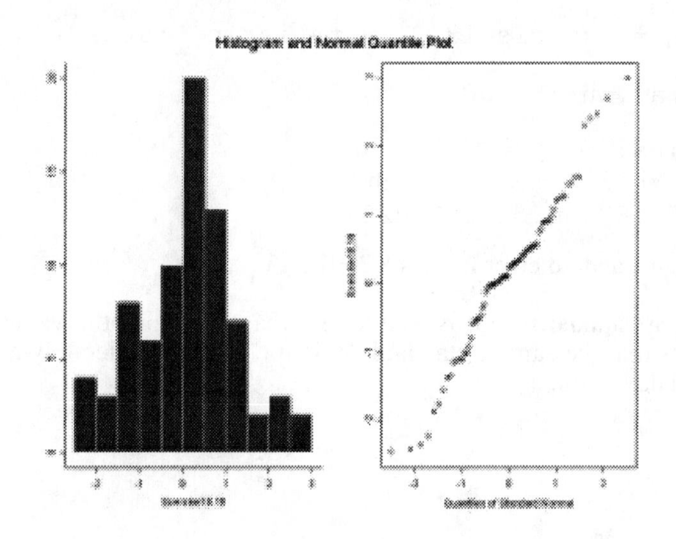

There do not appear to be any significant departures from Normality. The histogram is centered at about 0 and the spread is approximately what we would expect for the $N(0, 1)$ distribution.

b) S-PLUS commands to bootstrap the mean, and the resulting output are given below.

```
> boot19 <- bootstrap(Exercise18.19, mean)
```

```
> boot19
Call:
bootstrap(data = Exercise18.19, statistic = mean)
Number of Replications: 1000

Summary Statistics:
       Observed   Mean      Bias      SE
mean   0.1249 0.1209 -0.003978 0.1326
```

We see that the standard error of the bootstrap mean is 0.1326.

c) The bias is small (–0.003978) in part (b). The histogram of the data in part (a) looks approximately Normal, so the bootstrap distribution of the mean will also look Normal. When the bootstrap distribution is approximately Normal and the bias is small, it is safe to use the bootstrap t confidence interval.

From part (b), we see that the bootstrap mean is 0.1209 and the standard error is SE = 0.1326. The sample size is $n = 78$, and the upper 0.025 percentile of the t distribution with $n - 1 = 77$ df is found, using Table D (use the value that corresponds to 80 df) to be $t^* = 1.99$. Thus,

$$\text{bootstrap mean} \pm t^*\text{SE} = 0.1209 \pm (1.99)(0.1326) = 0.1209 \pm 0.2639 = (-0.1430, 0.3848)$$

18.21 a) Using computer software, we calculate the sample standard deviation to be

$$s = 7.705967$$

b) We bootstrap the standard deviation s. The S-PLUS commands and the corresponding output are given below.

```
> bootstrap(data = Exercise18.21, statistic = stdev)

Number of Replications: 1000

Summary Statistics:
        Observed  Mean    Bias    SE
stdev     7.706 7.187 -0.5189 2.226
```

We see that the bootstrap standard error for s is SE = 2.226.

c) The bootstrap standard error is a little less than one-third the value of the sample standard deviation. This suggests that the sample standard deviation is only moderately accurate as an estimate of the population standard deviation.

d) The S-PLUS commands to produce a plot of the bootstrap distribution and a Normal quantile plot are given below, along with the plots.

```
> plot(boot21)
> qqnorm(boot21)
```

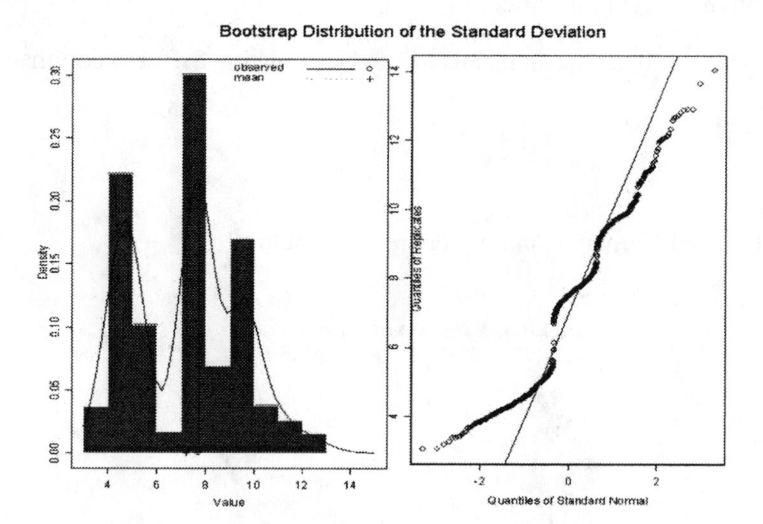

These plots show that the bootstrap distribution is not Normal. Thus, it would *not* be appropriate to give a bootstrap t interval for the population standard deviation.

18.23 a) Separate histograms and Normal quantile plots for the minority and white refusal rates are given below.

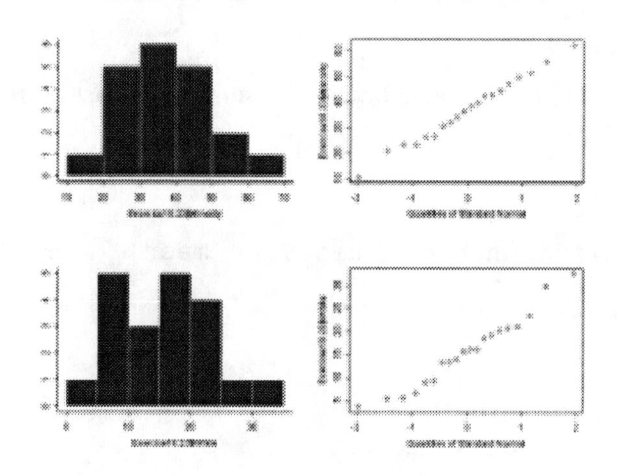

The plots do not show any significant departures from Normality, so there is nothing in the plots to suggest that the difference in means might be non-Normal.

b) Using statistical software, we compute

mean difference (minority refusal rate – white refusal rate) = 21.255
standard deviation of the differences = 8.296066

There are n = 20 banks, and the upper 0.025 percentile of the t distribution with n – 1 = 19 df is found, using Table D to be t^* = 2.093. Thus, a 95% paired t confidence interval for the difference in population means is mean difference $\pm$ t*(standard deviation of the differences)/ $\sqrt{n}$

$= 21.255 \pm 2.093(8.296066/\sqrt{20}) = 21.255 \pm 3.883 = (17.372, 25.138)$

The interval does not contain 0 and only includes positive values. This is evidence that the minority refusal rate is larger than the white refusal rate.

c) We bootstrapped the difference in means using the following S-PLUS commands.

```
> boot23 <- bootstrap(difference, mean)
> plot(boot23)
> qqnorm(boot23)
```

The bootstrap distribution and Normal quantile plot are given below.

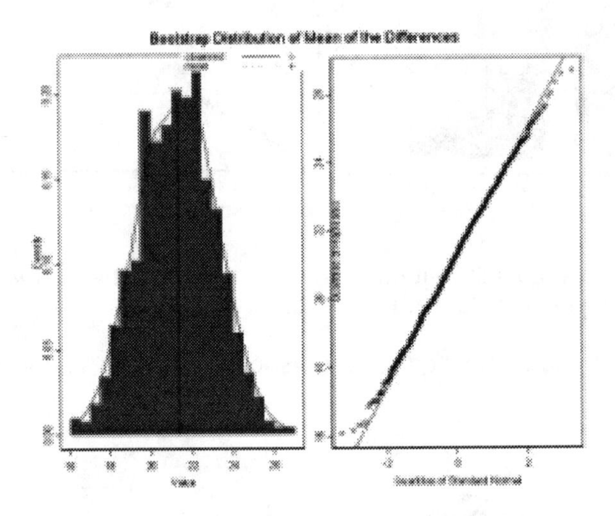

The bootstrap distribution looks reasonably Normal. The summary statistics from the bootstrap are given below.

```
> boot23
Call:
bootstrap(data = difference, statistic = mean)

Number of Replications: 1000

Summary Statistics:
     Observed  Mean    Bias     SE
mean    21.25 21.26 0.00874 1.836
```

The bias is small. Thus, a bootstrap t confidence interval is appropriate here. We see that the bootstrap mean is 21.26 and the standard error is SE = 1.836. The sample size is n = 20 and the upper 0.025 percentile of the t distribution with $n - 1 = 19$ df is found, using Table D to be $t^* = 2.093$. Thus, a 95% bootstrap t confidence interval is

bootstrap mean $\pm$ t*SE $= 21.26 \pm (2.093)(1.836) = 21.26 \pm 3.84 = (17.42, 25.10)$

This is very close to the traditional interval that we calculated in (b).

18.25 a) The S-PLUS commands we used to bootstrap the mean for the CLEC data are given below.

```
> boot25 <- bootstrap(CLEC, mean)
> plot(boot25)
> qqnorm(boot25)
> title("Bootstrap Distribution of Mean for CLEC data")
```

The resulting bootstrap distribution and Normal quantile plot are given below.

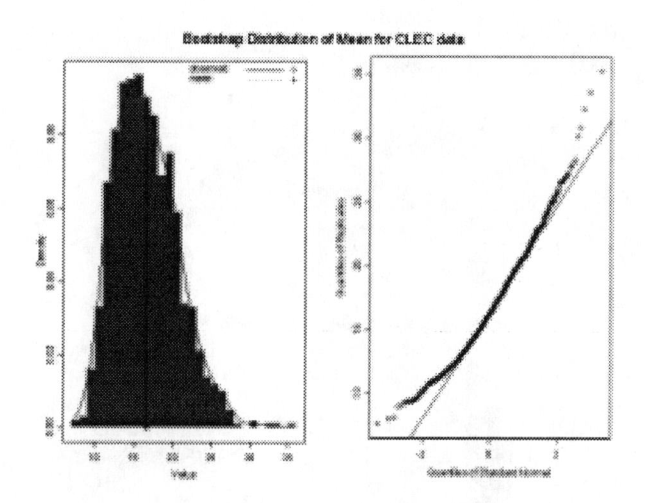

This bootstrap distribution is right-skewed. The bootstrap distribution of the Verizon repair times given in Figure 18.3 appears to be approximately Normal.

b) The source of the skew in the bootstrap distribution of the difference in means appears to be due to the skew in the CLEC data.

SECTION 18.4

OVERVIEW

The selection of the original random sample from the population is the source of almost all the variation in a bootstrap distribution. The resampling process introduces little additional variation.

Bootstrap distributions based on small samples can be quite variable. Their shape and spread reflect the characteristics of the sample. Small samples do not reflect the population from which they are drawn as well as large samples. Thus, bootstrap distributions based on small samples may not accurately estimate the shape and spread of the sampling distribution.

Bootstrapping is unreliable for statistics like the median and quartiles when the sample size is small. In this case, the bootstrap distributions tend to be broken up (discrete) and highly variable in shape.

SECTION 18.4 EXERCISES

18.27 a) In Section 4 of Chapter 4, we learned that the sampling distribution of the sample mean $\bar{x}$ has mean equal to the population mean μ and standard deviation equal to the population standard deviation σ divided by $\sqrt{n}$. Thus,

$$\text{mean of the sampling distribution of } \bar{x} = \mu = 8.4$$
$$\text{standard deviation of the sampling distribution of } \bar{x} = \sigma/\sqrt{n} = 14.7/\sqrt{n}$$

b) We drew a SRS of size $n = 10$ and bootstrapped the sample mean $\bar{x}$ using 1000 resamples from our sample. The S-PLUS commands to carry out the bootstrap, the summary statistics, and the resulting distribution and Normal quantile plot are given below.

```
> y10 <- sample(ILEC, size = 10)
> boot27a <- bootstrap(y10, mean)
> plot(boot26a)
> qqnorm(boot26a)
> boot26a
Call:
bootstrap(data = y10, statistic = mean)

Number of Replications: 1000

Summary Statistics:
     Observed  Mean      Bias    SE
mean    8.602 8.577 -0.02517 2.899
```

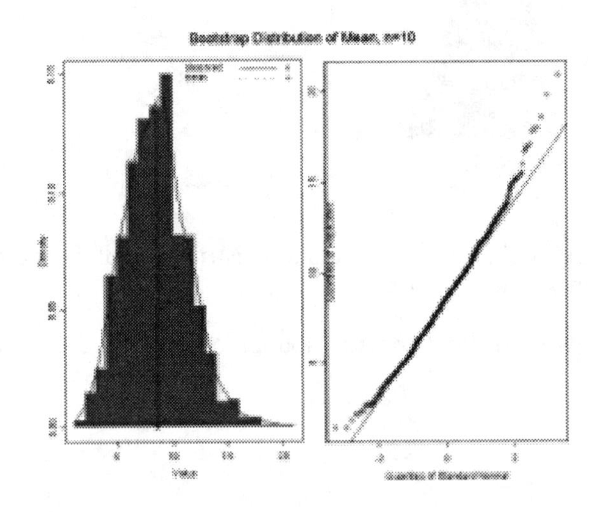

We see from the summary statistics that the bootstrap standard error is SE = 2.899.

c) We drew a SRS of size $n = 40$ and bootstrapped the sample mean $\bar{x}$ using 1000 resamples from our sample. The S-PLUS commands to carry out the bootstrap, the summary statistics, and the resulting distribution and Normal quantile plot are given below.

```
> y40 <- sample(ILEC, size = 40)
> boot26b <- bootstrap(y40, mean)
> plot(boot26b)
> qqnorm(boot26b)
> title("Bootstrap Distribution of Mean, n=40")
> boot26b
Call:
bootstrap(data = y40, statistic = mean)

Number of Replications: 1000
```

```
Summary Statistics:
      Observed  Mean     Bias    SE
mean     7.158 7.254 0.09551 1.516
```

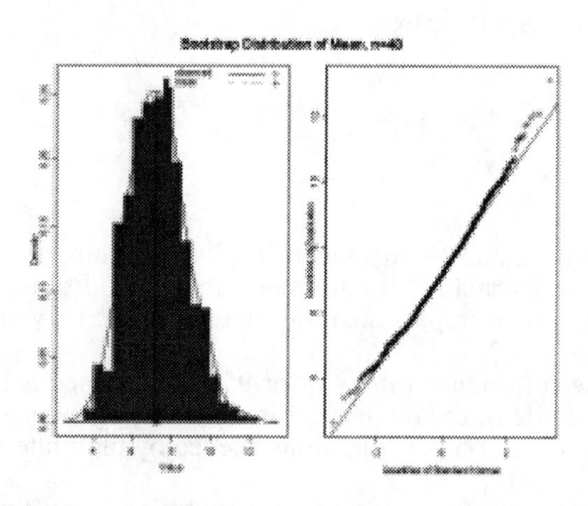

We see from the summary statistics that the bootstrap standard error when $n = 40$ is SE = 1.516.

Next, we drew a SRS of size $n = 160$ and bootstrapped the sample mean $\bar{x}$ using 1000 resamples from our sample. The S-PLUS commands to carry out the bootstrap, the summary statistics, and the resulting distribution and Normal quantile plot are given below.

```
> y160 <- sample(ILEC, size = 160)
> boot26c <- bootstrap(y160, mean)
> plot(boot26c)
> qqnorm(boot26c)
> title("Bootstrap Distribution of Mean, n=160")
> boot26c
Call:
bootstrap(data = y160, statistic = mean)

Number of Replications: 1000

Summary Statistics:
      Observed  Mean      Bias     SE
mean     6.863 6.859 -0.003907 0.7693
```

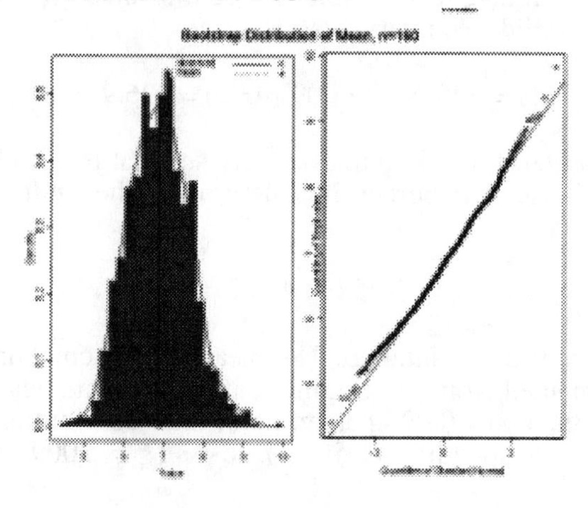

We see from the summary statistics that the bootstrap standard error when n = 160 is SE = 0.7693.

d) If we look at the distributions, and especially the Normal quantile plots, we see that as n increases the bootstrap distributions look more and more Normal. We also see that the standard errors SE decrease roughly by a factor of 2 as n increases.

SECTION 18.5

OVERVIEW

Both bootstrap t and traditional z and t confidence intervals require statistics with small biases and sampling distributions close to Normal. We can check these conditions by examining the bootstrap distribution (a histogram of the bootstrap, a quantile plot, and summary statistics) for bias and lack of Normality.

The **bootstrap percentile confidence interval** for 95% confidence is the interval between the 2.5% percentile and the 97.5% percentile of the bootstrap distribution. Agreement between the bootstrap t and percentile intervals is an added check on the conditions needed by the t interval. Do not use t or percentile intervals if these conditions are not met.

When bias or skewness is present in the bootstrap distribution, use either a **bootstrap tilting** or **BCa** interval. The t and percentile interval give inaccurate results under these circumstances unless the sample sizes are very large. The tilting and BCa confidence intervals adjust for bias and skewness and are generally very accurate except for small samples.

APPLY YOUR KNOWLEDGE

18.29 For a 90% bootstrap confidence interval, we would use the 5% and 95% percentiles as the endpoints. This is because the percentage of the distribution between the 5% and the 95% percentiles is 95% – 5% = 90%.

18.31 In Example 18.10, the 95% bootstrap t interval is (–0.144, 0.358) and the bootstrap percentile interval is (–0.128, 0.356). There is very close agreement between the upper endpoints of both intervals. However, the lower endpoints differ somewhat and this may indicate some skewness.

18.33 The confidence intervals requested in the statement of the problem are as follows.

Bootstrap t 95% confidence interval: From the software output in Figure 18.18, we see the bootstrap mean is 326.9 and the bootstrap SE is 43.9. The sample size is n = 50 and the upper 0.025 percentile of the t distribution with n – 1 = 49 df is found, using Table D (use the value corresponding to 50 df), to be t^* = 2.009. Thus, a 95% bootstrap t confidence interval is

$$\text{bootstrap mean} \pm t^*\text{SE} = 326.9 \pm (2.009)(43.9) = 326.9 \pm 88.2 = (238.7, 415.1)$$

Percentile 95% confidence interval: From Figure 18.18 we see that the 2.5% percentile is 252.5 and the 97.5% percentile is 433.1. These two percentiles determine the endpoints of the 95% percentile confidence interval, which is

$$(252.5, 433.2)$$

Traditional one-sample t 95% confidence interval: The mean of the actual data is $\bar{x}$ = 329.3. The sample standard deviation can be computed from the original data using statistical software and is s = 316.83. The sample size is n = 50 and the upper 0.025 percentile of the t distribution with n – 1 = 49 df is found, using Table D (use the value corresponding to 50 df), to be t^* = 2.009. Thus, a 95% traditional one-sample t confidence interval is

$$\bar{x} \pm t^*s/\sqrt{n} = 329.3 \pm (2.009)(316.83)/\sqrt{50} = 329.3 \pm 90.0 = (239.3, 419.3)$$

From Figure 18.18 we also see that the 95% BCa and Titling confidence intervals are (using the 2.5% percentile and 97.5% percentile as the lower and upper confidence limits)

BCa 95% confidence interval: (270, 455.7)

Tilting 95% confidence interval: (265, 458.7)

A picture that compares all five confidence intervals is given below.

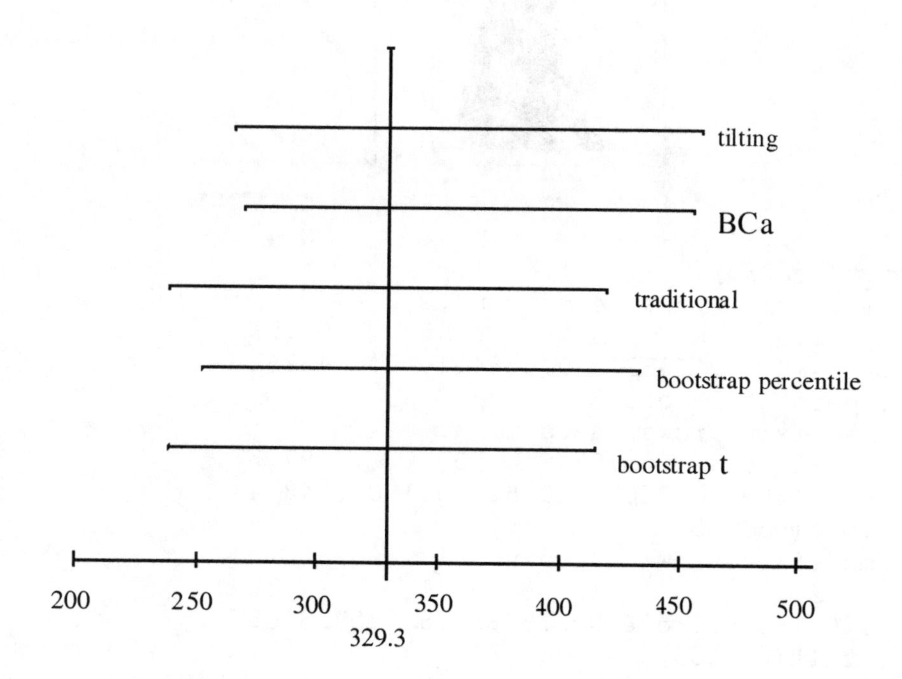

The bootstrap t and traditional intervals are centered on the sample mean. The bootstrap percentile interval is shifted to the right of these two. A BCa and tilting intervals are shifted even further to the right. The latter two better reflect the skewed nature of the data. Using a t interval or the bootstap percentile interval, we get a biased picture of what the value of the population mean is likely to be. In particular, we would underestimate its value. Any policy decisions based on these data, such as tax rates, would reflect this underestimate.

18.35 The S-PLUS commands needed to produce the desired intervals and the resulting output are given below.

```
> boot35 <- bootstrap(Exercise18.32, cor(Wages, LOS))

> boot35
Call:
bootstrap(data = Exercise18.32, statistic = cor(Wages, LOS))

Number of Replications: 1000

Summary Statistics:
       Observed   Mean     Bias     SE
Param    0.3535 0.3379 -0.01552 0.1154
> plot(boot35)
> qqnorm(boot35)
> title("Bootstrap Distribution of the Correlation Coefficient")
```

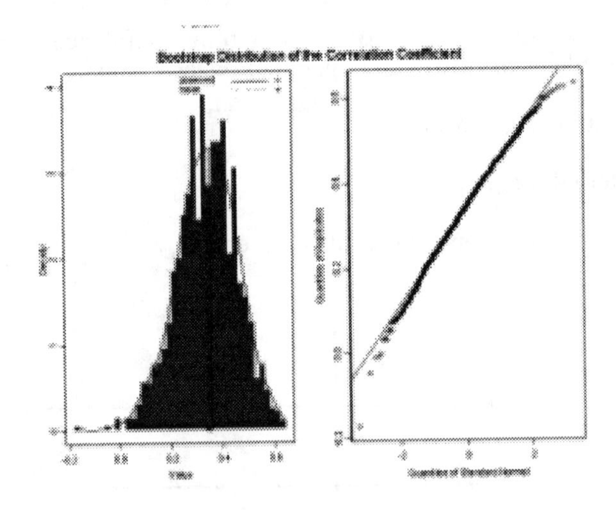

```
> limits.t(boot35)
numeric matrix: 1 rows, 4 columns.
           2.5%        5%        95%      97.5%
Param 0.1224106 0.1605213 0.5464144 0.5845251
> limits.percentile(boot35)
numeric matrix: 1 rows, 4 columns.
            2.5%      5.0%      95.0%      97.5%
Param 0.09481499 0.1360894 0.5170051 0.5548744
> limits.bca(boot35)
numeric matrix: 1 rows, 4 columns.
           2.5%        5%        95%      97.5%
Param 0.1203184 0.1650655 0.5468778 0.5723105
> limits.tilt(boot35)
Tilting Confidence Limits (exponential tilting):
numeric matrix: 1 rows, 4 columns.
           2.5%       5%        95%      97.5%
Param 0.1163342 0.154661 0.5270107 0.5542852

Tilting Confidence Limits (maximum-likelihood tilting):
numeric matrix: 1 rows, 4 columns.
           2.5%        5%     95%      97.5%
Param 0.1141616 0.1533321 0.52769 0.5550224
```

From the software output above, we see the bootstrap mean is 0.3379 and the bootstrap SE is 0.1154. The sample size is $n = 59$ and the upper 0.025 percentile of the t distribution with $n - 1 = 58$ df is found, using Table D (use the value corresponding to 60 df), to be $t^* = 2.00$. Thus, a 95% bootstrap t confidence interval is,

$$\text{bootstrap mean} \pm t^*\text{SE} = 0.3379 \pm (2.00)(0.1154) = 0.3379 \pm 0.2308 = (0.1071, 0.5687)$$

We use the 2.5% and 97.5% percentiles for the 95% bootstrap percentile, BCa, and tilting intervals. The intervals are

95% bootstrap percentile: (0.0948, 0.5549)

BCa: (0.1203, 0.5723)

tilting (exponential): (0.1163, 0.5542)

tilting (maximum likelihood): (0.1142, 0.5550)

A BCa and tilting intervals have a larger lower endpoint and are narrower than the bootstrap t and percentile intervals.

If you did Exercise 18.32, the bootstrap t and percentile intervals here will differ slightly from those in Exercise 18.32, because they are based on a different bootstrap sample.

SECTION 18.5 EXERCISES

18.37 a) The S-PLUS commands to bootstrap the difference in means for the repair time data are as follows.

```
> boot37 <- bootstrap(Verizon, mean(Time[Group == "ILEC"]) -
mean(Time[Group == "CLEC"]), group = Group)
```

b) The S-PLUS command to get the 95% BCa interval and the resulting output is as follows.

```
> limits.bca(boot37)
numeric matrix: 1 rows, 4 columns.
              2.5%        5%        95%       97.5%
Param -16.82863 -15.49855 -2.925063 -2.009733
```

The 95% tilting interval is similar.

Using the 2.5% and 97.5% percentiles as the lower and upper limits, the 95% BCa confidence interval is

$$(-16.82863, -2.009733)$$

This interval does not include 0, so we would conclude that the mean repair times for all Verizon customers are lower than the mean repair times for all CLEC customers.

c) Using a t or percentile interval, we would tend to understate the difference in mean repair times, and perhaps fail to recognize that the mean repair times for Verizon customers are significantly shorter than for CLEC customers.

18.39 a) Using software, we find the mean of the actual data is $\bar{x}$ = 63.012 and the sample standard deviation is s = 7.705967. The sample size is n = 25 and the upper 0.025 percentile of the t distribution with $n - 1$ = 24 df is found, using Table D to be t^* = 2.064. Thus, a 95% traditional one-sample t confidence interval is

$$\bar{x} \pm t^* s/\sqrt{n} = 63.012 \pm (2.064)(7.705967)/\sqrt{25} = 63.012 \pm 3.181 = (59.831, 66.193)$$

b) A histogram and Normal quantile plot of the data are given below.

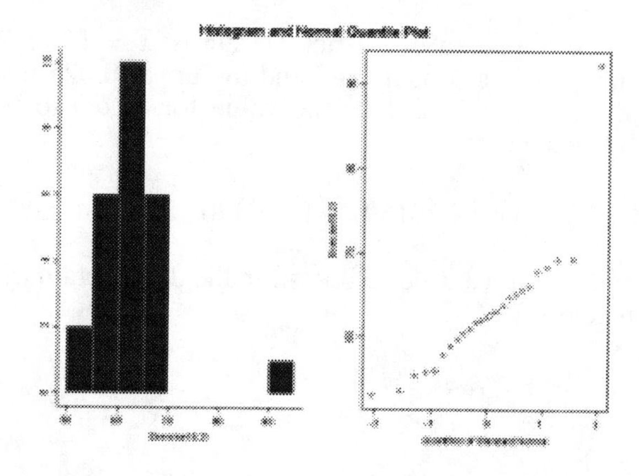

The value 92.3 is an outlier and might strongly influence the traditional confidence interval given in (a).

c) The S-PLUS command s we used to bootstrap the sample and obtain the bootstrap percentile interval, along with the output are given below.

```
> boot21 <- bootstrap(Exercise18.21, mean)

> limits.percentile(boot21)
numeric matrix: 1 rows, 4 columns.
        2.5%     5.0%    95.0%    97.5%
mean 60.368 60.6964 65.6542 66.1479
```

Using the 2.5% and 97.5% percentiles as the lower and upper limits, the 95% percentile interval is

(60.368, 66.1479)

This interval is narrower than the interval given in (a).

d) A 95% confidence interval for the mean weights of male runners (in kilograms) is (60.368, 66.1479).

18.41 a) A histogram and Normal quantile plot of the data are given below.

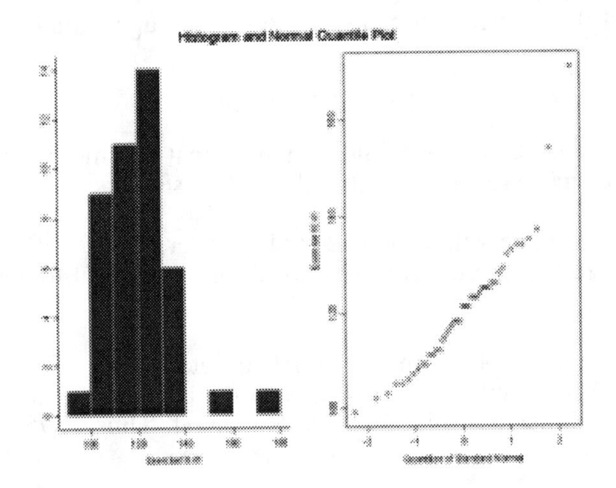

There are two large outliers present. In Section 1 of Chapter 7, we saw that t procedures can be used even for clearly skewed distributions when the sample size is large, roughly $n \geq 40$. In this example, $n = 43$, so one-sample t procedures may be safe even though there are two outliers present.

b) Using software, we find the mean of the actual data is $\bar{x} = 120.5814$ and the sample standard deviation is $s = 13.75159$. The sample size is $n = 43$ and the upper 0.025 percentile of the t distribution with $n - 1 = 42$ df is found, using Table D (use the value for 40 df) to be $t^* = 2.021$. Thus, a 95% traditional one-sample t confidence interval is

$$\bar{x} \pm t^* s/\sqrt{n} = 120.5814 \pm (2.021)(13.75159)/\sqrt{43} = 120.5814 \pm 4.2382 = (116.3432, 124.8196)$$

c) The S-PLUS commands that we used to bootstrap the data, obtain the bootstrap distribution, and the resulting output are given below.

```
> boot41 <- bootstrap(Exercise18.41, mean)
> plot(boot41)
> qqnorm(boot41)
> title("Bootstrap Distribution of the Mean")
```

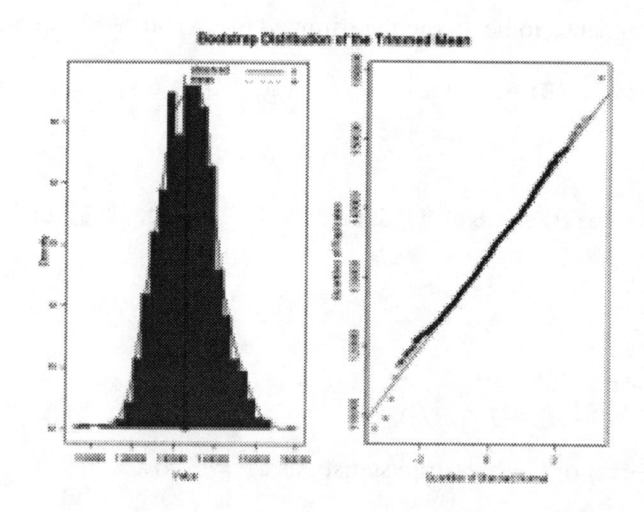

The bootstrap distribution shows moderate skewness to the right, so a bootstrap t interval should be moderately accurate.

d) The S-PLUS commands to produce the 95% bootstrap percentile interval and the resulting output are given below.

```
> limits.percentile(boot41)
numeric matrix: 1 rows, 4 columns.
        2.5%      5.0%     95.0%     97.5%
mean 116.907 117.3721 123.8837 124.5349
```

Using the 2.5% and 97.5% percentiles as the lower and upper limits, the 95% percentile interval is

(116.907, 124.5349)

This agrees closely with the interval found in (b), so we conclude that the one-sample t interval is reasonably accurate here.

18.43 a) A histogram and Normal quantile plot of the data are given below.

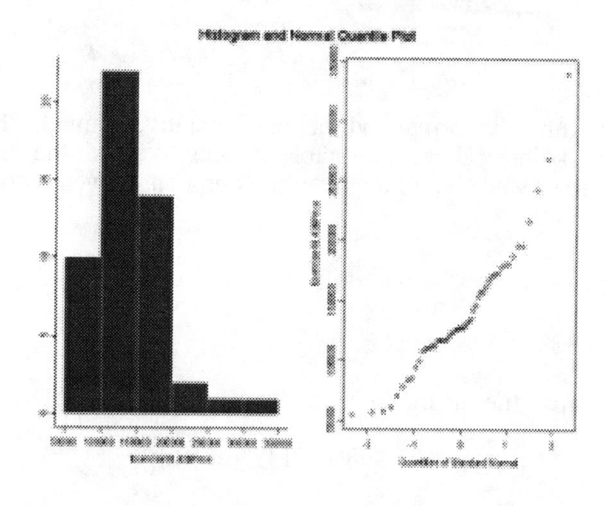

The data are clearly right-skewed. The mean would not be a useful measure of the price of a typical house in Ames. The trimmed mean or median might be more useful. We examine the trimmed mean.

b) The S-PLUS commands to bootstrap the trimmed mean and resulting output are given below.

```
> boot43 <- bootstrap(Exercise18.43, mean(Price, trim = 0.25))

> boot43
Call:
bootstrap(data = Exercise18.43, statistic = mean(Price, trim = 0.25))

Number of Replications: 1000

Summary Statistics:
      Observed   Mean  Bias   SE
Param   132717 133231 513.6 6731
```

We see that the standard error of our bootstrap statistic is SE = 6731.

c) The S-PLUS commands for the bootstrap distribution and the resulting histogram and Normal quantile plot are given below.

```
> plot(boot43)
> qqnorm(boot43)
> title("Bootstrap Distribution of the Trimmed Mean")
```

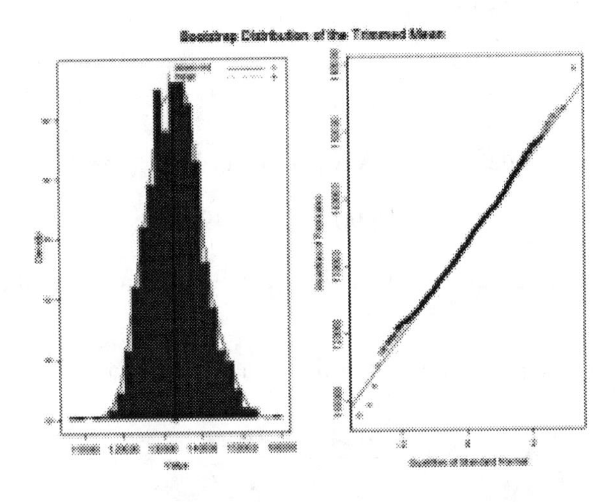

The distribution looks approximately normal and the bias is relatively small. The 95% bootstrap *t* interval or the 95% bootstrap percentile interval are reasonable choices. We give the 95% percentile interval. The S-PLUS commands needed to produce it and the resulting output are given below.

```
> limits.percentile(boot43)
numeric matrix: 1 rows, 4 columns.
          2.5%      5.0%     95.0%     97.5%
Param 121398.1 122633.1 144896.8 147127.4
```

Using the 2.5% and 97.5% percentiles as the lower and upper limits, the 95% percentile interval is

(121,398.10, 147,127.40)

Note: For comparison, we give the S-PLUS commands and output for the other possible intervals.

```
> limits.bca(boot43)
numeric matrix: 1 rows, 4 columns.
          2.5%        5%     95%    97.5%
Param 120654.8 121882.1 143829 146543.2
```

95% BCa interval is (120,654.80, 146,543.20)

```
> limits.tilt(boot43)
Tilting Confidence Limits (exponential tilting):

numeric matrix: 1 rows, 4 columns.
          2.5%        5%     95%    97.5%
Param 120822.5 122439.8 144099 146418.9
```

95% exponential tilting interval is (120,822.50, 146,418.90)

```
Tilting Confidence Limits (maximum-likelihood tilting):
numeric matrix: 1 rows, 4 columns.
          2.5%        5%       95%    97.5%
Param 120368.2 122063.6 144142.4 146516.5
```

95% maximum-likelihood tilting interval is (120,368.20, 146,516.50)

We can also compute the 95% bootstrap *t* interval, which is

$$\text{bootstrap mean} \pm t^*SE = 133{,}231 \pm (2.009)(6731) = 133{,}231 \pm 13{,}522 = (119{,}709,\ 146{,}753)$$

All give similar answers to the 95% percentile interval.

d) We are 95% confident that the mean selling price of all homes sold in Ames for the period represented by these data is in the interval ($121,398.10, $147,127.40).

18.45 a) A scatterplot of the data is given below.

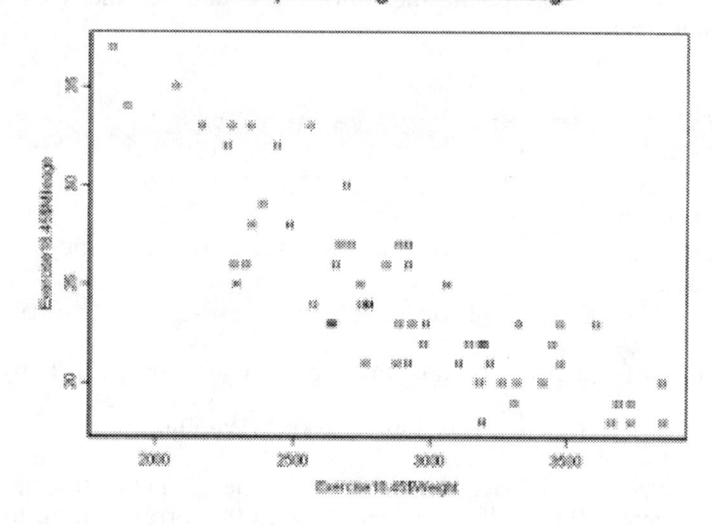

The relationship appears linear and the association is negative. Using software, we compute the sample correlation between weight and mileage to be

$$r = -0.848$$

b) The S-PLUS commands to bootstrap the correlation and to produce the 95% BCa and tilting confidence intervals, along with the output, are given below.

```
> boot45a <- bootstrap(Exercise18.45, cor(Weight, Mileage))

> limits.bca(boot45a)
numeric matrix: 1 rows, 4 columns.
            2.5%         5%          95%         97.5%
Param -0.9017411 -0.8928032 -0.7938861 -0.7819712
> limits.tilt(boot45a)
Tilting Confidence Limits (exponential tilting):
numeric matrix: 1 rows, 4 columns.
            2.5%         5%          95%         97.5%
Param -0.8995068 -0.8923711 -0.7890916 -0.7767903

Tilting Confidence Limits (maximum-likelihood tilting):
numeric matrix: 1 rows, 4 columns.
            2.5%         5%          95%         97.5%
Param -0.8998629 -0.8926798 -0.7886543 -0.7760803
```

The 95% confidence intervals are obtained by using the 2.5% and 97.5% percentiles, respectively, as the lower and upper confidence limits. The resulting intervals are as follows.

95% BCa interval: (–0.9017411, –0.7819712)

95% exponential tilting interval: (–0.8995068, –0.7767903)

95% maximum-likelihood tilting interval: (–0.8998629, –0.7760803)

All should be accurate intervals. They provide a 95% confidence interval for the population correlation between weight and gas mileage in miles per gallon for all 1990 model year cars.

c) Using statistical software, we obtain the following results for the least-squares regression to predict gas mileage from weight.

```
Coefficients:
              Value   Std. Error  t value   Pr(>|t|)
(Intercept)  48.3493    1.9794    24.4261    0.0000
     Weight  -0.0082    0.0007   -12.1779    0.0000

Residual standard error: 2.562 on 58 degrees of freedom
Multiple R-Squared: 0.7189
F-statistic: 148.3 on 1 and 58 degrees of freedom, the p-value is 0
```

From this, we see that the least-squares regression line to predict gas mileage from weight is

$$\text{Mileage} = 48.3493 - 0.0082\,(\text{Weight})$$

To compute the traditional 95% t confidence interval for the slope, we notice that there are 58 df for error. From Table D, there is no entry for 58 df, so we use $t^* = 2.000$ corresponding to 60 df. The resulting interval is est. of slope ± t^*(SE of slope) = –0.0082 ± (2.000)(0.0007) = –0.0082 ± 0.0014 = (–0.0096, –0.0068)

d) The S-PLUS commands to bootstrap the regression model and the results are given below.

```
> boot45b <- bootstrap(lm45, coef)
```

```
> boot45b <- bootstrap(lm45, coef)
> limits.percentile(boot45b)
numeric matrix: 2 rows, 4 columns.
                      2.5%           5.0%          95.0%          97.5%
(Intercept)  43.923778631   44.656189149   51.748587849   52.349064223
     Weight  -0.009475148  -0.009354222   -0.007004879   -0.006789247
```

From these results, a 95% percentile confidence interval for the slope is obtained by using the 2.5% and 97.5% percentiles as the lower and upper confidence limits. The interval is

(–0.009475148, –0.0067789247)

This is similar to the traditional interval computed in part (c).

18.47 a) A stemplot of the data is given below. The outlier is clearly seen in the plot.

```
Decimal point is 3 places to the right of the colon

   12 : 6
   13 :
   14 :
   15 :
   16 : 0069
   17 : 1358
   18 : 24
   19 : 133
   20 : 8
```

We first bootstrap the mean with the outlier present. The S-PLUS commands and results are given below.

```
> boot47a <- bootstrap(Exercise18.47, mean)
> boot47a
Call:
bootstrap(data = Exercise18.47, statistic = mean)

Number of Replications: 1000

Summary Statistics:
     Observed  Mean  Bias  SE
mean    17529 17532 3.161 482


> boot47b <- bootstrap(Exercise18.47[-10], mean)
> boot47b
Call:
bootstrap(data = Exercise18.47[-10], statistic = mean)

Number of Replications: 1000
```

```
Summary Statistics:
      Observed  Mean   Bias    SE
mean     17878 17879 0.5846 371.7
```

Notice that the bootstrap bias is 3.161 with the outlier included and 0.5846 with the outlier excluded. The presence of the outlier increases the bias.

```
> plot(boot47a)
> qqnorm(boot47a)
> boot47b <- bootstrap(Exercise18.47[-10], mean)
> boot47b
```

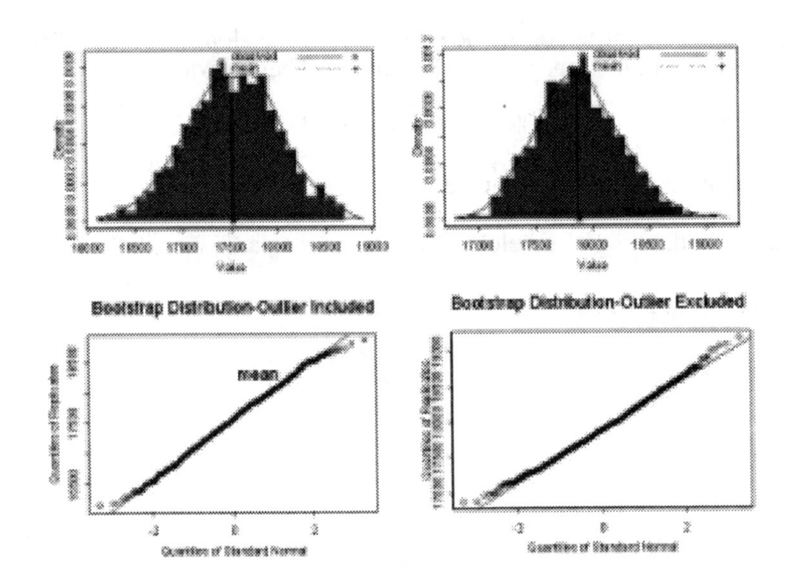

Examining the plots, we see that the bootstrap distribution with the outlier included is shifted significantly to the left (centered at a smaller value) of the bootstrap distribution with the outlier excluded. Also, the bootstrap distribution with the outlier included appears to be slightly left-skewed. However, there is little bias in either case (in fact, any apparent bias is due to random resampling, because the sample mean has no true bias).

b) The S-PLUS commands and 95% BCa interval for the mean with the outlier included are given below.

```
> limits.bca(boot47a)
numeric matrix: 1 rows, 4 columns.
          2.5%       5%      95%    97.5%
mean 16481.83 16655.21 18255.51 18431.11
```

From these results, a 95% BCa confidence interval for the mean is obtained by using the 2.5% and 97.5% percentiles as the lower and upper confidence limits. The interval is

$$(16481.83, 18431.11)$$

The S-PLUS commands and 95% BCa interval for the mean with the outlier removed are given below.

```
> limits.bca(boot47b)
numeric matrix: 1 rows, 4 columns.
          2.5%       5%      95%    97.5%
mean 17230.28 17307.24 18552.32 18697.67
```

From these results, a 95% BCa confidence interval for the mean is obtained by using the 2.5% and 97.5% percentiles as the lower and upper confidence limits. The interval is

(17230.28, 18698.67)

The lower confidence limits are quite different. The lower confidence limit for the interval based on the data that includes the outlier is much smaller than the lower confidence limit for the interval based on the data with the outlier excluded. The outlier was an unusually small value, so it appears that the effect of the outlier is to pull the lower limit down. The upper confidence limits for both intervals are more nearly equal, but the upper confident limit for the interval based on the data that includes the outlier is smaller than the lower confidence limit for the interval based on the data with the outlier excluded. Thus, the small value of the outlier also pulls the upper confidence limit down somewhat.

SECTION 18.6

OVERVIEW

Permutation tests are significance tests based on **permutation resamples** drawn at random from the original data. Permutation resamples are drawn **without replacement** in contrast to bootstrap samples, which are drawn with replacement. Permutation resamples must be drawn in a way that is consistent with the null hypothesis and with the study design. In a **two-sample design,** the null hypothesis says that the two populations are identical. Resampling randomly assigns observations to the two groups. In a **matched pairs** design, randomly permute the two observations within each pair separately. To test the hypothesis of **no relationship** between two variables, randomly reassign values of one of the two variables.

The **permutation distribution** of a suitable statistic is formed by the values of the statistic in a large number of resamples. Find the *P*-value of the test by locating the original value of the statistic on the permutation distribution.

When they can be used, permutation tests have great advantages. They do not require specific population shapes such as Normality. They apply to a variety of statistics, not just to statistics that have a simple distribution under the null hypothesis. They can give very accurate *P*-values, regardless of the shape and size of the population (if enough permutations are used).

It is often useful to give a confidence interval along with a test. To create a confidence interval we no longer assume the null hypothesis is true so we use a bootstrap resampling rather than a permutation resampling.

APPLY YOUR KNOWLEDGE

18.49 a) Let μ_1 denote the mean selling price for all Seattle real estate transactions in 2001 and μ_2 the mean selling price for all Seattle real estate transactions in 2002. We wish to test whether these means are significantly different. To do this, we test the hypotheses

$$H_0: \mu_1 = \mu_2$$
$$H_a: \mu_1 \neq \mu_2$$

b) To carry out a significance test for $H_0: \mu_1 = \mu_2$, we use the two-sample *t* statistic

$$t = \frac{(\bar{x}_1 - \bar{x}_2)}{\sqrt{\frac{s_1^2}{n_1} + \frac{s_2^2}{n_2}}}$$

We know that $n_1 = n_2 = 50$, and using statistical software, we find

$$\bar{x}_1 = 288.9265,\ \bar{x}_2 = 329.2571,\ s_1 = 157.7778\ s_2 = 316.83$$

Thus,

$$t = \frac{288.9265 - 329.2571}{\sqrt{\frac{(157.7778)^2}{50} + \frac{(316.83)^2}{50}}} = \frac{-40.3306}{\sqrt{2505.5017}} = -0.8057$$

The P-value is found by using the approximate distribution $t(k)$, where k taken to be the smaller of $n_1 - 1 = 49$ and $n_2 - 1 = 49$ for a conservative procedure. Using the row labeled 40 df in Table D (rounding 49 down to 40 to be additionally conservative), we see that $|-0.8057| = 0.8057$ is between the values in the columns labeled 0.25 and 0.20. Thus the upper tail probability that a $t(40)$ random variable is larger than 0.8507 is between 0.20 and 0.25. Because we are interested in a two-sided test, we double the upper tail probabilities to get the P-value. Hence

$$0.40 < P\text{-value} < 0.50$$

There is little evidence that the population means μ_1 and μ_2 differ.

c) To perform the permutation test, we use S-PLUS. The commands we used and the resulting output are given below.

```
> perm49 <- permutationTestMeans(Exercise18.49, treatment = Year)
> perm49
Call:
permutationTestMeans(data = Exercise18.49, treatment = Year)

Number of Replications: 999

Summary Statistics:
       Observed      Mean     SE   alternative   p.value
Price     40.33  -0.01678  49.68     two.sided     0.438
```

From the output, we see the P-value is 0.438. This is consistent with the P-value we computed in (b).

We again conclude that there is little evidence that the population means μ_1 and μ_2 differ.

d) The S-PLUS commands to produce a BCa 95% confidence interval and the resulting output are given below.

```
> boot49 <- bootstrap(Exercise18.49, mean(Price[Year == 2001]) -
mean(Price[Year ==2002]))

> limits.bca(boot49)
numeric matrix: 1 rows, 4 columns.
             2.5%        5%       95%     97.5%
Param  -175.5956  -154.2974  24.53738  35.87395
```

From these results, a 95% BCa confidence interval for the mean change from 2001 to 2002 is obtained by using the 2.5% and 97.5% percentiles as the lower and upper confidence limits. The interval is

$$(-175.5956, 35.87395)$$

This interval includes 0 and suggests that the two means are not significantly different at the 0.05 level. Again, this is consistent with the conclusions in parts (b) and (c).

18.51 The guideline given in the Chapter states "If the true (one-sided) P-value is p, the standard deviation of the estimated P-value is approximately

$$\sqrt{\frac{p(1-p)}{B}}$$

where B is the number of resamples." Using this guideline we find the following.

The standard deviation for the estimated P-value of 0.015 for the DRP study, based on B = 999 resamples is

$$\sqrt{\frac{p(1-p)}{B}} = \sqrt{\frac{0.015(1-0.015)}{999}} = 0.003846$$

The standard deviation for the estimated P-value of 0.0183 based on the 500,000 resamples in the Verizon study is

$$\sqrt{\frac{p(1-p)}{B}} = \sqrt{\frac{0.0183(1-0.0183)}{500{,}000}} = 0.0001896$$

18.53 a) Let p_1 denote the probability of success of new franchise firms with exclusive territory clauses and p_2 the probability of success of new franchise firms with no exclusive territory clause. Because we conjecture that exclusive territory clauses increase the chance of success, we test the hypotheses

$$H_0\colon p_1 = p_2$$
$$H_a\colon p_1 > p_2$$

b) Let X_1 = 108 be the number of success among the n_1 = 142 firms in the sample that had an exclusive territory clause, X_2 = 15 the number of success among the n_2 = 28 firms in the sample that had no exclusive territory clause. To perform the z test, we first compute the pooled estimate of the common (under H_0) value of p_1 and p_2,

$$\hat{p} = \frac{X_1 + X_2}{n_1 + n_2} = \frac{108+15}{142+28} = 0.724$$

The test uses the z statistic

$$z = \frac{\hat{p}_1 - \hat{p}_2}{\sqrt{\hat{p}(1-\hat{p})\left(\frac{1}{n_1} + \frac{1}{n_2}\right)}} = \frac{0.761 - 0.536}{\sqrt{0.724(1-0.724)\left(\frac{1}{142} + \frac{1}{28}\right)}} = 2.434$$

where we have used $\hat{p}_1 = X_1/n_1 = 0.761$, $\hat{p}_2 = X_2/n_2 = 0.536$. The P-value is the probability that a standard normal random variable would be larger than 2.434. According to Table A, this probability is approximately 0.0075. Thus,

$$P\text{-value} = 0.0075$$

and we conclude that there is strong evidence that exclusive territory clauses increase the chance of success.

c) Under the null hypothesis, all 170 firms are equally likely to be a success. In this case, successes occur for reasons that have nothing to do with whether the firm has an exclusive territory clause. We can resample in a way consistent with the null hypothesis choosing an ordinary SRS of 142 of the firms without replacement and assigning them to the exclusive territory clause group.

We use S-PLUS to perform the permutation test. The commands and ouptut are given below.

```
> perm53 <- permutationTestMeans(Exercise18.53, treatment = Exclusive)
> perm53
Call:
permutationTestMeans(data = Exercise18.53, treatment = Exclusive)

Number of Replications: 999

Summary Statistics:
        Observed     Mean      SE alternative p.value
Success   0.2248 0.004133 0.09306   two.sided    0.04
```

The output gives the P-value for a two-sided alternative. We divide this by 2 in order to get the P-value for our one-sided alternative. The resulting P-value is

$$P\text{-value} = 0.02$$

This P-value is larger than that found in part (b).

d) There is evidence at the 0.05 level that exclusive territory clauses increase the chance of success. There is not evidence that exclusive territory clauses increase the chance of success at the 0.01 level.

e) The S-PLUS commands needed to produce a BCa 95% confidence interval for the difference between the two population proportions, and the resulting output are given below.

```
> boot53 <- bootstrap(Exercise18.53, mean(Success[Exclusive == "Yes"])
- mean(Success[Exclusive == "No"]))

> limits.bca(boot53)
numeric matrix: 1 rows, 4 columns.
            2.5%         5%       95%     97.5%
Param 0.02536757 0.05681727 0.3917654 0.4316245
```

From these results, a 95% BCa confidence interval for the difference between the two population proportions is obtained by using the 2.5% and 97.5% percentiles as the lower and upper confidence limits. The interval is

$$(0.02536757, 0.4316245)$$

This interval does not include 0 and lies to the positive side of 0. This suggests that we have evidence that exclusive territory clauses increase the chance of success, which is consistent with the results of the permutation test in part (d) (recall that this test would reject the null hypothesis at the 0.05 level).

18.55 a) Let ρ denote the correlation between the salaries and batting averages of all Major League Baseball players. In this case, if there is correlation, we expect it to be positive, so we test the hypotheses

$$H_0: \rho = 0$$
$$H_a: \rho > 0$$

b) The S-PLUS commands that we used and the resulting output are given below. Only one of the variables needs to be permuted. In our case, we permuted the variable salary.

```
> perm55 <- permutationTest(Exercise18.55$Salary, cor(data,
Exercise18.55$Average))
```

```
> perm55
Call:
permutationTest(data = Exercise18.55$Salary, statistic = cor(data,
Exercise18.55$Average))

Number of Replications: 999

Summary Statistics:
      Observed        Mean     SE alternative p-value
Param   0.1068 -0.0007451 0.1493   two.sided   0.488
```

We want the one-sided P-value, so we divide the two-sided value given in the output by 2. We then see that the *P*-value is

$$P\text{-value} = 0.244$$

and we conclude that there is not strong evidence that salaries and batting averages are correlated in the population of all major-league players.

SECTION 18.6 EXERCISES

18.57 a) For the median, we test the hypotheses

H_0: median time for right hand = median time for left hand
H_a: median time for right hand ≠ median time for left hand

For the 25% trimmed mean, we test the hypotheses

H_0: 25% trimmed mean time for right hand = 25% trimmed mean time for left hand
H_a: 25% trimmed mean time for right hand ≠ 25% trimmed mean time for left hand

b) To perform the permutation test, we must either permute the variable hand, holding time fixed, or permute the times, holding hand fixed. We permuted the variable time. The S-PLUS commands that we used and the resulting ouptut are given below.

```
> perm57a <- permutationTest(Exercise18.57$Time,
median(data[Exercise18.57$Hand =="right"]) -
median(data[Exercise18.57$Hand == "left"]))

> perm57a
Call:
permutationTest(data = Exercise18.57$Time, statistic =
median(data[Exercise18.57$Hand == "right"]) -
median(data[Exercise18.57$Hand == "left"]))

Number of Replications: 999

Summary Statistics:
      Observed  Mean    SE alternative p-value
Param   -101.5 1.715 36.56   two.sided   0.002

> plot(perm57a)
> title("Permutation Distribution - Difference in Medians")
```

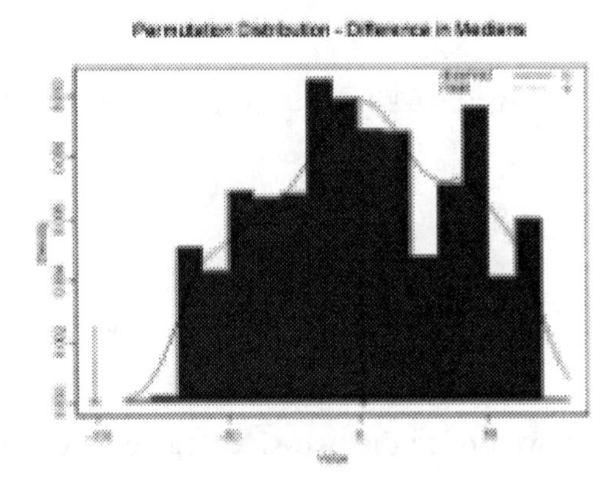

The permutation distribution is clearly not Normal.

The P-value for the permutation test is given in the above output and is

$$P\text{-value} = 0.002$$

There is strong evidence that there is a difference in the population median times when using the right hand versus when using the left hand.

c) To perform the permutation test, we must either permute the variable hand, holding time fixed, or permute the times, holding hand fixed. We permuted the variable time. The S-PLUS commands that we used and the resulting ouptut are given below.

```
> perm57b <- permutationTest(Exercise18.57$Time,
mean(data[Exercise18.57$Hand == "right"], trim = 0.25) -
mean(data[Exercise18.57$Hand == "left"], trim = 0.25))

> perm57b
Call:
permutationTest(data = Exercise18.57$Time, statistic =
mean(data[Exercise18.57$Hand == "right"], trim = 0.25) -
mean(data[Exercise18.57$Hand == "left"],
     trim = 0.25))

Number of Replications: 999

Summary Statistics:
      Observed   Mean    SE alternative p-value
Param      -99 0.3766 26.07   two.sided   0.002

> plot(perm57b)
> title("Permutation Distribution - Difference in Trimmed Means")
```

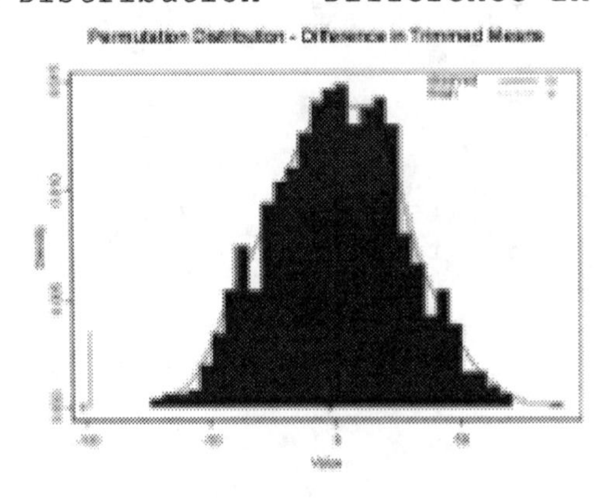

This permutation distribution looks much more like a Normal distribution than the permutation distribution in (b) for the difference in medians.

The P-value for the permutation test is given in the above output and is

$$P\text{-value} = 0.002$$

There is strong evidence that there is a difference in the population 25% trimmed mean times when using the right hand versus when using the left hand.

d) We stated the conclusions in parts (b) and (c).

18.59 a) Let p_1 denote the proportion of women in the population who pay attention to a "No Sweat" label when buying a garment and p_2 denote the proportion of men in the population who pay attention to a "No Sweat" label when buying a garment. We test the following hypotheses.

$$H_0: p_1 = p_2$$
$$H_a: p_1 \neq p_2$$

b) To perform the permutation test, we use S-PLUS. The commands we used and the resulting output are given below.

```
> perm59 <- permutationTestMeans(Exercise18.59, treatment = Sex)
> perm59
Call:
permutationTestMeans(data = Exercise18.59, treatment = Sex)

Number of Replications: 999

Summary Statistics:
          Observed       Mean      SE alternative p.value
LabelUser   0.1053 -0.0009235 0.03134   two.sided   0.002

> plot(perm59)
> title("Permutation Distribution")
```

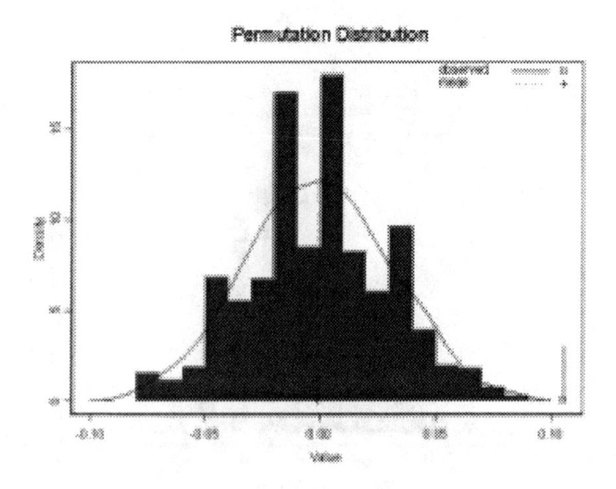

The P-value for the permutation test is given in the output above and is

$$P\text{-value} = 0.002$$

There is strong evidence that there is a difference between the proportion of women in the population who pay attention to a "No Sweat" label when buying a garment and the proportion of men in the population who pay attention to a "No Sweat" label when buying a garment.

c) The permutation distribution is given in part (b). The shape is approximately Normal (except that it is discrete; you can see this using a normal quantile plot or by observing spikes in the histogram) and thus it is not surprising that the permutation test agrees closely with the z test in Example 8.6.

18.61 On each house, we record two variables: One is the square footage and the other is the age of the house. To perform the permutation test, we permute the square footage of the houses, keeping the age fixed. The S-PLUS commands we used to conduct the tw0-sided permutation test and the results are given below.

```
> perm61 <- permutationTest(Exercise18.61$SquareFootage, cor(data,
Exercise18.61$Age))

> perm61
Call:
permutationTest(data = Exercise18.61$SquareFootage, statistic =
cor(data, Exercise18.61$Age))

Number of Replications: 999

Summary Statistics:
       Observed     Mean     SE alternative p-value
Param   -0.4065 0.008022 0.1415   two.sided   0.002
```

The P-value is 0.002, and we conclude that there is strong evidence that there is a correlation between square footage and age of a house in Ames, Iowa.

18.63 To bootstrap the difference in proportions, we used the following S-PLUS commands. Output from these commands is included below.

```
> boot63 <- bootstrap(Exercise18.62, mean(Success[Treatment ==
"calcium"]) - mean(Success[Treatment == "placebo"]), group =
Treatment)

> plot(boot63)
> title("Bootstrap Distribution")
```

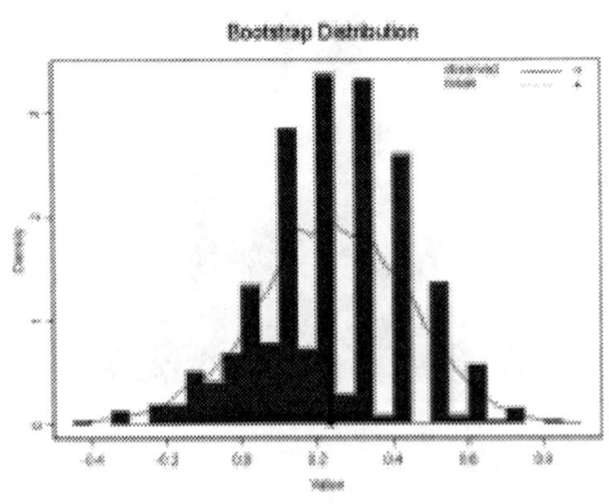

```
> boot63
Call:
bootstrap(data = Exercise18.62, statistic = mean(Success[Treatment ==
"calcium"]) -
     mean(Success[Treatment == "placebo"]), group = Treatment)
```

```
Number of Replications: 1000

Summary Statistics:
      Observed  Mean      Bias      SE
Param   0.2364 0.239 0.002673 0.2028
```

We construct a 95% z interval using the observed difference in proportions (0.2364) and the bootstrap standard error (SE = 0.2028). For 95% confidence, we use $z^* = 1.96$, and the resulting interval is

$$\text{observed difference } \pm z^*(\text{SE}) = 0.2364 \pm 1.96(0.2028) = 0.2364 \pm 0.3975 = (-0.1611, 0.6339)$$

18.65 a) To perform a two-sided permutation test on the ratio of standard deviations, we used the following S-PLUS commands. Output from these commands is included below.

```
> perm65 <- permutationTest(Exercise18.64$Time,
stdev(data[Exercise18.57$Group
=="CLEC"])/stdev(data[Exercise18.57$Group == "ILEC"]))

> perm65
Call:
permutationTest(data = Exercise18.64$Time, statistic =
stdev(data[Exercise18.64$ Group ==
"ILEC"])/stdev(data[Exercise18.64$Group == "CLEC"]))

Number of Replications: 999

Summary Statistics:
      Observed Mean SE alternative p-value
Param   0.9321  Inf NA   two.sided   0.408
```

Some of the ratios were infinite because the permutation test produced a standard deviation of 0 in the denominator. The mean and SE are not given, but we are still able to give the P-value = 0.408 based on the permutation distribution. This P-value tells us that there is not strong evidence that the variability in the repair times for ILEC and CLEC customers differ.

b) The P-value for the permutation test differs from that obtained by the F statistic. This suggests that the test based on the F statistic is not accurate.

18.67 a) Because of the way in which S-PLUS carries out the permutation test, we will formulate our hypotheses as follows. Let μ be the mean change (pretest – posttest) that would be achieved if the entire population of executives received similar instruction. Because we hope to show that the course improves comprehension scores, i.e., the posttest scores are higher than the prestest scores, we test the hypotheses

$$H_0\colon \mu = 0$$
$$H_a\colon \mu < 0$$

Note that negative values of μ indicate that mean posttest scores are higher than mean pretest scores, and hence that test scores have improved.

b) To perform a paired-sample permutation test, we used S-PLUS. First, we rearranged the data as follows to allow us to carry out the test in S-PLUS.

```
   Subject Score     Exam
1        1    30  Pretest
2        2    28  Pretest
3        3    31  Pretest
4        4    26  Pretest
5        5    20  Pretest
6        6    30  Pretest
7        7    34  Pretest
8        8    15  Pretest
9        9    28  Pretest
10      10    20  Pretest
11      11    30  Pretest
12      12    29  Pretest
13      13    31  Pretest
14      14    29  Pretest
15      15    34  Pretest
16      16    20  Pretest
17      17    26  Pretest
18      18    25  Pretest
19      19    31  Pretest
20      20    29  Pretest
21       1    29 Posttest
22       2    30 Posttest
23       3    32 Posttest
24       4    30 Posttest
25       5    16 Posttest
26       6    25 Posttest
27       7    31 Posttest
28       8    18 Posttest
29       9    33 Posttest
30      10    25 Posttest
31      11    32 Posttest
32      12    28 Posttest
33      13    34 Posttest
34      14    32 Posttest
35      15    32 Posttest
36      16    27 Posttest
37      17    28 Posttest
38      18    29 Posttest
39      19    32 Posttest
40      20    32 Posttest
```

The commands we used and the resulting output are given below.

```
> newData67 <- data.frame(Subject = rep(1:20, 2), Score =
c(Exercise18.67$Pretest, Exercise18.67$Posttest), Exam =
rep(c("Pretest", "Posttest"), each = 20))

> perm67 <- permutationTestMeans(newData67, treatment = Exam, group =
Subject,alternative = "less")
> perm67
Call:
permutationTestMeans(data = newData67, treatment = Exam, alternative =
"less",group = Subject)

Number of Replications: 999

Summary Statistics:
      Observed    Mean     SE alternative p.value
Score    -1.45 0.02628 0.7569        less   0.036
```

The *P*-value is 0.036, so there is evidence (significant at the 0.05 level but not at the 0.01 level) that the mean change (pretest – posttest) is negative and hence posttest scores are higher, on average, than pretest scores.

c) A graph of the permutation distribution is given below.

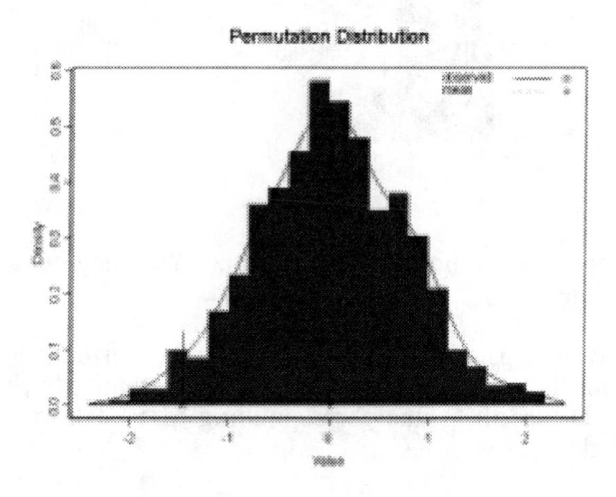

The observed mean change in scores is –1.45 and this is marked on the graph with a solid line at –1.45. The area to the left of –1.45 is the *P*-value.

CHAPTER 18 REVIEW EXERCISES

18.69 a) For a Uniform distribution on 0 to 1, we know that the population median is 0.5 because half of the population is below 0.5 and half is above 0.5. We used the following commands in S-PLUS to generate a sample of 50 observations from the Uniform distribution on 0 to 1 and to bootstrap the sample median. The commands used and the resulting output are given below.

```
> unif <- runif(50)
> boot69 <- bootstrap(unif, median)
```

```
> boot69
Call:
bootstrap(data = unif, statistic = median)

Number of Replications: 1000

Summary Statistics:
        Observed  Mean       Bias       SE
median    0.6037 0.601 -0.002662 0.04706
```

The commands to generate the bootstrap distribution and the resulting plot, along with a Normal quantile plot, are given below.

```
> plot(boot69)
> qqnorm(boot69)
```

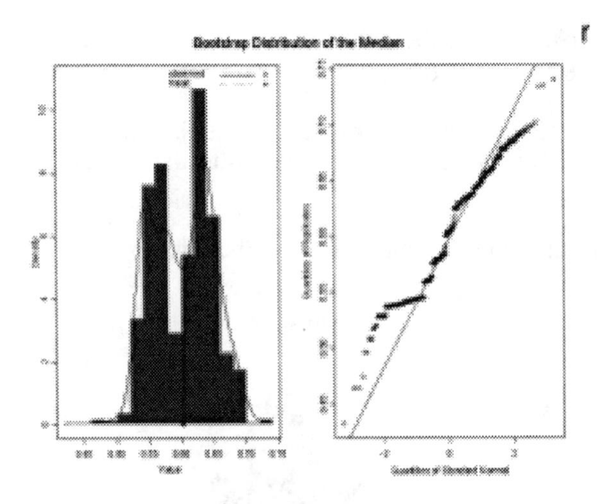

The bootstrap distribution appears to be bimodal, not Normal. You may get a different picture depending on the random data you generate.

b) From the output given in part (a), the bootstrap standard error is SE = 0.04706. The bootstrap mean is 0.601. The sample size is $n = 50$ and the upper 0.025 percentile of the t distribution with $n - 1 =$ 49 df is found, using Table D (use the value corresponding to 50 df), to be $t^* = 2.009$. Thus, a 95% bootstrap t confidence interval is

$$\text{bootstrap mean} \pm t^*\text{SE} = 0.601 \pm (2.009)(0.04706) = 0.601 \pm 0.095 = (0.506, 0.696)$$

c) The bootstrap BCa 95% confidence interval is obtained using the following S-PLUS commands. The resulting output is also given.

```
> limits.bca(boot69)
numeric matrix: 1 rows, 4 columns.
            2.5%       5%       95%     97.5%
median 0.5285407 0.537198 0.6779015 0.6845293
```

From these results, a 95% BCa confidence interval for the mean is obtained by using the 2.5% and 97.5% percentiles as the lower and upper confidence limits. The interval is

$$(0.5285407, 0.6845293)$$

The bootstrap t 95% confidence interval is a little wider than the 95% BCa interval. Although the bootstrap distribution for the median is not Normal, the bootstrap t interval is not too unreliable here.

18.71 a) We used S-PLUS to bootstrap the correlation between overseas and US stocks. The commands we used, including those to produce the bootstrap distribution and Normal quantile plot, and the resulting output are given below.

```
> boot71 <- bootstrap(Exercise18.71, cor(Overseas, U.S.))

> boot71
Call:
bootstrap(data = Exercise18.71, statistic = cor(Overseas, U.S.))
Number of Replications: 1000
Summary Statistics:
    Observed  Mean        Bias     SE
Param 0.5034 0.4941 -0.009342 0.139

> plot(boot71)
> qqnorm(boot71)
> title("Bootstrap Distribution of the Correlation")
```

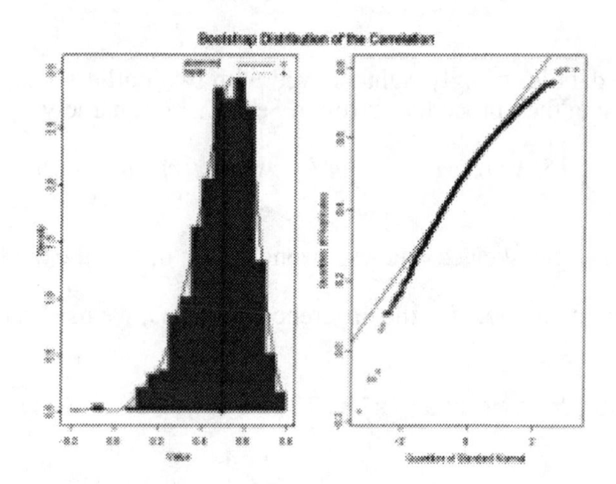

The bootstrap distribution is left-skewed and does not appear to be approximately Normal. From the output, we see that the bootstrap standard error is

$$SE = 0.139$$

b) The bootstrap t confidence interval is not appropriate here because the bootstrap distribution is not approximately Normal.

c) We use S-PLUS to produce a 95% BCa confidence interval. The commands we used and the resulting output are given below.

```
> limits.bca(boot71)
numeric matrix: 1 rows, 4 columns.
           2.5%        5%       95%     97.5%
Param 0.1854328 0.2390449 0.6873278 0.7166615
```

From these results, a 95% BCa confidence interval for the mean is obtained by using the 2.5% and 97.5% percentiles as the lower and upper confidence limits. The interval is

$$(0.1854328, 0.7166615)$$

18.73 a) Below are histograms of the 2000 and 2001 data.

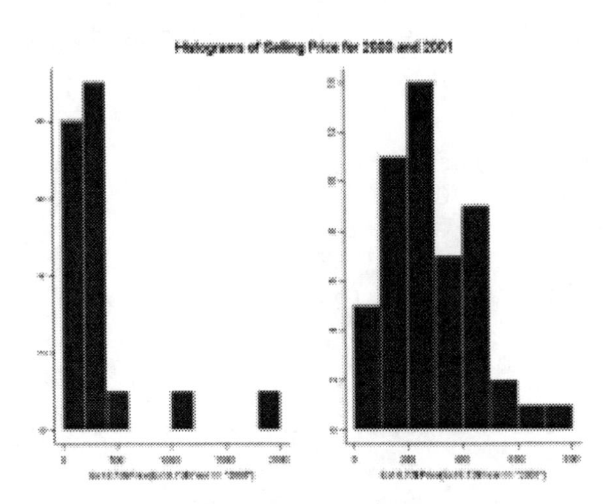

The histogram of the 2000 data is strongly right-skewed with two outliers, one of which is extreme. This violates the guideline for using the *t* procedure given in Section 17.1, namely

For a sample size of at least 15, t procedures can be used except in the presence of outliers or strong skewness.

The histogram of the 2001 is right-skewed, but less strongly than that of the 2000 data.

b) To perform a permutation test for the difference in means, we used the S-PLUS. The commands and output are given below.

```
> perm73_permutationTestMeans(Ex18.73, treatment=Year)

> perm73
Call:
permutationTest(data = Ex18.73$Price, statistic =
mean(data[Ex18.73$Year ==
     "2001"]) - mean(data[Ex18.73$Year == "2000"]))

Number of Replications: 999

Summary Statistics:
      Observed   Mean    SE alternative p-value
Param   -80.02 0.7614 67.49   two.sided   0.302
```

From the output, we see that the *P*-value for the permutation test for the difference in means is

$$P\text{-value} = 0.302$$

We conclude that there is not strong evidence that the mean selling prices for all Seattle real estate in 2000 and in 2001 are different.

18.75 We can use permutation tests when we can see how to resample in a way that is consistent with the study design and with the null hypothesis. In Section 18.6, we saw how to do this for two-sample problems, matched pairs designs, and relationships between two quantitative variables. The study described in Exercise 18.74 is a one-sample problem. We have no methods for carrying out a permutation test in such one-sample problems (there is no obvious way to resample that is consistent with a one-sample test for a mean).

18.77 To give a bootstrap confidence interval for the mean difference within pairs, first create the differences and then bootstrap. The S-PLUS commands to do this and the resulting output are given below.

```
> Exercise18.77 <- Exercise18.76$After - Exercise18.76$Before
> boot77 <- bootstrap(Exercise18.77, mean)

> boot77
Call:
bootstrap(data = Exercise18.77, statistic = mean)

Number of Replications: 1000

Summary Statistics:
     Observed   Mean      Bias      SE
mean   0.6386 0.6335 -0.00506 0.1064
```

The S-PLUS commands to produce the bootstrap distribution and Normal quantile plot are given below, along with the output.

```
> plot(boot77)
> qqnorm(boot77)
> title("Bootstrap Distribution of the Mean of the Differences")
```

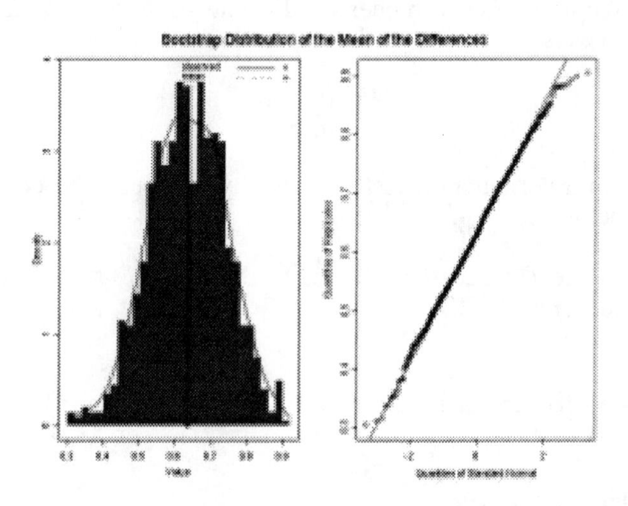

The distribution looks approximately normal, so we compute the 95% bootstrap t interval.

From the output, the bootstrap standard error is SE = 0.1064. The bootstrap mean is 0.6335. The sample size is $n = 14$ and the upper 0.025 percentile of the t distribution with $n - 1 = 13$ df is found, using Table D to be $t^* = 2.160$. Thus, a 95% bootstrap t confidence interval for the mean after minus the mean before is

$$\text{bootstrap mean} \pm t^*\text{SE} = 0.6335 \pm (2.160)(0.1064) = 0.6335 \pm 0.2298 = (0.4037, 0.8633)$$

Zero is outside this interval, so the result is significant at the 0.05 level. We conclude that there is strong evidence that the mean difference is different from 0.

18.79 a) From the Table, we compute

$$\text{proportion of girls who like chocolate ice cream} = \frac{\text{Number of girls who said yes}}{\text{Total number of girls}} = \frac{40}{50} = 0.80$$

$$\text{proportion of boys who like chocolate ice cream} = \frac{\text{Number of boys who said yes}}{\text{Total number of boys}} = \frac{30}{45} = 0.667$$

b) The S-PLUS commands to perform the permutation test and the resulting output are given below.

```
> perm79 <- permutationTestMeans(Exercise18.79, treatment = Sex)
> perm79
Call:
permutationTestMeans(data = Exercise18.79, treatment = Sex)

Number of Replications: 999

Summary Statistics:
     Observed      Mean      SE alternative p.value
Like   0.1333 0.0008764 0.08926   two.sided   0.222
```

From the output, the P-value for a two-sided test is 0.222. There is not strong evidence that there is a difference in the proportion of boys and girls who like chocolate ice cream.

18.81 Let μ_1 denote the mean word count of all ads placed in magazines aimed at people with high education levels, and μ_2 denote the mean word count of all ads placed in magazines aimed at people with medium education levels. We wish to see if higher word counts occur in ads placed in magazines aimed at people with high education levels. Thus, we test the hypotheses

$$H_0: \mu_1 = \mu_2$$
$$H_a: \mu_1 > \mu_2$$

We use S-PLUS to perform a permutation test of these hypotheses. The commands we used and the resulting output are given below.

```
> perm81 <- permutationTestMeans(Exercise18.80, treatment = Education,
alternative = "greater")
> perm81
Call:
permutationTestMeans(data = Exercise18.80, treatment = Education,
alternative = "greater")

Number of Replications: 999

Summary Statistics:
          Observed    Mean    SE alternative p.value
WordCount    18.61 -0.1402 22.98     greater   0.209
```

From the output, we see that the P-value is 0.209. Thus, there is not strong evidence that the mean word count is higher for ads placed in magazines aimed at people with high education levels than for ads placed in magazines aimed at people with medium education levels.

The 95% confidence interval in Exercise 18.80 (d) for the difference in means contained 0. This suggests that there is not strong evidence of a *difference* in mean word counts. Here we conclude that there is not strong evidence that the mean word counts is *higher* for ads placed in magazines aimed at people with high education levels than for ads placed in magazines aimed at people with medium education levels.

18.83 a) The S-PLUS commands we used to bootstrap the difference in mean monthly burglary counts (after – before) and to produce a histogram and Normal quantile plot of the bootstrap distribution are given below, along with the output.

```
> boot83 <- bootstrap(Exercise18.82, mean(Burglaries[When == "After"])
- mean(Burglaries[When == "Before"]))
> plot(boot83)
> qqnorm(boot83)
> title("Bootstrap Distribution")
```

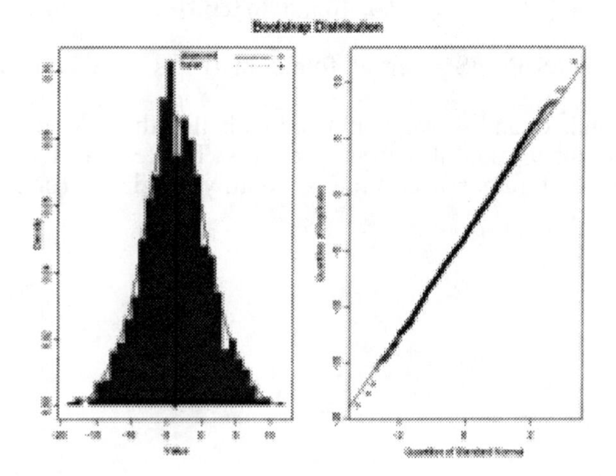


```
> boot83
Call:
bootstrap(data = Exercise18.82, statistic = mean(Burglaries[When ==
"After"]) - mean(Burglaries[When == "Before"]))

Number of Replications: 1000

Summary Statistics:
      Observed   Mean      Bias    SE
Param    -3.67 -3.676 -0.006333 4.566
```

The bootstrap distribution appears to be approximately Normal.

b) From the output in (a), we see that the bootstrap standard error is

$$SE = 4.566$$

The bootstrap mean is –3.676. The sample sizes are $n = 41$ for the period before and $n = 17$ for the period after the commencement of a citizen-police program. We use the conservative method and use the smaller sample size to determine the degrees of freedom. Thus, we take the degrees of freedom to be $n - 1 = 17 - 1 = 16$. The upper 0.025 percentile of the t distribution with 16 df is found, using Table D, to be $t^* = 2.12$. Thus, a 95% bootstrap t confidence interval is

$$\text{bootstrap mean} \pm t^*SE = -3.676 \pm (2.12)(4.566) = -3.676 \pm 9.680 = (-13.356, 6.004)$$

Using S-PLUS rather than the conservative method yields the t interval

$$(-12.81314, 5.473112)$$

c) The S-PLUS commands to produce the 95% bootstrap percentile interval and the resulting output are given below.

```
> limits.percentile(boot83)
numeric matrix: 1 rows, 4 columns.
            2.5%       5.0%     95.0%     97.5%
Param -12.43483 -11.53113 4.278903 5.763863
```

Using the 2.5% and 97.5% percentiles as the lower and upper limits, the 95% percentile interval is

$$(-12.43483, 5.763863)$$

This agrees closely with the S-PLUS interval found in (b), so we conclude that the intervals are reasonably accurate here.

These intervals include 0, and so we would conclude that there is not strong evidence (at the 0.05 level) of a *difference* in the mean monthly burglary counts. The tests in Exercise 18.82 were one-sided tests and showed no evidence of a *decrease* in mean monthly burglaries (or of an *increase* in the case of part [d] of Exercise 18.82).